Biological Wastewater Treatment Processes
Mass and Heat Balances

T0220815

Biological Wastewater Treatment Processes
Mass and Heat Balances

Davide Dionisi

CRC Press
Taylor & Francis Group
Boca Raton London New York

CRC Press is an imprint of the
Taylor & Francis Group, an **informa** business

CRC Press
Taylor & Francis Group
6000 Broken Sound Parkway NW, Suite 300
Boca Raton, FL 33487-2742

First issued in paperback 2020

© 2017 by Taylor & Francis Group, LLC
CRC Press is an imprint of Taylor & Francis Group, an Informa business

No claim to original U.S. Government works

ISBN 13: 978-0-367-57393-5 (pbk)
ISBN 13: 978-1-4822-2926-4 (hbk)

Library of Congress Cataloging-in-Publication Data

Names: Dionisi, Davide, author.
Title: Biological wastewater treatment processes : mass and heat balances / Davide Dionisi.
Description: Boca Raton : CRC Press, 2017.
Identifiers: LCCN 2016030054 | ISBN 9781482229264 (hardback)
Subjects: LCSH: Sewage--Purification--Biological treatment--Mathematical models.
Classification: LCC TD755 .D567 2017 | DDC 628.3/5--dc23
LC record available at https://lccn.loc.gov/2016030054

Visit the Taylor & Francis Web site at
http://www.taylorandfrancis.com

and the CRC Press Web site at
http://www.crcpress.com

To Luca and Valerio

Contents

Preface

WHY ANOTHER BOOK ON biological wastewater treatment, considering that dozens of very good ones are already available on the market? I am afraid I have to tell my readers that even if you read all of this book, it will not exempt you from reading other good books published on this topic. However, I believe the opposite is also true: Even though you manage to go through, for example, the almost 1000 pages of the excellent resource *Biological Wastewater Treatment* (Grady et al., CRC Press, 3rd ed.), you will still find value in reading this book. Here I do not aim to provide the same comprehensive description of all the types of biological wastewater treatment given by the book by Grady et al. or by other resources. My aim with this book is to show how the principles of reaction stoichiometry and kinetics, of mass, and of heat balances can be used to study, design and optimise biological processes for wastewater treatment. In the end, the overall aim of this book is to show how the application of chemical engineering principles can give significant benefits to this sector, where they are not always applied.

In this book, the design of biological wastewater treatment processes is carried out by using mass, and, when they are relevant, heat balances, using relatively simple models for reaction kinetics and stoichiometry. In the end, the steady state mass balances give a system of equations that, having chosen some values for the design parameters, can be solved to calculate all the variables that characterise the system. A possible obvious criticism to this approach is that the processes that occur in biological wastewater treatment processes are very complex, that the organic substrate is composed of many different substances and that many species of microorganisms are involved. Therefore, it is very difficult to build accurate kinetic models of these processes, let alone to find reliable values for the kinetic parameters. However, my reply to these possible comments is that the value of reaction kinetics and of mass and heat balances is not to give accurate predictions of the variable values. In this book, you will find hundreds of design calculations, and I am not afraid to say that nobody should trust the numerical values obtained. For example, in one of the examples, we calculate that in the activated sludge process for a solid's residence time of five days, the effluent concentration of biodegradable COD is 24.98 mg COD/L. Do we give any importance to this particular value or do we have any trust in its accuracy? Of course not. The processes occurring in biological wastewater treatment are too complex for us to generate models that give an accurate quantitative prediction of their behaviour. Yet, in spite of these considerations, I believe that even simple mass balances and kinetic models, such as the ones used in this book, are of tremendous importance in the study of biological processes. The reason is that these models help us understand which are the design parameters that affect the performance of these processes and allow us to calculate the trends (not the accurate values) of the process variables as a function of the design variables. For example, the approach used in this book allows us to understand what is the effect of the solid's residence time on the treatment efficiency, what is the effect of the alkalinity of the feed on the process pH, what is the effect of the internal recycle on nitrogen removal and so on. Mathematical modelling is not (only) a way to fit a curve to some experimental data, but it is a way to understand biological wastewater treatment processes and to design them so to satisfy our desire of an efficient and sustainable performance.

For the reasons just mentioned, what this book aims to give you is the methodology to perform mass and heat balances, not the numerical results. This is the reason I put plenty of focus in showing all the steps in the derivation of the equations used in this book. It will be the responsibility of the reader to use this approach with the kinetic models, mass and heat balances that are appropriate for their particular systems.

This book can be read by different audiences at different levels. It can be useful for undergraduate students or postgraduate students in courses that include biological wastewater treatment. I teach some parts of Chapters 1 through 5 as part of the course 'Air and Water Pollution Control' for the MEng undergraduate programme in Chemical Engineering at University of Aberdeen, Scotland, United Kingdom and some parts of Chapters 1 and 5 as part of the course 'Energy from Biomass' for the MSc programme in Renewable Energy Engineering at the same institution. It can be useful for specialised courses on this subject or for PhD students, but it can also be useful for plant operators who might not be interested in the maths but are interested in understanding the effect of the design parameters on the performance of the plant.

Chapter 1 presents a general overview of biological wastewater treatment processes, describing the main concepts that will be at the basis of the mathematical modelling described and used in the subsequent chapters. Chapter 2 presents the modelling fundamentals about reaction stoichiometry and kinetics, mass transfer and other phenomena relevant in biological wastewater treatment. Chapter 3 presents the general methodology to write mass and heat balances and describes some simple experiments that can be used for parameter estimation. Chapter 4 applies the concepts of reaction kinetics and mass balances, introduced in Chapters 2 and 3, to the activated sludge process, and Chapter 5 does the same for anaerobic digestion processes. Chapters 6 and 7 apply the same concepts to the study of sequencing batch reactors and of attached growth processes.

I thank the many people who, directly or indirectly, have contributed to my professional formation on biological wastewater treatment and have therefore made this book possible. Naming them all here is not possible, so I will mention just very few of them. First of all I wish to thank my main mentor and teacher, Prof. Mauro Majone of Sapienza University, Rome, Italy, from whom I have learnt most of what I know today on this subject, and Prof. Mario Beccari for all his support during and after my PhD thesis. I have also been greatly influenced by Prof. Maria Cristina Annesini, who supervised my undergraduate thesis in chemical engineering at Sapienza University, and therefore I am sincerely indebted to her. If I look at the approach I use in Chapter 2 to write the stoichiometry of chemical reactions, I realise that I have not learnt it at the university, but earlier on when I was at secondary school, thanks to the brilliant chemistry classes of Prof. Anna Maria Murchio, of Liceo Classico Norberto Turriziani, Frosinone, Italy. Therefore I wish to thank her sincerely. Finally I wish to thank all my students of the courses of Advanced Chemical Engineering (academic year 2013/14), Air and Water Pollution Control and Renewable Energy 2 (Biomass) (academic year 2015/16) who, maybe without even knowing it, have helped me with the revision of this book having the patience to study on the draft chapters. I also wish to give a big thank to my PhD students Igor Silva, Ifeoluwa Bolaji, Chinedu Casmir Etteh, Chukwuemeka Uzukwu, and Adamu Rasheed who have helped me with the revision of this book. In particular, I am greatly indebted to Adamu for all his help in the final formatting of the manuscript.

Finally I am most grateful to my wife Federica for all her patience, encouragement and support and to my parents, and I wish to finish with a quotation from my father: 'Some days you feel that you have given all that is possible to give. In reality, you have not done even half of what you could have done.'

About the Author

Dr. Davide Dionisi is a Senior Lecturer in Chemical Engineering at University of Aberdeen, Scotland, United Kingdom, where he teaches wastewater treatment, biochemical engineering and renewable energy from biomass in undergraduate and postgraduate programmes. He has more than 15 years experience in biological wastewater treatment and process modelling and simulation. Dr. Dionisi obtained his degree in Chemical Engineering at Sapienza University, Rome, Italy, where he also obtained his PhD. Before starting his current position at University of Aberdeen, Dr. Dionisi was a research associate and then a lecturer at the School of Mathematical, Physical and Natural Sciences at Sapienza University and worked for the agrochemical company Syngenta as process engineer and principal process engineer in the Process Studies Group. Dr. Dionisi is the author of more than 80 contributions to international journals and conferences, a member of the AIDIC (Italian Association of Chemical Engineering) and a chartered member of the IChemE and the UK Engineering Council.

CHAPTER 1

Biological Wastewater Treatment Processes

1.1 POLLUTING PARAMETERS IN WASTEWATERS

Raw wastewaters from urban or industrial discharges may contain many substances which can cause pollution to the environment and can cause in the end very negative effects on human life and health. Therefore, wastewaters cannot in general be discharged to the receiving water body, which may be a river, lake or sea, unless they have been treated to remove the polluting substances or to reduce the concentration of these substances below some safe levels (Figure 1.1).

In general, the main parameters or substances which need to be removed from wastewaters are suspended solids, soluble organic matter, heavy metals, toxic organic chemicals, nitrogen and phosphorus.

Suspended solids are usually measured as total suspended solids (TSS) and volatile suspended solids (VSS). The TSS are the total solids which are deposited on a filter of specified pore size (typically around 1 μm) and are composed of both organic and inorganic solids. The VSS are the fraction of the TSS which volatilise at a specified temperature, typically 550°C. The VSS are in general considered to represent the fraction of the TSS which is organic in nature. The difference between TSS and VSS is called fixed suspended solids (FSS), which is considered the inorganic fraction of the TSS. As an example, starch, cellulose or microorganisms contribute to the VSS, whereas calcium carbonate ($CaCO_3$) is part of the FSS.

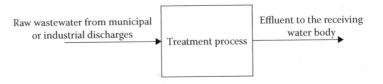

FIGURE 1.1   Wastewaters and treatment processes.

Soluble organic matter has to be removed from wastewaters because if large amounts of organic matter are discharged into water bodies they will cause growth of microorganisms. Microorganisms' growth will have a negative effect on aquatic life since it will cause a depletion in dissolved oxygen in water causing death of many aquatic species. Also, uncontrolled growth of microorganisms in water bodies may originate pathogen bacteria which may spread diseases among fish, animals and humans. Soluble organic matter is composed of many different organic species which cannot be measured individually. Therefore, concentration of soluble organic matter in wastewaters is generally expressed via two lump parameters, the biochemical oxygen demand (BOD) and the chemical oxygen demand (COD). The BOD measures the amount of oxygen that an inoculum of microorganisms consumes when is in contact with the considered wastewater for a prescribed period of time. This time is often taken as five days, and the BOD measured with five-day contact time between microorganisms and wastewater is referred to as BOD_5. So the BOD of a wastewater is considered an indicator of the biodegradable organic matter in that wastewater. The COD measures the amount of oxygen that is necessary for the complete oxidation of all the organic matter contained in the sample. The COD test is carried out heating a sample of the considered wastewater in the presence of a strong acid and of a strong chemical oxidant, typically potassium dichromate ($K_2Cr_2O_7$). The ratio BOD:COD is taken as an indicator of the biodegradability of the organic matter contained in the wastewater. The higher the BOD:COD ratio, the higher is the biodegradability of the organic matter. This is correct; however, it is only an approximate indication. Indeed, even for a completely biodegradable substance, for example, glucose, the BOD will be lower than the COD because in the BOD test part of the substance is assimilated into new microorganisms and therefore is not oxidised by oxygen, even though it is biodegraded.

Heavy metals such as chromium, copper, palladium and nickel are usually not present at harmful concentrations in urban wastewaters, although they may be present in industrial wastewaters, depending on the nature of the industrial discharge. Heavy metals are toxic, above certain concentrations, to living organisms and, if present in the wastewater above certain limits, they must be removed before discharge into the receiving water body.

Toxic organic chemicals are substances which can be toxic or harmful to living organisms even when present in wastewaters at very low concentrations. Many different categories of chemicals can fit into this definition, for example, pharmaceuticals, detergents, polycyclic aromatic hydrocarbons (PAHs), polychlorinated biphenyls (PCBs), and others. These substance are also called xenobiotics, micropollutants or priority pollutants. They can be present in municipal wastewaters (e.g. pharmaceuticals and detergents) and in industrial wastewaters (e.g. PAHs and PCBs).

Nitrogen and phosphorus compounds can be present in municipal and industrial wastewaters. Nitrogen is often present as ammonia (NH_3 or NH_4^+) or as organic nitrogen, for example, in amino acids. Phosphorus is mainly present as phosphates (PO_4^{3-}). In municipal wastewaters, nitrogen is mainly present as a product of human metabolism, whereas phosphorus may be present due to its presence in laundry liquids, even though more stringent regulations and improved formulations by the detergents industry have greatly decreased the presence of phosphorus in municipal wastewaters.

Table 1.1 summarises the typical concentration of the main polluting parameters in municipal wastewaters.

1.2 COD AND BOD

COD and BOD are the two key parameters used to characterise the organic content of wastewaters and the efficiency of biological wastewater treatment plants.

TABLE 1.1   Typical Range of Polluting Parameters in Raw (before Treatment) Municipal Wastewaters

Parameter	Value
TSS	100–350 mg/l
BOD_5	100–400 mg/l
COD	250–1000 mg/l
Total nitrogen	20–100 mgN/l
Total phosphorus	4–15 mgP/l

Note: TSS: total suspended solids; BOD: biochemical oxygen demand; COD: chemical oxygen demand.

1.2.1 COD

The COD of an organic compound represents the amount of oxygen that is required to oxidise the substance to carbon dioxide and water. For a wastewater, where many organic substances may be present, the COD is the amount of oxygen that is required to oxidise all the organic substances present in the wastewater. Examples of oxidation reactions for some organic species are shown below.

$$C_6H_{12}O_6 \text{(glucose)} + 6O_2 \rightarrow 6CO_2 + 6H_2O \qquad (1.1)$$

From this stoichiometry it can be calculated that 1 g of glucose corresponds to 1.067 g of COD.

$$CH_4 \text{(methane)} + 2O_2 \rightarrow CO_2 + 2H_2O \qquad (1.2)$$

1 g of methane corresponds to 4 g of COD.

$$C_5H_9O_4N \text{(glutamic acid)} + 4.5O_2 \rightarrow 5CO_2 + 3H_2O + NH_3 \qquad (1.3)$$

1 g of glutamic acid corresponds to 0.98 g of COD. Note that in the COD reaction nitrogen is not oxidised and remains at the oxidation state at which is present in the organic compound, in this case NH_3.

The COD can be calculated, and measured, not only for soluble species but also for suspended solids. For example, the following is the COD reaction for microorganisms, which can be represented by the empirical formula $C_5H_7O_2N$.

$$C_5H_7O_2N \text{(microrganisms)} + 5O_2 \rightarrow 5CO_2 + 2H_2O + NH_3 \qquad (1.4)$$

1 g of microorganisms corresponds to 1.42 g of COD.

In practice measurement of COD does not involve oxygen, but it is usually done by mixing the wastewater sample with a hot sulphuric acid solution, containing potassium dichromate ($K_2Cr_2O_7$). Dichromate is the oxidant that oxidises the organic matter. The sulphuric acid solution also usually contains silver sulphate as catalyst. The COD measurement involves measurement of the amount of consumed dichromate (or of the amount of dichromate which is left unreacted at the end of the test) from which the COD of the sample can be calculated taking into account the stoichiometry of the reduction reactions of dichromate and oxygen:

$$K_2Cr_2O_7 + 6e^- + 6H_3O^+ \rightarrow Cr_2O_3 + 2KOH + 8H_2O \qquad (1.5)$$

$$O_2 + 4e^- + 4H_3O^+ \rightarrow 6H_2O \tag{1.6}$$

Therefore, 1 mol of dichromate consumed corresponds to 1.5 mol of oxygen, that is, of COD.

Example 1.1

A COD test is performed on a wastewater. In the test tube, there are 2 ml of a 4-mM solution of potassium dichromate $(K_2Cr_2O_7)$ and 2 ml of a wastewater sample are added. At the end of the test, the residual concentration of dichromate is 0.2 mM. Calculate the COD of the wastewater (as mg COD/l).

Solution

According to the reaction stoichiometries (1.5) and (1.6) 1 mol of dichromate consumed correspond to 1.5 mol of COD. The initial dichromate solution is present in a 2-ml volume, while the volume at the end of the test is 4 ml; therefore, the amount of dichromate consumed is:

$$4 \cdot 10^{-3} \frac{mol}{l} \cdot 2 \cdot 10^{-3} l - 0.2 \cdot 10^{-3} \frac{mol}{l} \cdot 4 \cdot 10^{-3} l = 7.2 \cdot 10^{-6} mol$$

This corresponds to $10.8 \cdot 10^{-6}$ mol COD. This COD is present in 2 ml of wastewater, so the COD concentration of the wastewater is:

$$\frac{10.8 \cdot 10^{-6} \, mol\,COD}{2 \cdot 10^{-3} l} = 5.4 \cdot 10^{-3} \frac{mol\,COD}{l} = 172.8 \frac{mg\,COD}{l}$$

1.2.2 BOD

The BOD is the amount of oxygen that microorganisms require to grow on a certain organic compound or on the organic species contained in a certain wastewater. BOD is measured in bottles containing the wastewater under consideration and a small inoculum of microorganisms. Clearly, the value of the BOD depends on the length of the test. The typical profile of oxygen consumed by microorganisms as a function of time during a BOD test is shown in Figure 1.2. Initially, microorganisms might need an adaptation (lag) phase before starting consuming oxygen.

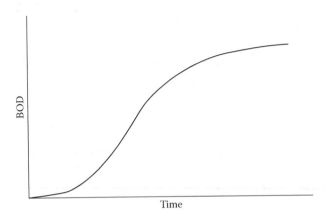

FIGURE 1.2   Typical profile of the time profile of BOD versus time.

This happens when the microorganisms are not acclimated to the waste-water before the start of the test. After the initial acclimation, BOD increases rapidly, because microorganisms are actively growing on the organic substrates and removing oxygen. When all the biodegradable substances in the wastewater have been removed, the BOD curves plateaus out to the final value. Often, BOD measurement is taken after a conventional length of the test of five days, in which cases the BOD values are referred to as BOD_5.

For a given wastewater containing organic substances, the BOD is always lower than the COD for two reasons:

- Some organic substances might not be biodegradable, at least under the conditions (temperature, pH, type of inoculum, etc.) used in the BOD test. Organic substances which are not biodegradable contribute to the COD but not to the BOD.

- Even for totally biodegradable substances the BOD will be lower than the COD, because the COD is proportional to all the electrons that an organic compound can donate to oxygen, whereas the BOD measures only the electrons that have actually been donated to oxygen during microbial growth. The difference is that during microbial growth some of the electrons are not donated to oxygen but are used to form new microorganisms. This will be discussed in the next section on the COD balance.

1.2.3 The COD Balance

The COD of an organic species is proportional to the maximum number of electrons that can be removed from that substance. For example, for the oxidation of acetic acid we have:

$$CH_3COOH + 2O_2 \rightarrow 2CO_2 + 2H_2O \qquad (1.7)$$

In this reaction, oxygen removes electrons from glucose, according to the half-reaction shown previously, Equation 1.6:

$$O_2 + 4e^- + 4H_3O^+ \rightarrow 6H_2O \qquad (1.6)$$

Therefore, since the oxidation of acetic acid requires 2 mol of oxygen and each mol of oxygen removes 4 mol of electrons, the COD stoichiometry (1.7) indicates that 8 mol of electrons can be removed from acetic acid. Therefore, the COD of a substance, that is, the amount of oxygen that is required to oxidise it, is proportional to the number of electrons that can be removed from that substance.

Now, let us consider a biological reaction where microorganisms grow on a certain carbon source (substrate) under aerobic conditions. Under these conditions, as will be explained in Section 1.4 and in Chapter 2, microorganisms use the carbon source to produce other microorganisms and oxidise part of it using oxygen to obtain the energy necessary for growth. Therefore, in such biological process, the 'removable' electrons of the substrate, which are represented by the COD, have two possible destinations: they can be removed by oxygen or they can be still present in the microorganisms which have been produced using that carbon source.

This concept is the basis for the COD balance, which can be written as follows:

$$\text{Total 'removable' electrons present in the removed}$$
$$\text{substrate} = \text{'removable' electrons present in the produced} \qquad (1.8)$$
$$\text{microorganisms} + \text{electrons removed by oxygen}$$

Note that if any inert products or metabolic intermediates are generated during the biological process, the 'removable' electrons contained in them have also to be accounted for in the right hand side of Equation 1.8.

In symbols:

$$\left(-\Delta S_{COD}\right) = \Delta X_{COD} + \left(-\Delta O_2\right) \qquad (1.9)$$

where:

$(-\Delta S_{COD})$ is the substrate removed (as COD) (in concentration units, e.g. kg COD/m³). This term gives the total removable electrons in the removed substrate.

ΔX_{COD} is the microorganisms produced (as COD) (kg COD/m³ or other concentration units). This term gives the removable electrons present in the produced microorganisms.

$(-\Delta O_2)$ is the oxygen consumed (kg O_2/m³ or other concentration units). This term gives the electrons that have been removed from the system.

The COD balance, expressed by Equation 1.9, is the reason why the BOD is always lower than the COD even for totally biodegradable substrates. The COD gives the total number of removable electrons contained in the substrate, whereas the BOD measures only the electrons that are actually removed from the system. In other words, the COD gives the term $(-\Delta S_{COD})$, whereas the BOD gives the term $(-\Delta O_2)$ in Equation 1.9, and therefore the COD is always higher than the BOD, because the production of microorganisms ΔX_{COD} is not accounted for in the BOD.

Note that the COD balance is always valid and always applicable in biological processes, even though the process might not involve oxygen at all, for example, under anaerobic conditions or when a different electron acceptor is present (e.g. nitrate or sulphate).

For example, if nitrate is the electron acceptor instead of oxygen the COD balance can be expressed as:

$$\text{Total 'removable' electrons present in the removed}$$
$$\text{substrate} = \text{'removable' electrons present in the produced} \quad (1.10)$$
$$\text{microorganisms} + \text{electrons removed by nitrate}$$

To use the COD balance with nitrate as electron acceptor, Equation 1.10, we need to express the equivalence between nitrate and oxygen as electron acceptors. This can be obtained by writing and comparing the reduction reactions for oxygen and nitrate (assuming that nitrate is reduced to molecular nitrogen):

$$O_2 + 4e^- + 4H_3O^+ \rightarrow 6H_2O \quad (1.6)$$

$$2HNO_3 + 10e^- + 10H_3O^+ \rightarrow N_2 + 16H_2O \quad (1.11)$$

By making equal the number of electrons accepted in the two reactions (1.6) and (1.11), we see that 1 g of HNO_3 (as nitrogen) corresponds to 2.86 g of O_2. In other words, when 1 g of nitrate (as nitrogen) is reduced to molecular nitrogen, it has removed from the substrate the same number of electrons as 2.86 g oxygen. This allows us to write COD balance with nitrate as electron acceptor as

$$(-\Delta S_{COD}) = \Delta X_{COD} + (-\Delta NO_3) \cdot 2.86 \qquad (1.12)$$

where $(-\Delta NO_3)$ is the nitrate consumed (as nitrogen in concentration units, e.g. kg $N-NO_3/m^3$).

Under anaerobic fermentative conditions there is no electron acceptor and the COD balance can be written as

$$
\begin{aligned}
&\text{Total 'removable' electrons present in the removed} \\
&\text{substrate} = \text{'removable' electrons present in the produced} \qquad (1.13)\\
&\text{microorganisms} + \text{'removable' electrons present in the products}
\end{aligned}
$$

which corresponds to

$$(-\Delta S_{COD}) = \Delta X_{COD} + \Delta P_{COD} \qquad (1.14)$$

where ΔP_{COD} is the total COD of the products (in the liquid and gas phase) in concentration units, where the concentrations are referred to the liquid phase.

Note that the various forms of the COD balance, Equations 1.9, 1.12 and 1.14, express all the same concept that the removable electrons are conserved in the system and are all based on the fact that carbon dioxide, which is a common product in most biological reactions, has no removable electrons (carbon dioxide cannot be oxidised further) and therefore does not contribute to the COD balance.

The COD balance will be further discussed in Chapter 2 and used extensively in all the following chapters.

Example 1.2

The BOD of a wastewater is being measured. The wastewater has a COD of 300 mg/l. At the start of the BOD test, an inoculum of 10 mg/l of microorganisms ($C_5H_7O_2N$) is added to the BOD bottle (assume that the volume of microorganisms and mineral salts added is negligible compared to the volume of wastewater). A value of BOD_5 of 180 mg/l is obtained. Which is the residual COD in the BOD bottle at the end of the test? What could be the nature of this residual COD?

Solution

At the beginning of the test, the total COD comes from the waste-water (300 mg COD/l) and from the inoculum. As shown in previous sections, the conversion factor of microorganisms into COD is 1.42 mg COD/mg microorganism, so the COD due to the inoculum is approx. 14 mg COD/l. Therefore, the total COD at the start of the test is 314 mg COD/l.

During the test 180 mg/l of oxygen are consumed so, from the COD balance, the residual COD at the end of the test is 134 mg COD/l. There can be several contributions to the residual COD:

- Microorganisms. These will be the microorganisms in the initial inoculum, plus the microorganisms generated during the test. This contribution will certainly be present, since microbial growth inevitably generates new microorganisms.
- Non-biodegradable COD present in the wastewater sample. This contribution might or might not be present in this case, and this test does not necessarily indicate that a fraction of the COD in the sample is not biodegradable.
- Intermediate metabolic products which are not further biodegradable or inert products generated by microorganisms decay. This contribution might or might not be present.

Example 1.3

Sometimes, nitrate is not reduced to molecular nitrogen but to nitrite (NO_2). Calculate the equivalence factor between nitrate removed and oxygen consumed if nitrate is reduced to nitrite.

Solution

If nitrate is reduced to nitrite, instead than to molecular nitrogen, the reduction reaction for nitrate is:

$$HNO_3 + 2e^- + 2H_3O^+ \rightarrow HNO_2 + 3H_2O \qquad (1.15)$$

Comparing with the reduction reaction for oxygen, Equation 1.6, we see that 1 g of nitrate (as nitrogen) corresponds to 1.14 g of oxygen, that is, when 1 g of nitrate (as nitrogen) is reduced to nitrite it removes from the substrate the same number of electrons as 1.14 g of oxygen.

Note that if nitrite is used as electron acceptor and is converted to molecular nitrogen it corresponds to a certain amount of oxygen removed, and the equivalence factor between nitrite and oxygen can be calculated in the same way by writing the reduction reaction of nitrite to molecular nitrogen:

$$2HNO_2 + 6e^- + 6H_3O^+ \rightarrow N_2 + 10H_2O \tag{1.16}$$

Comparing again with the reduction reaction for oxygen, Equation 1.6, we see that 1 g of nitrite (as nitrogen) corresponds to 1.71 g of oxygen, that is, when 1 g of nitrite (as nitrogen) is reduced to molecular nitrogen it removes from the substrate the same number of electrons as 1.71 g of oxygen.

Note that, as expected, the conversion factor of nitrate reduction to molecular nitrogen can also be obtained as the sum of the conversion factors for nitrate reduction to nitrite and for nitrite reduction to molecular nitrogen (any small difference is due to rounding).

Example 1.4

In an anaerobic process (no electron acceptor) a substrate is present at an initial concentration of 1.5 g COD/l and it is totally removed from the medium. We can assume that the products in the liquid phase contain only 0.1 g/l of microorganisms ($C_5H_7O_2N$) and 0.1 g/l of acetic acid (CH_3COOH). The produced gas is made only of methane and carbon dioxide. How much methane is produced in this process, per unit volume of the liquid phase?

Solution

The COD balance under anaerobic conditions is given by Equation 1.14. In this case

$$(-\Delta S_{COD}) = 1.5 \frac{gCOD}{l}$$

$$\Delta X_{COD} = 0.1 \cdot 1.42 \frac{gCOD}{l} = 0.14 \frac{gCOD}{l}$$

where 1.42 is the conversion factor between microorganisms and COD, according to Equation 1.4.

The total COD of the products is therefore, according to Equation 1.14:

$$\Delta P_{COD} = 1.5 - 0.14 = 1.36 \frac{gCOD}{l}$$

The products are acetic acid, methane and carbon dioxide, but carbon dioxide does not contribute to the COD balance (it contains no removable electrons). From the oxidation reaction of acetic acid, it can be calculated that the COD conversion factor for this species is 1.067 g COD/g acetic acid. Therefore, 0.1 g/l of acetic acid corresponds to 0.107 g COD/l. It follows that the COD of the produced methane is equal to 1.36 – 0.107 = 1.253 g COD/l, where the concentration is referred to the liquid phase (and not to the gas phase). From the oxidation reaction of methane we calculate that the conversion factor for methane into COD is 4 g COD/g methane. Therefore, the concentration of produced methane is 1.253/4 = 0.31 g methane/l (concentration referred to the liquid phase).

1.3 THE ROLE OF BIOLOGICAL PROCESSES IN WASTEWATER TREATMENT

Broadly speaking, municipal wastewaters usually have similar compositions, and therefore, the sequence of treatment used for them is often very similar. A wastewater treatment plant for municipal wastewaters typically consists of a sequence of three types of treatment (Figure 1.3): primary, secondary, and tertiary treatment. On the other hand, the nature and composition of industrial wastewaters is typically very dependent on the type of factory (e.g. chemical, energy, food and drink), and therefore, the sequence of treatments used is typically site specific.

The aim of the primary treatments is to remove most of the suspended solids contained in the raw wastewaters. Primary treatment may consist of screening, degritting and sedimentation, called primary sedimentation to

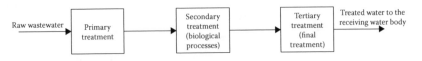

FIGURE 1.3 Typical treatment sequence for a municipal wastewater.

distinguish from secondary sedimentation processes occurring later in the treatment. After primary treatments, most of the suspended solids have been removed and the wastewater is sent to secondary treatments which are aimed to remove the soluble organic matter. Secondary treatments are typically biological processes which remove the biodegradable organic matter. Finally, the effluent from secondary treatment is sent to the final treatment processes (tertiary treatments) which may include disinfection and other processes aimed at removing any residual microorganisms which have not been separated at the end of the secondary treatment processes.

This book will only focus on secondary treatment, that is, on biological processes, which will be introduced in the rest of this chapter.

1.4 MICROORGANISMS' GROWTH ON SUBSTRATES

In biological processes microorganisms grow on the biodegradable matter contained in the wastewaters. Since microorganisms, like all living organisms, are mainly composed of carbon, hydrogen, oxygen and nitrogen, microbial growth removes these elements from the wastewater. In addition, microbial growth requires many other substances, such as phosphorus, metals and many mineral elements. The reactions involved in microbial growth are oxidation–reduction reactions. In some cases, microorganisms use an external oxidant (electron acceptor), typically oxygen (aerobic processes) or nitrate (anoxic processes). Sulphate may also be used as electron acceptor if present in the absence of oxygen or nitrate. In other cases, there is no external electron acceptor and the organic substrate itself is used as both oxidant and reducing agent. This is the case for anaerobic fermentation reactions.

A general scheme for microorganisms' growth is shown in Figure 1.4. Microorganisms consume carbon, hydrogen, nitrogen and oxygen sources, plus other nutrients and possibly an electron acceptor and generate new microorganisms and products. The carbon source can be organic or inorganic, that is, carbon dioxide. If they use an organic carbon source, microorganisms are called heterotrophs, and if they use an inorganic carbon source, they are called autotrophs. In general, if microorganisms use oxygen as external electron acceptor the only product of their metabolism are, in addition to new microorganisms, carbon dioxide and water. There are exceptions to this, however, when microorganisms remove one substance to generate some intermediates which might not be further biodegradable, this sometimes happens with the metabolism of synthetic chemicals (xenobiotics). If microorganisms use nitrate as external electron acceptor, then in addition to carbon dioxide and water molecular nitrogen is also produced

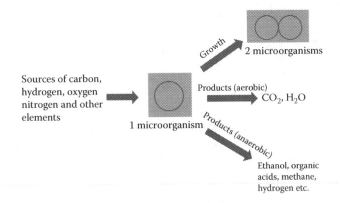

FIGURE 1.4   Conceptual scheme of microbial growth.

from nitrate reduction. When no external electron acceptors are present (anaerobic conditions), the metabolism of organic substances generates organic products, that is, acetic acid, ethanol, lactic acid and many others. Therefore, in general the process of microbial growth can be schematised as follows (components in brackets are not always utilised/produced):

$$\text{Microorganisms} + \text{carbon source} + \text{other elements}$$

$$+ (\text{electron acceptor}) \rightarrow \text{new microorganisms} \qquad (1.17)$$

$$+ (\text{carbon products}) + (CO_2) + (H_2O)$$

In this book, we will focus mainly on the following categories of microorganisms, which are the most important in biological treatment processes.

Oxygen- or nitrate-consuming heterotrophs: These microorganisms use organic carbon as substrate and use oxygen or nitrate as external electron acceptor. They use nitrate only in the absence of oxygen. The growth of heterotrophic microorganisms can be schematised as follows:

$$\text{Microorganisms} + \text{organic carbon source}$$

$$+ \text{other elements} + O_2 \ (\text{or } NO_3^-) \rightarrow \qquad (1.18)$$

$$\text{new microorganisms} + CO_2 + H_2O + (N_2)$$

Ammonia-consuming autotrophs (nitrifiers): These microorganisms use CO_2 as carbon source for growth and oxidise ammonia to nitrate obtain energy. Their metabolism can be schematised as follows:

$$\text{Microorganisms} + CO_2 + \text{other elements} + NH_3 + O_2 \rightarrow$$
$$\text{new microorganisms} + NO_3^- + H_2O \tag{1.19}$$

Fermentative: These microorganisms are active under anaerobic conditions and use organic carbon as carbon source without external electron acceptor. Depending on the particular organic substrate and microorganism species, they produce various organic products and might produce or utilise hydrogen. Carbon dioxide is usually produced but sometimes is utilised. Their metabolism can be schematised as follows:

$$\text{Microorganisms} + \text{organic carbon} + \text{other elements} + (H_2) \rightarrow$$
$$\text{new microorganisms} + \text{organic products} + CO_2 + H_2O + (H_2) \tag{1.20}$$

Hydrogen-consuming methanogens (hydrogenotrophic methanogens): These microorganisms are active under anaerobic conditions, use CO_2 as carbon source, oxidise hydrogen for energy needs and produce methane. Their metabolism can be schematised as follows:

$$\text{Microorganisms} + H_2 + CO_2 + \text{other elements} \rightarrow$$
$$\text{new microorganisms} + H_2O + CH_4 \tag{1.21}$$

Acetate-consuming methanogens (acetoclastic methanogens): These microorganisms are active under anaerobic conditions and utilise acetic acid as carbon source, producing methane as product. Their metabolism can be schematised as follows:

$$\text{Microorganisms} + CH_3COOH + \text{other elements} \rightarrow$$
$$\text{new microorganisms} + H_2O + CH_4 + CO_2 \tag{1.22}$$

1.5 OTHER PHENOMENA OCCURRING IN BIOLOGICAL WASTEWATER TREATMENT PROCESSES

1.5.1 Hydrolysis of Slowly Biodegradable Substrates

Microorganisms can only grow on soluble substrates which can pass through the cell membrane. However, wastewaters often contain large fractions of high-molecular weight substances, either soluble or insoluble, which need to be hydrolysed before they can be metabolised. This is the case, for example, for cellulose and starch, which are polymers of glucose. They need to be hydrolysed to glucose outside the cell so that glucose can be transported into the cell and metabolised. Similarly, proteins cannot be metabolised as such but need to be hydrolysed to amino acids, which can then be metabolised. Hydrolysis is usually considered to occur due to extracellular enzymes, either released by the microorganisms on the liquid medium, or attached to the cell membrane. Hydrolysis can therefore be schematised as follows:

$$\text{high molecular weight substances}$$

$$+ \text{water} \xrightarrow{\text{hydrolysis}} \text{low molecular weight substrates}$$

In this book, we will assume, as it is reasonable to do, that the microorganisms which carry out the hydrolysis products are the same microorganisms which utilise the hydrolysis products, that is, heterotrophs under aerobic or anoxic (nitrate used as electron acceptor) conditions and fermentative microorganisms under anaerobic conditions.

1.5.2 Endogenous Metabolism and Maintenance

In addition to growth, other metabolisms are also important in biological wastewater treatment processes.

Endogenous metabolism is the conversion of the active biomass into inert products or carbon dioxide. Endogenous metabolism is assumed to occur both in the presence and in the absence of external substrates. When there are no external substrates present, endogenous metabolism accounts for the fact that microorganisms utilise part of the internal macromolecules, such as enzymes or storage polymers, as energy source, converting them to carbon dioxide or to inert products. When external substrate is present, endogenous metabolism still accounts for the fact that in a mixed culture not all the microorganisms are actively growing on the substrate but some of them will be inactive and decaying.

Maintenance metabolism occurs only in the presence of the external substrate. Maintenance accounts for the fact that not all the substrate is used for growth but part of it is used to maintain basic cellular functions.

Even though maintenance and endogenous metabolism are entirely different phenomena, in reality their effect on biological processes is similar, that is, they contribute to reducing the amount of biomass produced per unit of substrate consumed. Therefore, usually only endogenous metabolism is used in mathematical models for biological wastewater treatment plants, and this approach will be used here.

1.6 ANAEROBIC DIGESTION MODEL

We have seen in Section 1.4 the conceptual scheme of some reactions that occur under anaerobic conditions (Equations 1.20–1.22). Anaerobic digestion is a complex process where different microbial communities interact to the final aim of producing the desired product, methane. A conceptual scheme which is frequently used to describe anaerobic digestion is shown in Figure 1.5. Complex organic substrates are first hydrolysed into their macromolecular constituents, mainly carbohydrates, proteins and lipids (inerts may also be generated in this process, but they are not considered here). Then macromolecules are hydrolysed to their respective monomers, sugars, amino acids and long-chain fatty acids, which are the substrates for anaerobic microorganisms. Many different species of microorganisms are active on these substrates. Considering sugars, certain species convert them to acetic acid and hydrogen, other microorganisms instead generate other products such as other volatile fatty acids, for example, propionic and butyric acid. Similarly, other microorganisms convert amino acids to

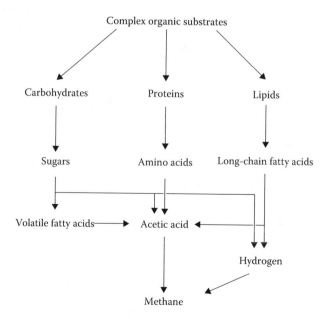

FIGURE 1.5 Scheme of the main processes that occur during anaerobic digestion of biodegradable organic matter.

acetic acid and other volatile fatty acids. Long-chain fatty acids are usually converted to acetic acid and hydrogen. The substrates for methane production are acetic acid and hydrogen.

1.7 PROCESS SCHEMES FOR BIOLOGICAL WASTEWATER TREATMENT

In this section, we will introduce some of the main processes used for biological treatment of wastewaters. The modelling of most of these processes will be presented later in this book.

1.7.1 Activated Sludge Processes

The main process used in biological wastewater treatment is the activated sludge process for carbon removal (Figure 1.6). In this process, heterotrophic microorganisms grow in the biological reactor where they remove the organic carbon sources in the influent wastewater. The biological reactor is aerated to provide the oxygen which is required for microorganisms' growth. In the effluent from the biological reactor, most or all of the biodegradable organic matters contained in the influent wastewater have been removed. However, the effluent from the biological reactor is a solid–liquid mixture which includes microorganisms, which need to be separated before the effluent can be discharged or directed to the final (tertiary treatment). Separation of the microorganisms from the liquid phase is very often achieved by settling in a settling tank, called 'secondary' settling tank to distinguish it from the primary settling tank at the end of primary treatments. A stream with the concentrated microorganisms is collected at the bottom of the settling tank and is recycled back to the biological reactor. This recycle stream has the function of increasing the concentration of microorganisms in the biological reactor. Part of the bottom stream of the settling tank stream (waste sludge) is removed from the system in order to

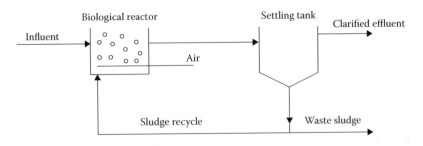

FIGURE 1.6   Activated sludge process for carbon removal.

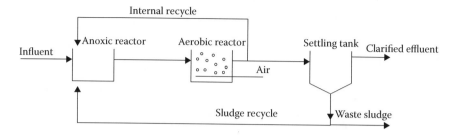

FIGURE 1.7   Activated sludge process for nitrogen removal with pre-denitrification.

control the residence time of the microorganisms in the reactor. The clarified effluent from the top of the settling tank is sent to the final treatments or directly to the receiving water body.

Figure 1.7 shows a typical scheme for the activated sludge process for carbon and nitrogen removal. This scheme is usually referred to as 'pre-denitrification'. The first tank is the anoxic reactor where, in the absence of oxygen, the organic material is removed by heterotrophic microorganisms using nitrate, instead of oxygen, as electron acceptor. In this reactor, nitrate is converted to molecular nitrogen, achieving the aim of nitrogen removal. Typically, most or all of the biodegradable organic matter is removed in the anoxic reactor. The main purpose of the subsequent aerobic reactor is to allow nitrification to occur, that is, the growth of nitrifying microorganisms which convert ammonia to nitrate. The internal recycle stream has the purpose to provide the anoxic reactor with nitrate. Note that nitrate is also provided to the anoxic reactor with the sludge recycle stream, where, assuming ideally no nitrate consumption or production in the settling tank, nitrate concentration is the same as in the internal recycle stream. However, the sludge recycle stream has a higher concentration of microorganisms than the internal recycle, and therefore, its flow rate cannot exceed certain values, otherwise the microorganisms concentration in the reactors will be too high. Therefore, it is not usually possible or wise to regulate the amount of nitrate recycled by adjusting the sludge recycle flow rate. It is usually better to use the internal recycle stream to control the nitrate flow to the anoxic reactor and to use the sludge recycle stream to control the microorganisms' concentration in the reactors.

Another scheme for nitrogen removal using the activated sludge process is the one in Figure 1.8. In this configuration, the influent wastewater is fed to the aerobic reactor where carbon removal and nitrification take

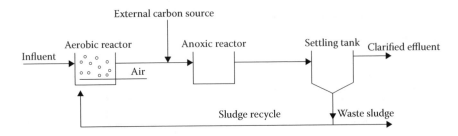

FIGURE 1.8 Activated sludge process for nitrogen removal with the influent wastewater fed to the aerobic reactor.

place. Nitrate conversion to molecular nitrogen occurs in the anoxic reactor; however, most of the organic carbon has been removed in the aerobic reactor, so an external carbon source, often methanol, has to be added to the anoxic reactor. The requirement of an external carbon source is the main disadvantage of this process scheme; however, this scheme gives the benefit of not requiring an internal recycle.

An important contribution to the operating costs of activated sludge processes, and of all aerobic biological wastewater treatment processes, is the energy cost required for aeration. Also, an operating cost can also be associated with the treatment and disposal of the waste sludge. However, often the waste sludge is sent to an anaerobic digestion process where it is used as feedstock to produce methane, and therefore energy, and in this case it may represent a profit rather than a cost. An issue associated with the anaerobic digestion of waste sludge is its relatively slow digestion rate.

To minimise the volume of the biological reactor(s), it is convenient to design the process with a high biomass concentration, so that the reaction rate per unit of reactor volume is maximised and the reactor volume can be minimised, with advantages in terms of capital costs and plant footprint. However, the maximum biomass concentration in the biological reactor is determined by the capacity of the settling tank. Indeed, the settling velocity decreases as the biomass concentration increases, and if the biomass concentration is too high, settling becomes virtually impossible and the activated sludge process will not be able to operate. To overcome this limitation of the conventional activated sludge process, activated sludge processes with membranes have been developed. In these processes, separation is carried out by membranes, rather than using a settling tank. The microorganisms which come out of the reactor are sent to crossflow membranes, which are permeable only to soluble species but

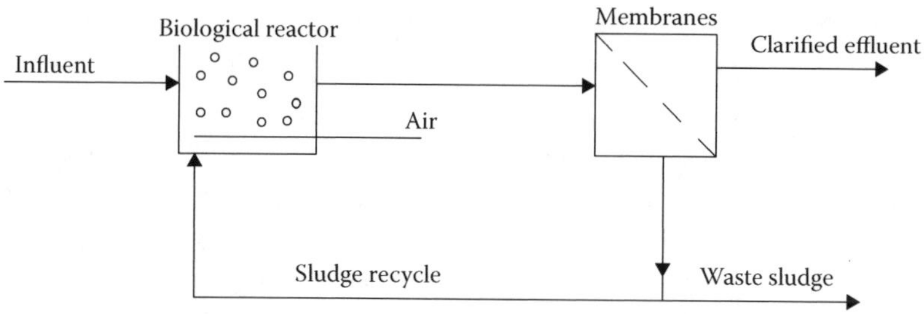

FIGURE 1.9 Scheme of an activated sludge process with membranes for solid–liquid separation.

not to suspended solids such as microorganisms, which are retained in the system and recycled to the reactor. Membranes have also the advantage of generating a higher quality effluent due to the better clarification and the lower presence of residual suspended solids than with settling tanks. The scheme of an activated sludge process with membranes (Figure 1.9) is conceptually the same as the conventional activated sludge process; the only difference is the way the solid–liquid separation is obtained. Instead of being placed outside the reactor, in some cases membranes can also be placed inside it. Membrane separation processes have the disadvantages of high capital and operating costs and of the requirement of frequent cleaning due to fouling. In addition to solid–liquid separation, another limitation on the maximum microorganisms' concentration that it is possible to obtain in aerobic biological processes is related to aeration. Indeed, high biomass concentrations can reduce the mass transfer coefficient for oxygen in water because of the effect on fluid viscosity and potential clogging of the aerators. Therefore, too high biomass concentrations in the biological reactor need to be avoided in any cases, even with solid–liquid separation provided by membranes.

The performance of activated sludge processes, and of most other biological wastewater treatment processes, is usually characterised in terms of the hydraulic residence time (HRT), the solids residence time (SRT) and the organic load rate (OLR). The HRT is a nominal residence time in the biological reactor and is expressed as the ratio between the volume of the reactor and the influent flow rate. It is a nominal residence time because in reality the residence time in the biological reactor is shorter than the HRT because of the recycle flow rate. The SRT, also called sludge age, is the residence time of the microorganisms (and indeed of any settleable or separable solids) in the reactor and is calculated as the ratio between the mass of solids present in the biological reactor and the mass flow rate of solids leaving the process with the waste sludge stream. The SRT is the most important parameter in the design of biological processes, because it represents the average time that microorganisms spend in the system and therefore determines the treatment efficiency. In particular, it is important to observe that activated sludge processes for carbon and nitrogen removal typically require a longer SRT than activated sludge processes for carbon removal only, because nitrifying microorganisms are slow growers, and therefore a longer residence time is required for nitrification than for carbon removal. The OLR is the ratio between the mass flow rate of biodegradable COD in the influent to the process and the volume of the biological reactor. Compatibly with the requirements of solid–liquid separation and of aeration described earlier, it is advantageous to have processes with the highest possible OLR, because this will correspond to the lowest volume of the reactor. In the design chapters on the activated sludge process and other biological processes, we will see that the HRT and the SRT are design parameters which need to be chosen by the process designer, whereas the OLR is an output of the design procedure and it can be calculated from the design results. Table 1.2 gives typical values for the HRT, SRT and OLR for activated sludge processes. The concepts of HRT, SRT and OLR are not limited to the activated sludge process but are applicable to any biological process.

Activated sludge processes are operated in a wide range of temperatures, from less than 5°C to more than 30°C, and can operate in a relatively large range of pH which is typically between 6 and 8.5.

1.7.2 Sequencing Batch Reactor

The sequencing batch reactor (SBR, Figure 1.10) is a suspended growth process which is conceptually the same as the activated sludge process. The only important difference is that in the SBR the process is operated in

TABLE 1.2 Typical Values of Activated Sludge Parameters. Values Outside These Ranges Are Also Common

Parameter	Values
HRT	0.2–2 day
SRT	2–20 day
OLR	0.1–2 kg COD/m³.day

Note: HRT: hydraulic residence time; SRT: solids residence time; OLR: organic load rate; COD: chemical oxygen demand.

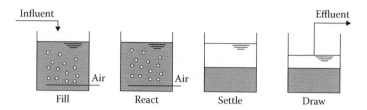

FIGURE 1.10   Sequencing batch reactor for carbon removal.

one single vessel, instead than in two vessels, reactor and settling tank, as in the activated sludge process. The SBR operates as a temporal sequence of phases, rather than as spatial sequence of vessels. A typical cycle of the SBR consists of fill, react, settle and draw. In the fill phase, the influent is fed to the reactor and the volume increases. After feed is completed, the reactor is left aerated, typically for several hours, until most or all of the biodegradable carbon is removed. At the end of the reaction phase, the aeration and mixing are stopped and the microorganisms are allowed to settle. When settling is completed the clarified effluent is removed and the reactor is ready for a new cycle to start.

Similarly as for the activated sludge process, the SBR cycle can also be adapted for nitrogen removal. A typical SBR cycle for nitrogen removal is shown in Figure 1.11. In this case, the fill and the first part of the reaction phase are not aerated. Therefore, in these phases the microorganisms consume the influent organic material using nitrate as electron acceptor. The second part of the reaction phase is aerated, so the microorganisms can oxidise ammonia to nitrate (nitrification), which is removed during the fill and reaction phase of the next cycle.

1.7.3 Attached Growth Processes

The main difference between attached growth (Figure 1.12) and activated sludge processes is that in the former the microorganisms are attached to support materials, instead of being suspended in the reactor mixture.

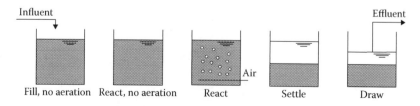

FIGURE 1.11   Sequencing batch reactor for carbon and nitrogen removal.

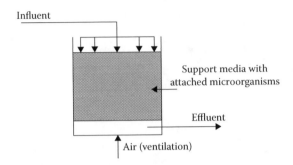

FIGURE 1.12   General scheme for an aerobic attached growth process.

Support materials can be made of many materials, for example, plastic, sand or activated carbon. Compared to activated sludge process, this allows potentially higher retention times of the microorganisms in the system and eliminates the need for a secondary settling tank. The influent wastewater can be fed from the top or from the bottom of the filter, and air can be provided by ventilation. Attached growth typically allows for higher biomass concentrations than suspended growth processes; therefore, they can in principle obtain the same efficiency of treatment with a reduced volume of the reactor. However, since large flocs are typically produced in these systems, they may suffer from diffusion limitation for oxygen and substrate, therefore decreasing their effectiveness.

A particular type of attached growth process is the rotating biological reactors (RBRs, also called rotating disc reactors or with similar names) (Figure 1.13). In RBRs microorganisms are attached to support materials which are placed inside a cylinder which is partially immersed in the wastewater to be treated. The cylinder with the microorganisms rotate so that the microorganisms are alternatively exposed to the wastewater and to air, from which they obtain the oxygen required for the removal of the substrate (RBRs are aerobic processes). RBRs have the advantage, common

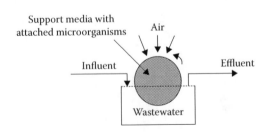

FIGURE 1.13   Scheme of a rotating biological reactor (RBR) process.

with other attached growth processes, of high volumetric reaction rates because of the high biomass concentration; however, for mechanical reasons these units are typically small and are therefore only suited for relatively low flow rates of the wastewater to be treated. Typical dimensions of the discs used for RBRs are 3.5 m in diameter and 8 m in length.

While RBRs are aerobic processes, in general attached growth processes can be used for aerobic or anaerobic processes.

1.7.4 Anaerobic Digestion

Anaerobic digestion is often carried out in single tank processes with no liquid–solid separation. In this case, the reactor can be often assumed to be perfectly mixed. The influent feed can be a concentrated wastewater, which is mainly liquid, or solid waste, for example, food waste, agricultural waste, etc. The biogas generated in the process, which is usually mainly composed of methane and carbon dioxide, leaves from the top of the vessel and collected for energy utilisation. The slurry in the reactor is composed of water and solids, which are made of microorganisms and of any solids in the feed which have not been converted. There are two main reasons why, differently from aerobic processes such as activated sludge, for anaerobic digestion there is often no settling and recirculation of the microorganisms. The first reason is that usually the influent of anaerobic digesters is very concentrated; therefore, it is possible to achieve high microorganisms' concentrations and relatively high volumetric reaction rates even without concentration and recirculation of the biomass. The second reason is that the effluent of anaerobic digesters is not usually to be discharged into a water body. Instead, it is often spread on land as fertiliser or sent to further aerobic biological treatment for removal of the residual COD. Therefore, there is often no need for biomass recirculation.

Anaerobic digestion processes can also be carried out with biomass recycle, analogously to the conventional activated sludge process. Biomass separation can be obtained using settling tanks or, especially if the biomass concentration is very high, using membranes (and the schemes are analogous to the aerobic processes shown in Figures 1.7 and 1.9). Also, anaerobic process can be carried out using attached growth processes, Section 1.7.3, with a scheme similar to the one in Figure 1.12 but without aeration. Attached growth processes are particularly interesting for anaerobic processes, because the biomass concentration can be quite high due to the high concentration of the feed, and therefore solid–liquid separation using sedimentation or even membranes could be problematic. Attached

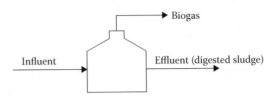

FIGURE 1.14 Anaerobic digestion process with completely mixed tank and no solid–liquid separation.

growth processes have the advantage of allowing a high biomass concentration without the need for solid–liquid separation and therefore are particularly suitable for anaerobic treatment of wastewater with high COD loading (i.e. with high mass of COD to be removed per day) (Figure 1.14).

Another type of anaerobic digestion process is the Upflow Anaerobic Sludge Blanket (UASB) (Figure 1.15). Differently than for the previous case, in this case the solid and the liquid phase are separated before the liquid and the gas exit the reactor. Therefore, in this case the residence time of the liquid and of the solids will not be the same. In the UASB reactor, the feed enters from the bottom and flows upwards. In the reactor, a dense blanket of microorganisms develops, which degrades the organic materials converting it to biogas. At the top of the reactor, appropriate devices retain the solids in the reactor and allow the clarified liquid and the gas to exit. To control the growth of the microorganisms, sludge is removed from the blanket. In the sludge blanket, the microorganisms form granules, which are typically large, dense and readily settleable. The formation of granules and the retention of the microorganisms inside the reactor allow higher concentration of biomass inside the reactor and

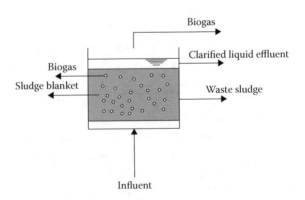

FIGURE 1.15 Upflow anaerobic sludge blanket (UASB) reactor.

therefore higher volumetric reaction rates. However, a main limitation of the UASB is that the mechanism of granules formation is not very well understood, and therefore lab and pilot scale studies need to be performed to evaluate the feasibility of this technology for each particular waste or wastewater.

Compared to aerobic processes, anaerobic processes give the important advantage of generating a valuable stream, biogas, which can be used to generate electricity or can be injected (previous removal of the carbon dioxide) to the gas grid to generate heat. In addition, they also have the important advantage of not requiring energy for aeration. However, they usually have the important disadvantage of a lower degradation rate of the organic substrate, therefore requiring larger vessels (larger values of the HRT and SRT than for aerobic processes) which consequent larger capital cost. Also, it is more difficult to achieve the same effluent quality in terms of COD removal with anaerobic digestion than with aerobic processes. To partially compensate for the slower reaction rates, anaerobic processes are usually operated at a temperature of at least 30°C–35°C, higher than in most activated sludge processes, and this also causes an operating cost for the anaerobic digestion process. Another important limitation of anaerobic processes is the higher sensitivity to pH and to inhibitors than aerobic processes. For efficient methanogenesis, the pH needs to be usually in the range 6.8–8, and the process is sensitive to high concentrations of many inhibiting species, for example, ammonia, organic acids, etc. Therefore, the operation of anaerobic digestion processes can be less straightforward than the operation of aerobic processes.

1.7.5 The SHARON® and Anammox Processes

The Single reactor system for High activity Ammonia Removal Over Nitrite (SHARON®) and Anaerobic ammonium oxidation (Anammox) processes (Figure 1.16) are innovative processes, currently under development but already installed in a number of full scale facilities, aimed at removing nitrogen from wastewaters with high ammonia concentration. Conventional nitrification denitrification processes, such as the

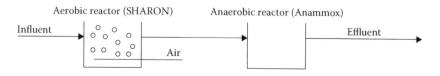

FIGURE 1.16 Scheme of the combined SHARON and Anammox processes for ammonia removal.

ones shown in Figures 1.7 and 1.8, have the drawbacks of high oxygen consumption for nitrification and of the possible requirement for external COD addition. External COD addition is particularly required if the wastewater has a low COD content but high ammonia concentration. In the SHARON process part of the ammonia is oxidised aerobically to nitrite rather than to nitrate (as in conventional nitrification process). Control of the ammonia oxidation to nitrite is achieved by control of the SRT, of the temperature and of the pH. Since only part of the ammonia is oxidised and the oxidation product is nitrite, rather than nitrate, the SHARON process requires much less oxygen than the conventional nitrification process. The outlet of the SHARON reactor contains ammonia and nitrite in approximately equal molar concentrations and is directed to the Anammox reactor. The Anammox process is based on the anaerobic oxidation of ammonia using nitrite (rather than oxygen) as electron acceptor, giving molecular nitrogen as final product. In summary, the process removes almost all the nitrogen from a wastewater with a fraction of the oxygen consumption of the conventional nitrification/denitrification process and with no need of organic carbon. Therefore, this process is particularly suitable for wastewaters of high nitrogen and no or low COD content. The main drawback of the SHARON process is the need of careful control of the reaction conditions to ensure that nitrification is not completed to the end product nitrate (which would make oxygen consumption higher and would prevent the Anammox process). For the Anammox process, the main disadvantage is that only very few species of microorganisms are able to carry out ammonia oxidation using nitrite and therefore requires very careful selection of the inoculum and of the reaction conditions. The Anammox process is also very sensitive to the presence of inhibitors in the wastewater.

1.8 GENERAL ASSUMPTIONS MADE IN THIS BOOK

In this book we will make several assumptions, with the aim of simplifying the notation, the modelling and the mass balances. The main assumptions are listed below:

- When we talk about COD, we always refer to biodegradable COD. In reality, part of the COD of a wastewater is biodegradable, part is not biodegradable, at least under the conditions of biological wastewater treatment processes. However, the non-biodegradable COD can be considered an inert and plays no role in biological wastewater treatment, and therefore, we will not consider it here.

- The liquid streams will always be considered to be dilute solutions. This assumption has two consequences: the physical properties, when they are needed, will be assumed to be the ones of pure water; we will assume that the flow rate of the liquid stream does not change because of the biological process. This means that in all cases the liquid flow rate in and out of the processes will be assumed to be the same value. This assumption of constant liquid flow rate will be used also for anaerobic digestion, when a gas flow rate (biogas) leaves the system, therefore obviously reducing the liquid flow rate. However, this reduction in the liquid flow rate will be ignored, because it is usually small. The assumption of constant liquid flow rate also means that we will ignore the amount of water generated by the fermentation reactions (metabolic water), There is only one case where we will consider the metabolic water, and this is when we will do heat balances. The reason is that in heat balances it is important to consider the heat of reaction, and this cannot be estimated with enough accuracy if all the reaction products, including the metabolic water, are considered.

- Unless stated otherwise, we will assume the processes to be at atmospheric pressure.

- In pH calculations, we will ignore any possible precipitation reactions.

1.9 KEY POINTS

- Biological wastewater treatment is necessary to remove the biodegradable organic matter from wastewaters. Without treatment the organic matter would end up in the receiving water bodies, where microorganisms would grow uncontrollably, causing death of many aquatic organisms (e.g. fish) and spreading diseases.

- COD is a measure of the total concentration of organic matter in a wastewater (or water in general). COD is also proportional to the total number of 'removable' electrons in the organic matter, and therefore, it is extremely useful in mass balances in biological wastewater treatment processes.

- BOD is the oxygen consumed by microorganisms when they remove the organic matter in a certain wastewater. BOD is always lower than the COD for two reasons: (a) some of the organic matter in the wastewater may not be biodegradable and (b) BOD only measures the oxygen consumption associated to the degradation of the substrate, while in reality part of the substrate is converted to new microorganisms.

- In general, microorganisms grow on organic matter by producing new microorganisms and generating products. Under aerobic conditions, the products are usually (at least if the substrate is completely metabolised) carbon dioxide and water. Under anaerobic conditions, many products are possible, for example, organic acids, hydrogen, methane and carbon dioxide.

- Many different types of biological wastewater treatment processes exist. They can be categorised in many ways, the most important distinction is between aerobic and anaerobic processes. Anaerobic processes have the advantages of generating a useful product, methane and of not requiring energy input for aeration. However, they have the drawbacks of slower reaction rate, which requires larger vessels, and of a lower treatment efficiency. Therefore, anaerobic processes are usually only preferred over aerobic processes if the influent COD loading is high enough to guarantee a high methane production rate

Questions and Problems

Atomic weights: $C = 12$; $H = 1$; $N = 14$; $O = 16$; $Cr = 52$; $K = 39$. Assume in all cases that microorganisms have the empirical formula $C_5H_7O_2N$.

1.1 Calculate the COD conversion factor (g COD/g substance) for the following species:

a) Propionic acid (CH_3CH_2COOH);

b) Benzene (C_6H_6);

c) Oleic acid ($C_{18}H_{34}O_2$);

d) Xylose ($C_5H_{10}O_5$);

e) Alanine ($C_3H_7O_2N$)

1.2 In a BOD test, the initial COD of the wastewater, after inoculation in the BOD bottle, is 150 mg COD/l. At the end of the test, the residual COD of the wastewater is 20 mg COD/l and 50 mg/l of microorganisms have been produced. Which is the BOD of the sample?

1.3 In a BOD experiment 30 mg/l of microorganisms are produced, and 60 mg/l of oxygen are consumed. The residual COD in the liquid phase at the end of the experiment is 10 mg COD/l. Which is the initial COD at the start of the BOD test?

1.4 In a COD experiment, initially potassium dichromate $(K_2Cr_2O_7)$ is present in a 5-ml sulphuric acid solution and 1 ml of a wastewater sample is added. At the end of the digestion, the concentration of produced chromium oxide (Cr_2O_3) is 100 mg/l. Which is the COD of the wastewater sample?

1.5 Let us assume, for argument's sake, that a BOD test is carried out using nitrate, instead than oxygen, as electron acceptor. At the start of the test the substrate concentration is 200 mg COD/l and at the end of the test the residual substrate concentration is 10 mg COD/l. During the test 40 mg/l of microorganisms are produced. How much nitrate has been consumed during this test? Assume that nitrate is all reduced to molecular nitrogen.

1.6 Compare two wastewaters, one having a low COD loading and one having a high COD loading. The COD loading is the mass flow rate of COD to be treated per day (kg COD/day). Which wastewater is better suited for aerobic treatment and which one for anaerobic treatment?

1.7 We have a wastewater with a flow rate of 10,000 m^3/day and a COD of 300 mg COD/l. This wastewater is treated in a process with a biological reactor having a volume of 4000 m^3, where the biomass concentration is 1500 mg/l. Microorganisms are removed from the system with the waste sludge stream, which has a flow rate of 200 m^3/day and biomass concentration of 3000 mg/l. Calculate the HRT, SRT and OLR for this process.

CHAPTER 2

Modelling Processes in Biological Wastewater Treatment

I N THIS CHAPTER, WE will present the theory and models that we will use in the rest of this book for the study of biological wastewater treatment processes.

2.1 MICROBIAL GROWTH

We have seen in Chapter 1 that growth of microorganisms on a substrate can be schematised as a chemical reaction where the substrate and nutrients (and possibly electron acceptors as oxygen or nitrate) are the reactants, and new microorganisms and other compounds (organic or inorganic) are the products. In this section, we want to learn how to write the stoichiometry and kinetics of fermentation reactions. Fermentation reactions, that is reactions which involve microbial growth on a substrate, can be described with the same principles that apply to any chemical reaction.

If we consider a generic chemical reaction (not necessarily a fermentation or biochemical reaction):

$$A + \alpha_1 B \rightarrow \alpha_2 C + \alpha_3 D \qquad (2.1)$$

once we know the stoichiometry (i.e. the coefficients α_1, α_2, α_3) and the value of, or an expression for, the rate of production or consumption of one of the species A, B, C, D, we can easily calculate the rate of production or consumption of all the species in the reaction. For example, assuming we know the value of, or an expression for, r_C (the rate of production of species C, in units of mass/volume.time), we can immediately calculate the rate of production or consumption of species A, B and D. If we use kmol to express the mass of the species and r_C is expressed as kmol/m^3.day the other rates will be:

$$r_A\left(\frac{kmol}{m^3 day}\right) = -\frac{1}{\alpha_2} r_C \quad r_B\left(\frac{kmol}{m^3 day}\right) = -\frac{\alpha_1}{\alpha_2} r_C \quad r_D\left(\frac{kmol}{m^3 day}\right) = \frac{\alpha_3}{\alpha_2} r_C \quad (2.2)$$

where we have written with the negative sign the rates of the species which are consumed, and with the positive sign the rates of the species which are produced.

Similarly, if we use kg to express the mass and r_C is expressed as kg/m^3.day the other rates will be:

$$r_A\left(\frac{kg}{m^3 day}\right) = -\frac{1}{\alpha_2} r_C \frac{MW_A}{MW_C} \quad r_B\left(\frac{kg}{m^3 day}\right)$$
$$= -\frac{\alpha_1}{\alpha_2} r_C \frac{MW_B}{MW_C} \quad r_D\left(\frac{kg}{m^3 day}\right) = \frac{\alpha_3}{\alpha_2} r_C \frac{MW_D}{MW_C} \quad (2.3)$$

So, in summary, from the knowledge of the rate of formation or consumption of one species taking part in a chemical reaction, we can calculate the rates of formation of all the species in the reaction, if the reaction stoichiometry is known. In the next sections, we will see how to apply this simple concept to the fermentation reaction that occurs in biological wastewater treatment processes.

Example 2.1

Consider the decomposition reaction of phosphine (this is not a fermentation or biochemical reaction, but the procedure is exactly the same for any reactions):

$$4PH_3 \rightarrow P_4 + 6H_2$$

1. Calculate the rate of production or consumption of the species PH_3 and H_2 in kmol/m^3.day if the rate of formation of the species P_4 is 100 kmol/m^3.day.

2. Calculate the rate of production or consumption of the species PH_3 and H_2 in kg/m³.day if the rate of formation of the species P_4 is 100 kg/m³.day.

Solution

1. Using kmol to express mass, we have:

$$r_{PH3} = -4 \cdot 100 \frac{kmol}{m^3 day} = -400 \frac{kmol}{m^3 day}$$

$$r_{H2} = 6 \cdot 100 \frac{kmol}{m^3 day} = 600 \frac{kmol}{m^3 day}$$

The rate of PH_3 is negative because the species is consumed, the rate of H_2 is positive because the species is produced.

2. Using kg to express mass, we have:

$$r_{PH3} = -\frac{100 \frac{kg}{m^3 day}}{124 \frac{kg}{kmol}} 4 \cdot 34 \frac{kg}{kmol} = -109.7 \frac{kg}{m^3 day}$$

$$r_{H2} = -\frac{100 \frac{kg}{m^3 day}}{124 \frac{kg}{kmol}} 6 \cdot 2 \frac{kg}{kmol} = 9.7 \frac{kg}{m^3 day}$$

where 34, 124 and 2 are the molecular weights of species PH_3, P_4 and H_2, respectively.

2.1.1 Stoichiometry

In order to obtain a mathematical description of microbial growth, we need to be able to quantify the relationship between microorganisms produced, substrate and nutrients consumed and products formed. In order words, we need to quantify the stoichiometry of the growth processes for the various microorganisms. In order to do this, we need to introduce the concepts of anabolism and catabolism.

Anabolism is the production of cellular material from the carbon source and the mineral elements. The cellular material formed in the anabolic reactions is constituted of proteins, lipids, DNA, RNA etc., that are all the polymers of which cells are made. Production of these polymers is not spontaneous and requires energy. Just to give an example the anabolic condensation reaction between the amino acids alanine and glycine to give the dipeptide alanine-glycine has a positive free energy change $\Delta G = 17.3$ kJ/mol at 37°C and pH 7:

$$\underset{\text{alanine}}{CH_3CHNH_2COOH} + \underset{\text{glycine}}{NH_2CH_2COOH} \leftrightarrow$$

$$\underset{\text{dipeptide alanine-glycine}}{CH_3CHNH_2COOCOCH_2NH_2} + H_2O \qquad (2.4)$$

The energy required for the anabolic reactions is provided by catabolism. Catabolic reactions are the intracellular reactions that generate the energy which is used in the anabolic processes to synthesise biomass. An example of a catabolic reaction is the oxidation of glucose in the presence of oxygen, which has a free energy change $\Delta G = -2879$ kJ/mol glucose:

$$C_6H_{12}O_6 + 6O_2 \leftrightarrow 6CO_2 + 6H_2O \qquad (2.5)$$

In cells, the energy vector is adenosine triphosphate (ATP). ATP is an energy rich molecule, which, when hydrolysed to Adenosine diphosphate (ADP), releases energy for the anabolic processes. ATP is generated from ADP in the catabolic reactions and is converted to ADP in the anabolic reactions. A conceptual scheme of anabolism and catabolism is shown in Figure 2.1.

2.1.1.1 Stoichiometry of Anabolism

The overall anabolic reaction for microorganisms can be schematised as follows:

$$\text{Carbon source} + \text{elements} \rightarrow \text{microorganisms} + \text{products}$$

In order to see how we can write a stoichiometry for the overall anabolic reaction, we can consider several examples. First of all, we specify that the

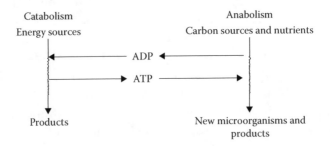

FIGURE 2.1 Conceptual scheme showing the coupling between anabolism and catabolism.

empirical formula for microorganisms used throughout this text will be $C_5H_7O_2N$. Let's consider now for example the aerobic metabolism of ethanol by heterotrophic microorganisms. The generic anabolic reaction for ethanol can be written as:

$$C_2H_5OH + aNH_3 + bO_2 \rightarrow cC_5H_7O_2N + dH_2O \qquad (2.6)$$

How do we determine the stoichiometric coefficients a, b, c, d?

One way of doing this is to write the half-reactions of oxidation and reduction that make up the overall anabolic reaction (Equation 2.6). In microorganisms the oxidation state of the main atoms is the following: $H = +1$, $O = -2$, $N = -3$. For carbon the oxidation state depends on the particular substance and it can be calculated by assuming that the molecule has an oxidation state of 0. Therefore, for the oxidation state of carbon in ethanol is: $2 \cdot Ox_C + 6 \cdot 1 + 1 \cdot (-2) = 0$ which $Ox_C = -2$. This is the average oxidation state for carbon in the ethanol molecule; the two carbons in this molecule may very well have different oxidation states. Using the same method it can be calculated immediately that the average oxidation state for carbon in biomass is 0. Therefore, the carbon atoms in ethanol need to be oxidised in the anabolic reactions to give the biomass components. To be precise, each carbon atom in the ethanol molecule needs to lose two electrons to be incorporated as biomass components.

The oxidation half-reaction can be written therefore as:

$$C_2H_5OH + \frac{2}{5}NH_3 + \frac{19}{5}H_2O \rightarrow \frac{2}{5}C_5H_7O_2N + 4e^- + 4H_3O^+ \qquad (2.7)$$

The stoichiometry above can be derived easily with the following steps:

1. Referring to one molecule of ethanol as a basis, balance the carbon atoms and so calculate the stoichiometric coefficient for biomass;

2. Balance the nitrogen atoms by calculating the stoichiometric coefficient for ammonia;

3. Write the number of electrons released by one molecule of ethanol to generate biomass and balance the charges using the appropriate number of H_3O^+ molecules;

4. Finally balance the number of hydrogen atoms finding the stoichiometric coefficient for H_2O.

Similarly we can write the reduction half-reaction, where molecular oxygen is reduced to water:

$$O_2 + 4e^- + 4H_3O^+ \rightarrow 6H_2O \tag{1.6}$$

The stoichiometry of the reduction half-reaction (Equation 1.6) has been obtained with a procedure absolutely analogous to the one used for the oxidation half-reaction (Equation 2.7).

The overall anabolic reaction can be written by combining the oxidation and reduction half-reactions so that both reactions generate and consume the same number of electrons. In the case of ethanol, we already have four electrons for each half-reaction so we can just combine the two reactions together to get the overall stoichiometry for anabolism:

$$C_2H_5OH + \frac{2}{5}NH_3 + O_2 \rightarrow \frac{2}{5}C_5H_7O_2N + \frac{11}{5}H_2O \tag{2.8}$$

As an alternative to using the method of half-reactions, another method to obtain the stoichiometry of the overall anabolic reaction (Equation 2.6) is to write the balances for the various elements. For example, the four coefficients a, b, c, d in the anabolic reaction for ethanol can be obtained by writing four balances for the elements C, O, H, N. We have, referring to reaction (Equation 2.6):

$$\text{C balance: } 2 = 5c \tag{2.9}$$
$$\text{H balance: } 6 + 3a = 7c + 2d \tag{2.10}$$
$$\text{O balance: } 1 + 2b = 2c + d \tag{2.11}$$
$$\text{N balance: } a = c \tag{2.12}$$

By solving Equations 2.9 through 2.12 simultaneously, for example by substitution, we can calculate the same a, b, c, d coefficients obtained above with the method of the half-reactions.

With the same approach, we can calculate the stoichiometry of anabolic reactions for the different types of metabolism occurring in biological wastewater treatment processes.

For the generic heterotrophic aerobic metabolism with an organic carbon source $C_wH_xO_yN_z$, the stoichiometry of the anabolic reaction can be calculated as follows. For heterotrophic metabolism to occur, the carbon in the organic substrate has to have an oxidation state lower than or equal to 0. This means that the carbon in the substrate has to be oxidised to be incorporated by the biomass and under aerobic conditions oxygen is the oxidant. If the carbon in the substrate has an oxidation state which is larger than 0, this means that the carbon needs to be reduced to be incorporated into biomass. Under aerobic conditions the reducing agent is ammonia which is oxidised to nitrate and this metabolism is called auto-trophic. So for heterotrophic microorganisms growing on the substrate $C_w H_x O_y N_z$ the oxidation state of the carbon in the substrate is equal to $(3z+2y-x)/w$ (which has to be negative or 0 for heterotrophic metabo-lism to occur) and each substrate molecule has to lose $x-2y-3z$ electrons to be converted to biomass. Therefore, the oxidation half-reaction for the anabolism of heterotrophic microorganisms, which can be calculated with the same steps described above for ethanol, is:

$$C_w H_x O_y N_z + \left(\frac{w}{5} - z\right) NH_3 + \left(x + \frac{2}{5}w - 3z - 3y\right) H_2 O \rightarrow$$

$$\frac{w}{5} C_5 H_7 O_2 N + (x - 3z - 2y) e^- + (x - 3z - 2y) H_3 O^+$$

(2.13)

The reduction half-reaction is the reduction of oxygen to water, Equation 1.6. Therefore, combining the oxidation and reduction half-reactions and making the appropriate rearrangements we obtain the overall anabolic reaction for aerobic heterotrophic metabolism on the generic organic substrate:

$$C_w H_x O_y N_z + \left(\frac{w}{5} - z\right) NH_3 + \left(\frac{x}{4} - \frac{y}{2} - \frac{3}{2}z\right) O_2 \rightarrow$$

$$\frac{w}{5} C_5 H_7 O_2 N + \left(\frac{x}{2} - \frac{2}{5}w - \frac{3}{2}z\right) H_2 O$$

(2.14)

In order to calculate the stoichiometry of other relevant anabolic reac-tions, we can use the same method with the following observations.

For anoxic metabolism of heterotrophic microorganisms the oxidant is nitrate instead than oxygen. The oxidation state of nitrate is +5 and it reduces to 0 as nitrate is reduced to molecular nitrogen. Apart from this important difference, the stoichiometry is the same as for aerobic heterotro-phic metabolism. The half-reaction corresponding to nitrate reduction is:

$$2HNO_3 + 10e^- + 10H_3O^+ \rightarrow N_2 + 16H_2O$$

(1.11)

The oxidation half-reaction is the same already earlier for aerobic metabo-lism, Equation 2.13, so the anabolic reaction is obtained by combining reactions (2.13) and (2.15) so that the number of electrons accepted and removed is the same:

$$C_wH_xO_yN_z + \left(\frac{w}{5} - z\right)NH_3 + \left(\frac{x}{5} - \frac{2}{5}y - \frac{3}{5}z\right)HNO_3 \rightarrow$$

$$\frac{w}{5}C_5H_7O_2N + \left(\frac{x}{10} - \frac{1}{5}y - \frac{3}{10}z\right)N_2 + \left(\frac{3}{5}x - \frac{1}{5}y - \frac{1}{5}w - \frac{9}{5}z\right)H_2O$$

(2.15)

For aerobic metabolism of autotrophic nitrifying microorganisms the carbon source is carbon dioxide (CO_2) where carbon has an oxidation state equal to +4. Therefore, carbon has to be reduced to be incorporated into the biomass and the reducing agent is NH_3. Nitrogen in NH_3 has an oxidation state equal to −3 and is oxidised to nitrate (oxidation state of $N = +5$). Therefore, the oxidation half-reaction for autotrophic nitrifiers growing on CO_2 is:

$$NH_3 + 11H_2O \rightarrow HNO_3 + 8e^- + 8H_3O^+ \qquad (2.16)$$

and the reduction half-reaction is:

$$CO_2 + \frac{1}{5}NH_3 + 4e^- + 4H_3O^+ \rightarrow \frac{1}{5}C_5H_7O_2N + \frac{28}{5}H_2O \qquad (2.17)$$

Combining the two half-reactions Equations 2.16 and 2.17 in the usual way, so that the number of electrons generated and removed in the two half-reactions is the same, we obtain:

$$CO_2 + \frac{7}{10}NH_3 \rightarrow \frac{1}{5}C_5H_7O_2N + \frac{1}{2}HNO_3 + \frac{1}{10}H_2O \qquad (2.18)$$

In fermentation reactions with an organic carbon source $C_wH_xO_yN_z$ as substrate, there is no external electron acceptor, that is oxidant, present. In this case, the electron acceptor is the H_3O^+ molecule, or better the H^+ ion, which is released in the oxidation of the carbon source. Therefore, it can be said that in anaerobic fermentation reactions the substrate itself is both oxidised and reduced. With the method of the half-reactions, the overall anabolic reaction for fermentative metabolism can be written. The oxidation half-reaction is the same we have seen for aerobic anabolism:

$$C_wH_xO_yN_z + \left(\frac{w}{5} - z\right)NH_3 + \left(x + \frac{2}{5}w - 3z - 3y\right)H_2O \rightarrow$$

$$\frac{w}{5}C_5H_7O_2N + (x - 3z - 2y)e^- + (x - 3z - 2y)H_3O^+$$

(2.13)

The reduction half-reaction is the reduction of the H^+ ion to H_2:

$$H_3O^+ + e^- \rightarrow H_2O + \frac{1}{2}H_2 \qquad (2.19)$$

By combining the two half-reactions in the usual way, we obtain the over-all anabolic reaction:

$$C_w H_x O_y N_z + \left(\frac{w}{5} - z\right)NH_3^+ \rightarrow$$
$$\frac{w}{5}C_5H_7O_2N^+ + \left(y - \frac{2}{5}w\right)H_2O + \left(\frac{x}{2} - y - \frac{3}{2}z\right)H_2 \qquad (2.20)$$

Other very important types of metabolism in anaerobic digesters are acetoclastic and hydrogenotrophic methanogenesis. In acetoclastic methanogenesis the substrate is acetic acid (CH_3COOH, i.e. $C_2H_4O_2$), which is produced with the catabolic reactions of many substrates, as shown in Chapter 1 and as will be discussed in the next sections. The ana-bolic reaction for acetoclastic methanogenesis is only a particular case of the general anabolic reaction for anaerobic fermentation of organic substrates and can be written immediately by substituting the coefficients $w = 2$, $x = 4$, $y = 2$ and $z = 0$ in Equation 2.20:

$$CH_3COOH + \frac{2}{5}NH_3 \rightarrow \frac{2}{5}C_5H_7O_2N + \frac{6}{5}H_2O \qquad (2.21)$$

Hydrogenotrophic methanogens use carbon dioxide as carbon source. For the anabolic reaction carbon dioxide is reduced to biomass and the reducing agent is hydrogen, which is oxidised to water. The reduction half-reaction is:

$$CO_2 + 4e^- + 4H_3O^+ + \frac{1}{5}NH_3 \rightarrow \frac{1}{5}C_5H_7O_2N + \frac{28}{5}H_2O \qquad (2.22)$$

The oxidation half-reaction is:

$$H_2 + 2H_2O \rightarrow 2H_3O^+ + 2e^- \qquad (2.23)$$

And combining them as previously described, we obtain the overall ana-bolic reaction for hydrogenotrophic methanogens:

$$2H_2 + CO_2 + \frac{1}{5}NH_3 \rightarrow \frac{1}{5}C_5H_7O_2N + \frac{8}{5}H_2O \qquad (2.24)$$

TABLE 2.1 Summary of Anabolic Reactions

Microorganisms	Anabolic Reaction
Heterotrophs (aerobic)	$C_wH_xO_yN_z + \left(\dfrac{w}{5} - z\right)NH_3 + \left(\dfrac{x}{4} - \dfrac{y}{2} - \dfrac{3}{2}z\right)O_2 \rightarrow$ $\dfrac{w}{5}C_5H_7O_2N + \left(\dfrac{x}{2} - \dfrac{2}{5}w - \dfrac{3}{2}z\right)H_2O$
Heterotrophs (anoxic)	$C_wH_xO_yN_z + \left(\dfrac{w}{5} - z\right)NH_3 + \left(\dfrac{x}{5} - \dfrac{2}{5}y - \dfrac{3}{5}z\right)HNO_3 \rightarrow$ $\dfrac{w}{5}C_5H_7O_2N + \left(\dfrac{x}{10} - \dfrac{1}{5}y - \dfrac{3}{10}z\right)N_2 + \left(\dfrac{3}{5}x - \dfrac{1}{5}y - \dfrac{1}{5}w - \dfrac{9}{5}z\right)H_2O$
Nitrifiers	$CO_2 + \dfrac{7}{10}NH_3 \rightarrow \dfrac{1}{5}C_5H_7O_2N + \dfrac{1}{2}HNO_3 + \dfrac{1}{10}H_2O$
Fermentative	$C_wH_xO_yN_z + \left(\dfrac{w}{5} - z\right)NH_3 \rightarrow \dfrac{w}{5}C_5H_7O_2N$ $+ \left(y - \dfrac{2}{5}w\right)H_2O + \left(\dfrac{x}{2} - y - \dfrac{3}{2}z\right)H_2$
Methanogens (acetoclastic)	$CH_3COOH + \dfrac{2}{5}NH_3 \rightarrow \dfrac{2}{5}C_5H_7O_2N + \dfrac{6}{5}H_2O$
Methanogens (hydrogenotrophic)	$2H_2 + CO_2 + \dfrac{1}{5}NH_3 \rightarrow \dfrac{1}{5}C_5H_7O_2N + \dfrac{8}{5}H_2O$

A summary of all the anabolic reactions discussed here is shown in Table 2.1.

So we have seen that from consideration of the oxidation and reduction reactions involved, we can write the full stoichiometry for all the anabolic reactions considered in this book. The same is true for catabolic reactions and is discussed in the next section.

2.1.1.2 Stoichiometry of Catabolism

As seen previously, catabolic reactions generate the energy that is used in anabolic reactions to generate new microorganisms. The stoichiometry of catabolic reactions can be derived with the same procedure used for anabolic reactions, that is by taking into consideration the oxidation and reduction reactions.

Considering the aerobic metabolism of heterotrophs on the organic substrate $C_wH_xO_yN_z$, the catabolic product is carbon dioxide and the oxidant is oxygen.

The oxidation half-reaction for catabolism can be written as:

$$C_wH_xO_yN_z + (6w - 3z - 3y)H_2O \rightarrow$$
$$wCO_2 + (4w + x - 3z - 2y)e^- + (4w + x - 3z - 2y)H_3O^+ + zNH_3 \tag{2.25}$$

The reduction half-reaction is the reduction of oxygen which we have already seen previously:

$$O_2 + 4e^- + 4H_3O^+ \rightarrow 6H_2O \tag{1.6}$$

So we obtain the overall catabolic reaction:

$$C_wH_xO_yN_z + \left(w + \frac{x}{4} - \frac{3}{4}z - \frac{y}{2}\right)O_2 \rightarrow$$
$$wCO_2 + \left(\frac{x}{2} - \frac{3}{2}z\right)H_2O + zNH_3 \tag{2.26}$$

Similarly we can obtain the overall catabolic reaction for anoxic metabolism where nitrate is the electron acceptor instead of oxygen.

For autotrophic nitrifying microorganisms, the catabolic reaction involves the oxidation of ammonia to nitrate using oxygen as electron acceptor. The stoichiometry of this reaction can be obtained in the usual way by writing the oxidation and reduction half-reactions, but in this case it is particularly easy and can be written immediately:

$$NH_3 + 2O_2 \rightarrow HNO_3 + H_2O \tag{2.27}$$

For anaerobic fermentative microorganisms, there is no electron acceptor and therefore, similarly as for the anabolic reaction, in the catabolic reaction the substrate itself is both oxidised and reduced. The final oxidation product for the catabolic reaction is very often carbon dioxide; however, the reduced products are different depending on the specific carbon source and on the microorganisms' type. For example, if glucose is the substrate, the reduced products from the catabolic reactions can be ethanol or hydrogen, hydrogen production being associated with the production of organic acids. We will consider here the case where glucose is converted to carbon dioxide (oxidation product), hydrogen (reduction product) and acetic acid. In this case the oxidation half-reaction can be written as:

$$C_6H_{12}O_6 + 10H_2O \rightarrow 2CO_2 + 8e^- + 8H_3O^+ + 2CH_3COOH \tag{2.28}$$

The reduction half-reaction is the reduction of the H^+ ion already seen before:

$$H_3O^+ + e^- \rightarrow H_2O + \frac{1}{2}H_2 \tag{2.19}$$

And combining the two half-reactions Equations 2.28 and 2.19 we obtain the full catabolic reaction:

$$C_6H_{12}O_6 + 2H_2O \rightarrow 2CH_3COOH + 4H_2 + 2CO_2 \tag{2.29}$$

For acetoclastic methanogens the catabolic reaction is the oxidation of acetic acid to carbon dioxide. The oxidant is acetic acid itself which reduces to methane (disproportion). This reaction can be split into the two half-reactions of oxidation:

$$CH_3COOH + 10H_2O \rightarrow 2CO_2 + 8e^- + 8H_3O^+ \tag{2.30}$$

and reduction:

$$CH_3COOH + 8e^- + 8H_3O^+ \rightarrow 2CH_4 + 10H_2O \tag{2.31}$$

From which the overall catabolic reaction can be obtained:

$$2CH_3COOH \rightarrow 2CH_4 + 2CO_2 \tag{2.32}$$

For hydrogenotrophic methanogens the catabolic reaction is the oxidation of hydrogen to water, with carbon dioxide being the oxidant. The oxidation half-reaction is:

$$H_2 + 2H_2O \rightarrow 2H_3O^+ + 2e^- \tag{2.33}$$

and the reduction half-reaction is:

$$CO_2 + 8e^- + 8H_3O^+ \rightarrow CH_4 + 10H_2O \tag{2.34}$$

So the overall catabolic reaction is:

$$CO_2 + 4H_2 \rightarrow CH_4 + 2H_2O \tag{2.35}$$

Table 2.2 summarises the catabolic reactions for the various types of microorganisms considered here.

2.1.1.3 Overall Growth Stoichiometry

We have seen in the previous sections that a stoichiometry can be written for the anabolic and catabolic reactions which are needed for microbial

TABLE 2.2   Summary of Catabolic Reactions

Microorganisms	Catabolic Reaction
Heterotrophs (aerobic)	$C_wH_xO_yN_z + \left(w + \dfrac{x}{4} - \dfrac{3}{4}z - \dfrac{y}{2}\right)O_2 \rightarrow$ $wCO_2 + \left(\dfrac{x}{2} - \dfrac{3}{2}z\right)H_2O + zNH_3$
Heterotrophs (anoxic)	$C_wH_xO_yN_z + \left(\dfrac{4}{5}w + \dfrac{1}{5}x - \dfrac{3}{5}z - \dfrac{2}{5}y\right)HNO_3 \rightarrow$ $wCO_2 + \left(\dfrac{2}{5}w + \dfrac{x}{10} - \dfrac{3}{10}z - \dfrac{1}{5}y\right)N_2$ $+ \left(\dfrac{2}{5}w + \dfrac{8}{5}x - \dfrac{9}{5}z - \dfrac{1}{5}y\right)H_2O + zNH_3$
Nitrifiers	$NH_3 + 2O_2 \rightarrow HNO_3 + H_2O$
Fermentative (glucose conversion to acetic acid)	$C_6H_{12}O_6 + 2H_2O \rightarrow 2CH_3COOH + 4H_2 + 2CO_2$
Methanogens (acetoclastic)	$2CH_3COOH \rightarrow 2CH_4 + 2CO_2$
Methanogens (hydrogenotrophic)	$CO_2 + 4H_2 \rightarrow CH_4 + 2H_2O$

growth. However, the question now is, can we write a stoichiometry for the overall process of microbial growth? In other words, can we combine the anabolic and catabolic reactions to obtain the overall growth stoichiometry? The answer is that this in general not possible, at least purely on paper. The reason why it is not possible to write an overall stoichiometry for microbial growth purely based on elemental balances and/or oxidation–reduction reactions is that it is in general not known how much of the energy generated in the catabolic reactions is actually transferred to the anabolic reactions. Energy is transferred from catabolism to anabolism via ATP and ADP and it is very difficult to predict on paper the amount of ATP that can be generated per unit of energy generated by the catabolic reactions. Even for the same substrate and products and for the same growth conditions, the efficiency of oxidative phosphorylation varies greatly among different microorganisms.

To clarify this further, let us consider the overall growth stoichiometry for aerobic heterotrophic microorganisms growing on a substrate $C_xH_yO_yN_z$. The overall growth reaction will be a combination of the anabolic and catabolic reactions, therefore, from Tables 2.1 and 2.2, the reactants will be the substrate, oxygen and ammonia (depending on the nitrogen content on the substrate nitrogen might be required as a reactant or produced as a product), while the products will be biomass, carbon dioxide and water. The overall growth stoichiometry therefore will be something like:

$$C_wH_xO_yN_z + aO_2 + bNH_3 \rightarrow cC_5H_7O_2N + dCO_2 + eH_2O \quad (2.36)$$

In the absence of any additional information the stoichiometric coefficients of reaction (Equation 2.36) cannot be calculated, because we have five unknown coefficients and only four equations, the elemental balances for C, N, O and H. Therefore, we need an additional equation to be able to write the stoichiometry of microbial growth. This additional equation needs to be obtained from experimental data and is represented by the growth yield. The growth yield is defined as the amount of biomass formed per unit amount of substrate consumed, that is:

$$Y_{X/S} \left(\frac{\text{kg biomass}}{\text{kg substrate}} \right) = \frac{\text{biomass produced due to growth}}{\text{substrate removed due to growth}} \quad (2.37)$$

In Equation 2.37 we have specified 'due to growth' because in biological processes there are other phenomena which may contribute to biomass production and substrate removal, for example endogenous metabolism and maintenance, which will be discussed later. However, the growth yield $Y_{X/S}$ only refers to the process of microbial growth.

The growth yield coefficient is usually obtained from experimental data. If the growth yield is known, this constitutes an additional equation relating the coefficients a, b, c, d, e of the growth Equation 2.36 and the overall stoichiometry can be calculated. For example, referring to the aerobic metabolism of the $C_wH_xO_yN_z$ substrate shown by Equation 2.36, knowledge of the growth yield $Y_{X/S}$ will give the following systems of five equations in five unknowns:

$$\begin{cases} \dfrac{c \cdot MW_{\text{biomass}}}{MW_{\text{substrate}}} = Y_{X/S} \\ 5c + d = w & \text{C balance} \\ 2c + 2d + e = 2a + y & \text{O balance} \\ 7c + 2e = 3b + x & \text{H balance} \\ 5c + d = w & \text{N balance} \end{cases} \quad (2.38)$$

Solving by substitution the system of Equations 2.38, we obtain the overall growth stoichiometry:

$$C_wH_xO_yN_z + \left(w + \frac{x}{4} - \frac{y}{2} - \frac{3}{4}z - 5\frac{Y_{X/S}MW_{substrate}}{MW_{biomass}}\right)O_2$$

$$+ \left(\frac{Y_{X/S}MW_{substrate}}{MW_{biomass}} - z\right)NH_3 \rightarrow \frac{Y_{X/S}MW_{substrate}}{MW_{biomass}}C_5H_7O_2N \qquad (2.39)$$

$$+ \left(w - 5\frac{Y_{X/S}MW_{substrate}}{MW_{biomass}}\right)CO_2 + \left(\frac{x}{2} - \frac{3}{2}z - 2\frac{Y_{X/S}MW_{substrate}}{MW_{biomass}}\right)H_2O$$

It is evident, therefore, that once the growth yield $Y_{X/S}$ is known, the stoichiometry of microbial growth is fully defined and can be calculated with the elemental balances described above.

Assuming, for example the substrate is glucose ($C_6H_{12}O_6$) and assuming a growth yield of 0.3 kg biomass/kg glucose (reasonable value for aerobic growth on many organic substrates), Equation 2.39 corresponds to the following overall growth stoichiometry:

$$C_6H_{12}O_6 + 3.61O_2 + 0.48NH_3 \rightarrow$$
$$0.48C_5H_7O_2N + 3.61CO_2 + 5.04H_2O \qquad (2.40)$$

The same approach can be applied to all the other types of microorganisms relevant to wastewater treatment. For example, for heterotrophs growing on organic substrate using nitrate as electron acceptor, the overall growth reaction will have the form:

$$C_wH_xO_yN_z + aHNO_3 + bNH_3 \rightarrow cC_5H_7O_2N + dCO_2 + eH_2O + fN_2 \quad (2.41)$$

The overall growth stoichiometry in this case can still be obtained introducing the growth yield $Y_{X/S}$, defined in the same way as for aerobic metabolism, and using the elemental balances for C, H, N and O. However, this will give us five equations, but there are six unknown coefficients here, due to the presence of molecular nitrogen. However, a sixth equation can easily be obtained by noticing that molecular nitrogen only comes from the reduction of nitrate, and that all the nitrate that reacts is converted to nitrogen. Therefore, we have the following additional equation: $f = (a/2)$.

By solving the system of equations we obtain the following stoichiometry for the overall growth of heterotrophic microorganisms on a carbon source using nitrate as electron acceptor:

$$
C_w H_x O_y N_z + \left(\frac{4}{5}w + \frac{1}{5}x - \frac{3}{5}z - \frac{2}{5}y - 4\frac{Y_{X/S}MW_{subs}}{MW_{biomass}} \right) HNO_3
$$

$$
+ \left(\frac{Y_{X/S}MW_{subs}}{MW_{biomass}} - z \right) NH_3 \rightarrow \frac{Y_{X/S}MW_{subs}}{MW_{biomass}} C_5 H_7 O_2 N \qquad (2.42)
$$

$$
+ \left(w - 5\frac{Y_{X/S}MW_{subs}}{MW_{biomass}} \right) CO_2
$$

$$
+ \left(\frac{2}{5}w + \frac{3}{5}x - \frac{9}{5}z - \frac{1}{5}y - 4\frac{Y_{X/S}MW_{subs}}{MW_{biomass}} \right) H_2 O
$$

$$
+ \left(\frac{2}{5}w + \frac{1}{10}x - \frac{3}{10}z - \frac{1}{5}y - 2\frac{Y_{X/S}MW_{subs}}{MW_{biomass}} \right) N_2
$$

For nitrifying microorganisms, combining the anabolic and catabolic reactions the general growth equation will have the form:

$$
NH_3 + aCO_2 + bO_2 \rightarrow cC_5 H_7 O_2 N + dHNO_3 + eH_2 O \qquad (2.43)
$$

For nitrifiers the growth yield is usually expressed as biomass produced per unit mass of nitrate (as nitrogen) produced, that is:

$$
Y_{XA/NO_3} \left(\frac{kg\,biomass}{kg\,N\text{-}NO_3} \right) = \frac{\left(\begin{array}{c} \text{Biomass produced} \\ \text{due to growth} \end{array} \right)}{\left(\begin{array}{c} \text{Nitrate (as nitrogen)} \\ \text{produced due to growth} \end{array} \right)} \qquad (2.44)
$$

In Equation 2.44, similarly to Equation 2.37, we have specified 'due to growth' to clarify that the growth yield refers only to the process of biomass growth, and not to other phenomena which may affect biomass production and substrate consumption, such as endogenous metabolism and maintenance. All the growth yield coefficients used in this book are only referred to biomass production due to growth.

Equation 2.44 can be written as, introducing the molecular weights of biomass and nitrogen and with reference to reaction (Equation 2.43):

$$
Y_{XA/NO_3} = \frac{c}{d} 8.07 \qquad (2.45)
$$

Combining Equation 2.45 with the elemental balances of C, H, O and N, referred to reaction Equation 2.43, we obtain the overall growth stoichiometry for nitrifying microorganisms:

$$NH_3 + 5\frac{Y_{XA/NO3}}{Y_{XA/NO3} + 8.07}CO_2 + \left(0.75 + \frac{10.1 - 5.75Y_{XA/NO3}}{Y_{XA/NO3} + 8.07}\right)O_2 \rightarrow$$

$$\frac{Y_{XA/NO3}}{Y_{XA/NO3} + 8.07}C_5H_7O_2N + \frac{8.07}{Y_{XA/NO3} + 8.07}HNO_3 \qquad (2.46)$$

$$+ \left(1.5 - \frac{4.035 + 3.5Y_{XA/NO3}}{Y_{XA/NO3} + 8.07}\right)H_2O$$

The same approach can be used to derive the stoichiometry of microbial growth under anaerobic conditions. For fermentative bacteria, the growth yield is always expressed exactly in the same way than for heterotrophic microorganisms, that is biomass produced due to growth/substrate removed due to growth. Due to the wide range of substrates and products that can be formed under fermentation conditions it is more practicable to refer, as an example, to a particular fermentation reaction. For example, with reference to the conversion of glucose to acetic acid and hydrogen, the general stoichiometry is:

$$C_6H_{12}O_6 + aH_2O + bNH_3 \rightarrow$$
$$cC_5H_7O_2N + dCH_3COOH + eH_2 + fCO_2 \qquad (2.47)$$

The six coefficients a, b, c, d, e, f in reaction (Equation 2.47) can be calculated from the growth yield

$$Y_{X/S}\left(\frac{\text{kg biomass}}{\text{kg glucose}}\right) = \frac{\left(\begin{array}{c}\text{biomass produced due}\\\text{to growth on glucose}\end{array}\right)}{\left(\begin{array}{c}\text{glucose removed due}\\\text{to growth on glucose}\end{array}\right)} \qquad (2.48)$$

from the four elemental balances and from the additional equation $d = f$ which comes from the fact that, for glucose fermentation to acetic acid, carbon dioxide is only produced in the catabolic reactions, which produce one mol of carbon dioxide per mol of acetic acid.

The equations can be solved in the usual way to give:

$$C_6H_{12}O_6 + 1.59Y_{X/S}NH_3 \rightarrow$$
$$1.59Y_{X/S}C_5H_7O_2N + (2-2.65Y_{X/S})$$
$$CH_3COOH + (4-5.3Y_{X/S})H_2 +$$
$$+(2-2.65Y_{X/S})CO_2 + (7.42Y_{X/S}-2)H_2O$$
(2.49)

Equation 2.49 represents the overall growth stoichiometry for the fermentative microorganisms which convert glucose to acetic acid and hydrogen.

For acetoclastic methanogens the general form of the growth stoichiometry is:

$$CH_3COOH + aNH_3 \rightarrow bC_5H_7O_2N + cCH_4 + dCO_2 + eH_2O \quad (2.50)$$

the growth yield is defined in the usual way as kg biomass/kg acetic acid, and, by solving the elemental balances we obtain:

$$CH_3COOH + 0.53Y_{X/S}NH_3 \rightarrow 0.53Y_{X/S}C_5H_7O_2N$$
$$+(1-1.325Y_{X/S})CH_4 + (1-1.325Y_{X/S})CO_2 + 1.59Y_{X/S}H_2O$$
(2.51)

Equation 2.51 represents the overall growth stoichiometry for acetoclastic methanogens.

For hydrogenotrophic methanogens, the overall growth stoichiometry has the form:

$$H_2 + aCO_2 + bNH_3 \rightarrow cC_5H_7O_2N + dH_2O + eCH_4 \quad (2.52)$$

The growth yield can be defined as kg biomass/kg hydrogen:

$$Y_{X/S}\left(\frac{kg\,biomass}{kg\,hydrogen}\right) = 56.5c \quad (2.53)$$

From Equation 2.53 and from the elemental balances the following stoichiometry can be obtained:

$$H_2 + (0.25+0.044Y_{X/S})CO_2 + 0.0177Y_{X/S}NH_3 \rightarrow$$
$$0.0177Y_{X/S}C_5H_7O_2N + (0.5+0.053Y_{X/S})H_2O$$
$$+(0.25-0.044Y_{X/S})CH_4$$
(2.54)

Equation 2.54 represents the overall growth stoichiometry for acetoclastic methanogens.

Table 2.3 summarises the overall reaction stoichiometry for the various microorganisms, expressed as a function of the growth yields.

TABLE 2.3 Summary of the Microbial Growth Reactions as a Function of the Growth Yield for Various Types of Microorganisms

Microorganisms	Growth Yield	Growth Stoichiometry
Heterotrophs (aerobic)	$Y_{X/S}\left(\dfrac{\text{kg biomass}}{\text{kg substrate}}\right)$	$$C_wH_xO_yN_z + \left(w+\frac{x}{4}-\frac{y}{2}-\frac{3}{4}z-5\frac{Y_{X/S}MW_{substrate}}{MW_{biomass}}\right)O_2 + \left(\frac{Y_{X/S}MW_{substrate}}{MW_{biomass}}-z\right)NH_3 \rightarrow$$ $$\frac{Y_{X/S}MW_{substrate}}{MW_{biomass}}C_5H_7O_2N + \left(w-5\frac{Y_{X/S}MW_{substrate}}{MW_{biomass}}\right)CO_2 + \left(\frac{x}{2}-\frac{3}{2}z-2\frac{Y_{X/S}MW_{substrate}}{MW_{biomass}}\right)H_2O$$
Heterotrophs (anoxic)	$Y_{X/S}\left(\dfrac{\text{kg biomass}}{\text{kg substrate}}\right)$	$$C_wH_xO_yN_z + \left(\frac{4}{5}w+\frac{1}{5}x-\frac{3}{5}z-\frac{2}{5}-4\frac{Y_{X/S}MW_{subs}}{MW_{biomass}}\right)HNO_3 + \left(\frac{Y_{X/S}MW_{subs}}{MW_{biomass}}-z\right)NH_3 \rightarrow$$ $$\frac{Y_{X/S}MW_{subs}}{MW_{biomass}}C_5H_7O_2N + \left(w-5\frac{Y_{X/S}MW_{subs}}{MW_{biomass}}\right)CO_2 + \left(\frac{2}{5}w+\frac{3}{5}x-\frac{9}{5}z-y-4\frac{1}{5}\frac{Y_{X/S}MW_{subs}}{MW_{biomass}}\right)H_2O$$ $$+\left(\frac{2}{5}w+\frac{1}{10}x-\frac{3}{10}z-\frac{1}{5}y-2\frac{Y_{X/S}MW_{subs}}{MW_{biomass}}\right)N_2$$
Nitrifiers	$Y_{XA/NO_3}\left(\dfrac{\text{kg biomass}}{\text{kg N}-NO_3}\right)$	$$NH_3 + 5\frac{Y_{XA/NO3}}{Y_{XA/NO3}+8.07}CO_2 + \left(0.75+\frac{10.1-5.75Y_{XA/NO3}}{Y_{XA/NO3}+8.07}\right)O_2 \rightarrow$$ $$\frac{Y_{XA/NO3}}{Y_{XA/NO3}+8.07}C_5H_7O_2N + \frac{8.07}{Y_{XA/NO3}+8.07}HNO_3 + \left(1.5-\frac{4.035+3.5Y_{XA/NO3}}{Y_{XA/NO3}+8.07}\right)H_2O$$
Fermentative (glucose conversion to acetic acid)	$Y_{X/S}\left(\dfrac{\text{kg biomass}}{\text{kg glucose}}\right)$	$$C_6H_{12}O_6 + 1.59Y_{X/S}NH_3 \rightarrow 1.59Y_{X/S}C_5H_7O_2N + (2-2.65Y_{X/S})CH_3COOH$$ $$+(4-5.3Y_{X/S})H_2 + (2-2.65Y_{X/S})CO_2 + (7.42Y_{X/S}-2)H_2O$$
Methanogens (acetoclastic)	$Y_{X/S}\left(\dfrac{\text{kg biomass}}{\text{kg acetic acid}}\right)$	$$CH_3COOH + 0.53Y_{X/S}NH_3 \rightarrow 0.53Y_{X/S}C_5H_7O_2N + (1-1.325Y_{X/S})CH_4 + (1-1.325Y_{X/S})CO_2$$ $$+1.59Y_{X/S}H_2O$$
Methanogens (hydrogenotrophic)	$Y_{X/S}\left(\dfrac{\text{kg biomass}}{\text{kg hydrogen}}\right)$	$$H_2 + (0.25+0.044Y_{X/S})CO_2 + 0.0177Y_{X/S}NH_3 \rightarrow$$ $$0.0177Y_{X/S}C_5H_7O_2N + (0.5+0.053Y_{X/S})H_2O + (0.25-0.044Y_{X/S})CH_4$$

Other examples of overall growth stoichiometry for other types of microorganisms and other substrates are shown in Examples 2.2 and 2.3.

Note that from the overall growth stoichiometry we can calculate some upper limits on the value of the growth yield $Y_{X/S}$. Indeed, for metabolism to occur both the anabolic and catabolic reactions need to take place, and the higher limit of $Y_{X/S}$ is the one for which no catabolism occurs. Let us consider, for example the aerobic heterotrophic metabolism of a carbon source, which is represented by Equation 2.39. In order for catabolism to occur, the stoichiometric coefficient for oxygen in Equation 2.39 needs to be higher than 0 (i.e. oxygen needs to be consumed), therefore it needs to be:

$$w + \frac{x}{4} - \frac{y}{2} - \frac{3}{4}z - 5\frac{Y_{X/S}MW_{substrate}}{MW_{biomass}} > 0$$

This means that:

$$Y_{X/S} < \left(w + \frac{x}{4} - \frac{y}{2} - \frac{3}{4}z \right)\frac{MW_{biomass}}{5MW_{substrate}}$$

For example, if the substrate is glucose ($C_6H_{12}O_6$), this condition means that it needs to be $Y_{X/S} < 0.75$ kg biomass/kg glucose.

We can use the same principle to calculate the upper limit for the growth yield for anaerobic reactions. For example for the growth of hydrogenotrophic methanogens, the need for catabolism to occur means that some methane needs to be produced and this translates into the condition:

$$0.25 - 0.044Y_{X/S} > 0$$

$$\text{that is } Y_{X/S} < 5.68\frac{\text{kg biomass}}{\text{kg hydrogen}}$$

It is important to observe, however, that the upper limit for the growth yield calculated in this way refers only to the chemistry of the reaction and not necessarily to its energetics. In practice, there needs to be some minimum energy generated from the catabolic reactions for the metabolism to occur, and this means that the maximum limit of the $Y_{X/S}$ can be significantly lower than the upper limit calculated here. However, it is always important to make a simple consistency check of the experimental data to make sure that the upper limit for $Y_{X/S}$ is not exceeded.

Example 2.2

Consider the anaerobic metabolism of the amino acid glycine ($C_2H_5O_2N$). One possible metabolism of this species under anaerobic conditions is its conversion to acetic acid. The microorganisms that carry out this reaction use glycine, ammonia and hydrogen and produce new microorganisms, acetic acid, carbon dioxide and water. We know from biochemistry that hydrogen and acetic acid are only involved in the catabolic reaction and that 1 mol of hydrogen is consumed per 1 mol of acetic acid produced. Assuming the growth yield $Y_{X/S}$ is defined as kg biomass/kg glycine, write the overall growth stoichiometry for anaerobic microbial growth on glycine.

Solution

From the information given we can write the overall growth reaction under anaerobic conditions on glycine as follows:

$$C_2H_5O_2N + aNH_3 + bH_2 \rightarrow cC_5H_7O_2N + dCH_3COOH + eCO_2 + fH_2O$$

From the biochemistry we know that $b = d$ and the growth yield is:

$$Y_{X/S}\left(\frac{\text{kg biomass}}{\text{kg glycine}}\right) = \frac{cMW_{\text{biomass}}}{MW_{\text{glycine}}} = 1.51c$$

The elemental balances are:

$$\begin{cases} 5c + 2d + e = 2 & \text{C balance} \\ 2c + 2d + 2e + f = 2 & \text{O balance} \\ 7c + 4d + 2f = 3a + 2b + 5 & \text{H balance} \\ c = 1 + a & \text{N balance} \end{cases}$$

Combining the elemental balances with the equation for the growth yield and with the equation $b = d$, we obtain the coefficients a, b, c, d, e, f, that is the overall growth stoichiometry:

$$C_2H_5O_2N + (0.66Y_{X/S} - 1)NH_3 + (1 - 2.2Y_{X/S})H_2 \rightarrow$$
$$0.66Y_{X/S}C_5H_7O_2N + (1 - 2.2Y_{X/S})CH_3COOH$$
$$+ 1.1Y_{X/S}CO_2 + 0.88Y_{X/S}H_2O$$

Example 2.3

Consider the anaerobic metabolism of palmitic acid $(C_{16}H_{32}O_2)$. Under anaerobic conditions palmitic acid is converted to acetic acid and hydrogen. The overall growth reaction includes the consumption of palmitic acid, ammonia, water and carbon dioxide and the production of microorganisms, acetic acid, hydrogen and carbon dioxide. We know from biochemistry that acetic acid and hydrogen are only produced during catabolism of palmitic acid and are not produced or consumed during anabolic reactions. The growth yield is defined as kg biomass/kg palmitic acid. Write the overall growth stoichiometry for the anaerobic metabolism of palmitic acid.

Solution

From the information given we can write the overall growth stoichiometry on palmitic acid under anaerobic conditions as:

$$C_{16}H_{32}O_2 + aNH_3 + bH_2O + cCO_2 \rightarrow dC_5H_7O_2N + eCH_3COOH + fH_2$$

The growth yield is defined as:

$$Y_{X/S}\left(\frac{kg\,biomass}{kg\,palmitic\,acid}\right) = \frac{dMW_{biomass}}{MW_{palmitic}} = 0.44d$$

Since we have six coefficients to determine, and we have four elemental balances plus the growth yield equation, we need an additional equation. This additional equation comes from the given information that acetic acid and hydrogen are only produced during the catabolic reaction and they don't play a role in the anabolic reaction. The catabolic reaction for palmitic acid can be written as:

$$C_{16}H_{32}O_2 + 14H_2O \rightarrow 8CH_3COOH + 14H_2$$

In this reaction the molar ratio between hydrogen and acetic acid is 14/8. This ratio does not change because of the anabolic reaction, because hydrogen and acetic acid are not involved in the anabolic reaction. Therefore, we have the additional equation that we were looking for and this is:

$$\frac{f}{e} = \frac{14}{8}$$

Now we have a system of six equations with the six unknown coefficients which can therefore be determined. The result is the following overall growth stoichiometry on palmitic acid under anaerobic conditions:

$$C_{16}H_{32}O_2 + 2.27Y_{X/S}NH_3 + (14 - 10.26Y_{X/S})H_2O + 3.45Y_{X/S}CO_2 \rightarrow$$

$$2.27Y_{X/S}C_5H_7O_2N + (8 - 3.95Y_{X/S})CH_3COOH + (14 - 6.91Y_{X/S})H_2$$

2.1.2 Kinetics

2.1.2.1 Microbial Growth

Once the stoichiometry of microbial growth is known, knowledge of the rate of microorganism growth will give the rate of consumption or production of all the species involved in the microbial metabolism.

Since microorganisms grow by duplicating themselves, it is to be expected that the rate of microbial growth, that is the rate of microorganisms production per unit volume and time, will be proportional to the number, or to the mass, of microorganisms present. Therefore, microbial growth rate can be expressed as:

$$r_X \left(\frac{kg \, biomass}{m^3 \times day} \right) = \mu \cdot X \qquad (2.55)$$

where:

X is the microorganisms (biomass) concentration (kg/m³)

μ is the specific growth rate (day⁻¹)

Many different models have been developed in the literature to express the specific growth rate as a function of the environmental conditions. The simplest and most widely used model is the Monod model, which relates the specific growth rate to the concentration of a limiting substrate, generically indicated as S in the equation below:

$$\mu = \frac{\mu_{max} S}{K_S + S} \qquad (2.56)$$

The unit of S are the ones of concentration, and typically S is expressed either as kg/m³, or as kg COD/m³. This kinetic equation is called the Monod equation and has two empirical parameters, μ_{max} and K_s, to be determined from experimental data. The Monod equation indicates that:

1. When the concentration of the limiting substrate S is very large $(S \gg K_S)$ the specific growth rate is independent of the substrate concentration S and is equal to μ_{max};

2. When the concentration of the limiting substrate S is very low ($S << K_S$), the specific growth rate is linearly dependent on S.

The typical shape of the curve $\mu = (\mu_{max} S)/(K_S + S)$ is shown in Figure 2.2. Note that the meaning of K_S is that when the substrate concentration S is equal to K_S, then the specific growth rate is equal to $\mu_{max}/2$.

The Monod equation assumes that there is only one limiting substrate and all the other substrates or mineral elements which are needed for microbial growth are in excess. This is often a realistic assumption for many wastewater treatment plants and therefore the Monod equation is widely used. For heterotrophic microorganisms, which feed on organic carbon, the limiting substrate is usually considered to be the carbon source. For autotrophic microorganisms, which use carbon dioxide as carbon source, the carbon source is usually considered to be in excess; therefore, other substrates are assumed to be rate limiting. For aerobic nitrifiers the limiting substrate is usually considered to be ammonia, while for hydrogenotrophic methanogens the rate limiting substrate is often considered to be the dissolved hydrogen.

If there is more than one limiting substrate, Monod equation can be easily adapted by adding more terms of the form $S_i/(K_{Si}+S_i)$. For example, for the growth of heterotrophic microorganisms in the absence of oxygen but with nitrate as electron acceptor we can assume that both the carbon

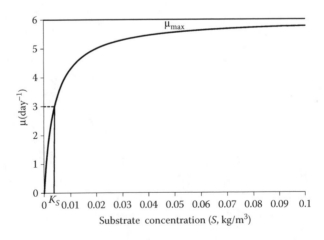

FIGURE 2.2 Typical shape of the rate equation for microbial growth on the limiting substrate S. This curve has been obtained with $\mu_{max} = 6$ day^{-1} and $K_S = 0.004$ kg/m^3.

source and nitrate can be rate limiting; therefore, the specific growth rate can be expressed as:

$$\mu = \frac{\mu_{max} S}{K_S + S} \times \frac{NO_3}{K_{NO_3} + NO_3} \qquad (2.57)$$

Equation 2.57 will be used to express the growth rate of heterotrophic microorganisms in the anoxic stage of the activated sludge process (or of the Sequencing Batch Reactor process) for carbon and nitrogen removal. A similar approach can be used to model the specific growth rate of aerobic heterotrophic microorganisms under conditions of low concentrations of dissolved oxygen. In this case both the carbon source and dissolved oxygen can be assumed to be rate limiting:

$$\mu = \frac{\mu_{max} S}{K_S + S} \times \frac{O_2}{K_{SO_2} + O_2} \qquad (2.58)$$

Table 2.4 summarises the limiting substrate and the form of the Monod equation for the various microbial populations considered in this book.

Many variations of the Monod model for the specific growth rate have been reported in the literature, to extend the validity of the equation beyond the simple case for which it was originally designed. One important modification of the Monod equation is the so-called Haldane or Andrews equation, which includes substrate inhibition. Indeed, many organic species which are substrates for microorganisms have an inhibiting effect on

TABLE 2.4 Limiting Substrate for the Monod Equation for the Main Types of Microorganisms Considered in This Book

Microorganisms	Limiting Substrate	Monod Equation
Heterotrophs (aerobic)	COD	$\mu = \dfrac{\mu_{max} S}{K_S + S}$
Heterotrophs (anoxic)	COD, nitrate	$\mu = \dfrac{\mu_{max} S}{K_S + S} \times \dfrac{NO_3}{K_{NO3} + NO_3}$
Autotrophs	NH_3	$\mu = \dfrac{\mu_{max\,A} NH_3}{K_{SA} + NH_3}$
Fermentative	Glucose	$\mu = \dfrac{\mu_{max\,GLU} GLU}{K_{SGlu} + GLU}$
Acetoclastic methanogens	Acetic acid	$\mu = \dfrac{\mu_{max\,AC} Ac}{K_{SAC} + Ac}$
Hydrogenotrophic methanogens	Hydrogen	$\mu = \dfrac{\mu_{maxH2} H_2}{K_{SH2} + H_2}$

microbial growth at high concentrations. In these cases the Haldane or Andrews equation can be used and it has the form:

$$\mu = \frac{\mu_{max} S}{K_S + S + (S^2/K_I)} \tag{2.59}$$

In Equation 2.59 in addition to μ_{max} and K_S, we have an additional parameter K_I. The lower the K_I value, the larger the inhibiting effect of the substrate or the lower the substrate concentration which is required to give a decrease in μ. A typical plot of the Haldane equation is shown in Figure 2.3.

Other modifications of the Monod equation have been proposed and used in the literature. For example, if microorganisms are growing on substrate S but are inhibited by another substance P, the specific growth rate can have the following form:

$$\mu = \frac{\mu_{max} S}{K_S + S} \frac{1}{1 + (P/K_P)} \tag{2.60}$$

Equations like 2.60 are widely used especially in anaerobic digestion. For example, this equation is often used to describe the inhibition of hydrogen on the growth of fermentative microorganisms on glucose and on the growth of acetogens microorganisms on volatile fatty acids.

Another important effect on the specific growth rate is pH. Microorganisms are typically only able to grow in a restricted pH range and,

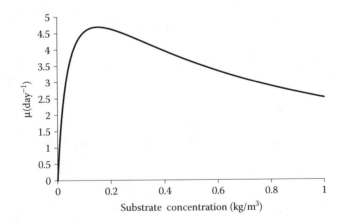

FIGURE 2.3 Typical shape of the rate equation for microbial growth on an inhibiting substrate S. This curve has been obtained with $\mu_{max} = 7.4$ day^{-1}, $K_S = 0.044$ kg/m3 and $K_I = 0.525$ kg/m^3, which are the parameters reported for microbial growth on phenol.

within this range; there is typically a pH value for which the specific growth rate is maximum. The effect of pH on the specific growth rate is often described by equations such as the following one:

$$\mu = \frac{\mu_{max} S}{K_S + S} \frac{1 + 2 \times 10^{0.5(a-b)}}{1 + 10^{(pH-b)} + 10^{(a-pH)}} \tag{2.61}$$

where a and b are two parameters which represent, respectively, the low and high extremes of pH for which there is still microbial activity. A plot of the pH inhibition factor is shown in Figure 2.4.

Another important effect on microbial growth rate is the effect of temperature. Typically the rate of microbial growth increases with temperature when the temperature increases from room temperature to approximately 35°C–40°C. At higher temperatures, denaturation of the cellular enzymes may start to occur, with a consequent decrease in microbial activity and in the growth rate. However, it is important to observe that temperature affects different microorganisms and different fermentation reactions in different ways, and for example anaerobic digestion is reported to occur even at temperatures higher than 60°C or 70°C, sometimes with rates higher than at 35°C–40°C. In the region where the reaction rate increases with temperature, it is usually assumed that the temperature dependence can be included in the μ_{max} parameter, for example with a conventional Arrhenius equation such as:

$$\mu_{max} = Ae^{-\frac{Ea}{RT}} \tag{2.62}$$

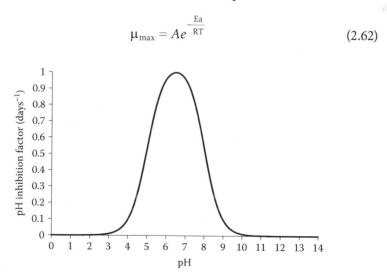

FIGURE 2.4 Typical shape of the pH inhibition factor. This curve has been obtained with $a = 5$, $b = 8$ which are parameters obtained for anaerobic growth on glucose.

2.1.2.2 Hydrolysis of Slowly Biodegradable Substrates

Not all the substrate in biological wastewater treatment processes is readily biodegradable. Some substances need to be hydrolysed before being transported inside the cell and metabolised. Typically high molecular weight species such as starch, lipids, proteins, polysaccharides have to be hydrolysed to their constituting monomers. The latter are readily biodegradable substrates that are metabolised by the microorganisms. The substances that require hydrolysis prior to metabolisation are called 'slowly biodegradable' substrates and indicated in this book with the symbol X_S. Figure 2.5 shows the general scheme of the hydrolysis process.

The kinetics of the hydrolysis process are usually expressed by the following rate equation:

$$r_{\text{hydr}} \left(\frac{\text{kg substrate}}{\text{m}^3 \times \text{day}} \right) = -k_{\text{h}} \frac{X_S/X}{K_X + (X_S/X)} X \qquad (2.63)$$

where the rate is taken with the negative sign because the substrate X_S is being removed. This equation indicates that the rate of hydrolysis depends on the biomass concentration and on the ratio between slowly biodegradable substrate and biomass concentration. The rationale for this dependence is that it is usually assumed that hydrolysis of high molecular weight species occurs on the external surface of the cells. The plot of the specific rate of hydrolysis

$$\frac{r_{\text{hydr}}}{X_H} \left(\frac{\text{kg substrate}}{\text{kg biomass} \times \text{day}} \right) = -k_{\text{h}} \frac{X_S/X_H}{K_X + (X_S/X_H)}$$

vs the (X_S/X_H) ratio is shown in Figure 2.6 (where for simplicity r_{hydr} is taken with the positive sign). The plot shows a profile which is similar to the μ vs S profile of the Monod equation for readily biodegradable substrates.

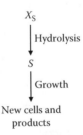

FIGURE 2.5   Conceptual scheme of the hydrolysis process.

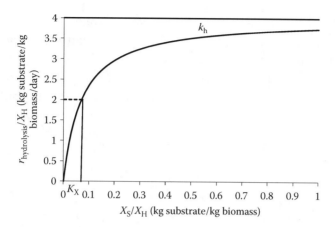

FIGURE 2.6 General profile of the specific rate of hydrolysis (taken with the positive sign) vs the ratio between slowly biodegradable substrate and biomass.

There is an important observation to make about the units of r_{hydr} and about the relationship between r_{hydr} and r_S, that is between the rate of hydrolysis of the slowly biodegradable substrate X_S and the rate of formation of the readily biodegradable substrate S. In the hydrolysis process one molecule of water is added to each monomer of the substrate that is hydrolysed. For example, if the substrate X_S is cellulose, its formula is $(C_6H_{10}O_5)_n$ and when cellulose is hydrolysed one water molecule is added to make glucose, $C_6H_{12}O_6$. Therefore, if the rate of cellulose hydrolysis r_{hydr} is expressed with the units of kg cellulose/m^3.day, then the rate of glucose formation r_S will be given by $r_S(kg\,glucose\,/\,m^3.day) = -(180\,/\,162)r_{hydr}$ where the ratio 180/162 is the ratio of the molecular weights of glucose and of the cellulose monomer. However, if the rate of cellulose hydrolysis and of glucose formation are both expressed in COD units, for example kg COD/m^3.day, then the rate of glucose formation is related to the rate of cellulose hydrolysis simply by: $r_S(kg\,COD\,/\,m^3.day) = -r_{hydr}$ because there is no loss or generation of COD during the hydrolysis process. So, in general, when calculating r_S from r_{hydr} we need to be careful about which units are being used for r_{hydr} and r_S. If X_S and S are expressed as mass of substrate, then the conversion between r_{hydr} and r_S needs to take into account the molecular formula of X_S and S, otherwise if X_S and S are both expressed as COD the numerical value of r_S will coincide with the numerical value of r_{hydr}, because there is no change in COD due to hydrolysis.

2.1.2.3 Endogenous Metabolism

Endogenous metabolism accounts for biomass auto oxidation, which occurs both in the absence and in the presence of external substrates. In the simplest model, under aerobic conditions endogenous metabolism converts biomass into carbon dioxide, water and ammonia and can be represented by the following stoichiometry:

$$C_5H_7O_2N + 5O_2 \rightarrow 5CO_2 + 2H_2O + NH_3 \qquad (2.64)$$

The stoichiometry of reaction (Equation 2.64) can be obtained, similarly to what was described for growth reactions previously, by using elemental balances. Similarly, for anoxic conditions, the stoichiometry of endogenous metabolism can be written as:

$$C_5H_7O_2N + 4HNO_3 \rightarrow 5CO_2 + 4H_2O + NH_3 + 2N_2 \qquad (2.65)$$

More detailed models of endogenous metabolism have also been reported. For example, it is possible to assume that not all the biomass is converted to carbon dioxide, but part of it is converted to inert products. In this case it is not possible to write an exact stoichiometry, since the formula of inert products is usually not known, and the generic stoichiometry can be written as:

$$C_5H_7O_2N + O_2 (\text{or nitrate}) \rightarrow CO_2 + H_2O + NH_3 (+N_2) + \text{products} \quad (2.66)$$

Endogenous metabolism under anaerobic conditions is not well known, but it is generally assumed that biomass is converted to inert solid products (which can be assimilated to dead microorganisms), which we will indicate as X_{inert} (kg/m^3).

In this book for simplicity we will assume that endogenous metabolism under aerobic and anoxic conditions generates only carbon dioxide, water and ammonia as product and we will ignore the possible formation of inert products. Under anaerobic conditions we will assume that endogenous metabolism produces inert solids, X_{inert}.

Endogenous metabolism is a very important phenomenon in biological wastewater treatment plants, and accounts for the fact that the net biomass production cannot be calculated simply from the growth yield, Equation 2.37. Indeed, endogenous metabolism causes a decrease in biomass concentration which is independent on the substrate concentration and on the growth rate. Therefore, due to endogenous metabolism in biological wastewater treatment processes the biomass concentration and the biomass production are lower than the values that could be calculated from the removed substrate using Equation 2.37. Also, it is evident from Equations 2.64 and 2.65 that endogenous metabolism under aerobic and anoxic conditions consumes the electron acceptor, oxygen or nitrate. The oxygen consumption due to endogenous metabolism is often an important contribution to the overall oxygen consumption in aerobic biological wastewater treatment processes.

The rate of endogenous metabolism is assumed to be dependent on the biomass concentration, that is:

$$r_{end}\left(\frac{\text{kg biomass}}{\text{m}^3 \times \text{day}}\right) = -bX \tag{2.67}$$

where b is an empirical coefficient. Like the growth rate, also the rate of endogenous metabolism is affected by inhibiting substances, pH and temperature.

In addition or in alternative to endogenous metabolism, other models have been developed to explain the biochemical phenomena which occur simultaneously to biomass growth in biological wastewater treatment plants (and in fermentation reactions in general). The most important of these models is based on the concept of 'maintenance'. Maintenance is the use of substrate for other metabolic functions not directly related to microorganisms growth, for example to generate energy for motility, to maintain the internal osmotic pressure or to repair damaged internal components. Even though maintenance and endogenous metabolism are different concepts, their effect on biological processes is similar (although not identical), that is they both cause the growth of microorganisms not to be entirely coupled with substrate removal and they both give an oxygen (or nitrate) consumption which is additional to the one due to biomass growth. For these reasons, it is usually difficult to distinguish between endogenous metabolism and maintenance. Therefore, usually either endogenous metabolism or maintenance is used in the modelling of biological wastewater treatment processes, but not both. In this book, we will only use the concept of endogenous metabolism.

2.1.2.4 Values of the Kinetic Parameters

Table 2.5 shows typical values or range of kinetic parameters from the literature. These values are only interesting for the order of magnitude of the parameters, because the reported values are very variable and values outside the given range are also common and reasonable. Typically,

TABLE 2.5   Typical Range of Values for the Main Parameters in
Kinetic Models of Biological Wastewater Treatment Processes.
Values Outside the Range Given Are Also Common

Parameter	Typical Values	Units
μ_{max}	0.1–10	day^{-1}
K_S	0.002–0.3	kg substrate/kg biomass
$Y_{X/S}$	0.05–0.5	kg biomass/kg substrate
b	0.05–0.4	day^{-1}
k_h	0.1–5	kg COD/kg biomass.day
K_X	0.07	kg COD/kg biomass

aerobic or anoxic growth of heterotrophic microorganisms show the highest
specific growth rate, that is the highest values of μ_{max} and the lowest values
of K_S. Nitrifying and anaerobic microorganisms, however, show lower values
of μ_{max} and, especially for anaerobic microorganisms, higher values of K_S. It
is important to remember that all the values depend on the nature of the substrate being metabolised or hydrolysed; therefore, it is reasonable that these
parameters might be different for different wastewaters. The value of the
growth yield $Y_{X/S}$ is particularly sensitive to the nature of the substrate and
in general is lower under anaerobic conditions than aerobic or anoxic ones,
because microorganisms obtain less energy from catabolic reactions per unit
of substrate being catabolised; therefore, a larger fraction of the substrate
needs to be used for catabolism, and this results in a lower growth yield.

2.1.3 Overall Rate Equations for Generation and Removal of Substrates and Products

In the previous sections we have seen how to write the stoichiometry and
kinetics of microbial growth on organic or inorganic substrates. The stoichiometry and kinetics expressions that we have written need to be combined together to obtain the rate expressions, or rate equations, for the
generation and removal of the substrates and products of the fermentation
reactions. For example let us consider the stoichiometry of heterotrophic
microorganisms' growth on soluble substrates under aerobic conditions
and use ethanol as an example substrate. From the general stoichiometry
for heterotrophic aerobic growth in Table 2.3, suing the molecular formula of ethanol, C_2H_6O, we obtain:

$$C_2H_5OH + \left(3 - 2.04Y_{X/S}\right)O_2 + 0.41Y_{X/S}NH_3 \rightarrow$$

$$0.41Y_{X/S}C_5H_7O_2N + \left(2 - 2.04Y_{X/S}\right)CO_2 + \left(1.5 - 0.82Y_{X/S}\right)H_2O \tag{2.68}$$

Equation 2.68 represents the stoichiometry of heterotrophic aerobic growth on ethanol, as a function of the growth yield $Y_{X/S}$, which is expressed as kg biomass/kg substrate. In the equations for the biomass growth rate reported hereafter, we will assume that the biomass concentration X is expressed as kg biomass/m³ and the specific growth rate μ as day⁻¹. We assume the Monod equation for the biomass growth rate (but the considerations we are doing here are valid whatever rate equation is chosen for biomass growth):

$$r_X\left(\frac{\text{kg biomass}}{\text{m}^3 \times \text{day}}\right) = \frac{\mu_{max}S}{K_S + S} \times X \tag{2.69}$$

From the growth stoichiometry (Equation 2.68) and the rate equation for biomass growth (Equation 2.69), the rate of generation of substrates and products can be calculated as follows, where a positive sign means that a species is being produced, negative sign means a species is being removed:

$$r_S\left(\frac{\text{kg substrate (ethanol)}}{\text{m}^3 \times \text{day}}\right) = -r_X\frac{1}{Y_{X/S}} \tag{2.70}$$

$$r_{NH3}\left(\frac{\text{kg ammonia (as nitrogen)}}{\text{m}^3 \times \text{day}}\right) = -r_X\frac{1}{8.07} \tag{2.71}$$

$$r_{O_2}\left(\frac{\text{kg oxygen}}{\text{m}^3 \times \text{day}}\right) = -\left(\frac{2.08}{Y_{X/S}} - 1.42\right)r_X \tag{2.72}$$

$$r_{CO_2}\left(\frac{\text{kg carbon dioxide}}{\text{m}^3 \times \text{day}}\right) = \left(\frac{1.90}{Y_{X/S}} - 1.94\right)r_X \tag{2.73}$$

$$r_{H_2O}\left(\frac{\text{kg carbon dioxide}}{\text{m}^3 \times \text{day}}\right) = \left(\frac{0.58}{Y_{X/S}} - 0.32\right)r_X \tag{2.74}$$

Equations 2.70 through 2.74 have been obtained with the same approach shown by Equation 2.3 and Example 2.1. For example, Equation 2.73 for the rate of carbon dioxide production has been obtained from:

$$r_{CO2} = \frac{(2 - 2.04Y_{X/S})}{0.41Y_{X/S}}\frac{MW_{CO2}}{MW_{biomass}}r_X \tag{2.75}$$

The same approach can be used to express the rate of generation and consumption of the species associated with the growth of other types of microorganisms. For nitrifiers the growth stoichiometry is, from Table 2.3:

$$NH_3 + 5\frac{Y_{XA/NO3}}{Y_{XA/NO3} + 8.07}CO_2 + \left(0.75 + \frac{10 - 5.75\,Y_{XA/NO3}}{Y_{XA/NO3} + 8.07}\right)O_2 \rightarrow$$

$$\frac{Y_{XA/NO3}}{Y_{XA/NO3} + 8.07}C_5H_7O_2N + \frac{8.07}{Y_{XA/NO3} + 8.07}HNO_3 + \tag{2.76}$$

$$\left(1.5 - \frac{4.035 + 3.5\,Y_{XA/NO3}}{Y_{XA/NO3} + 8.07}\right)H_2O$$

We assume that the growth rate of the autotrophic microorganisms is expressed as

$$r_{XA}\left(\frac{kg\,biomass}{m^3 \times day}\right) = \frac{\mu_{maxA}\,NH_3}{K_{SA} + NH_3} \cdot X_A \tag{2.77}$$

From the growth stoichiometry (Equation 2.76) and the rate equation for autotrophic microorganisms' growth (Equation 2.77) we can calculate the generation or consumption rate of all the other species involved in the metabolism of these microorganisms:

$$r_{NH3}\left(\frac{kg\,N-NH_3}{m^3 \times day}\right) = -\left(\frac{1}{Y_{XA/NO3}} + 0.12\right)r_{XA} \tag{2.78}$$

$$r_{CO_2}\left(\frac{kg\,carbon\,dioxide}{m^3 \times day}\right) = -1.95\,r_{XA} \tag{2.79}$$

$$r_{O_2}\left(\frac{kg\,oxygen}{m^3 \times day}\right) = -\left(\frac{4.57}{Y_{XA/NO3}} - 0.04\right)r_{XA} \tag{2.80}$$

$$r_{NO3}\left(\frac{kg\,N-NO_3}{m^3 \times day}\right) = \frac{1}{Y_{XA/NO3}}r_{XA} \tag{2.81}$$

With the same procedure we can obtain the rates of generation or consumption of all the substances involved in the process of microbial growth for all the species of microorganisms considered in this book.

In summary, we have seen that from the knowledge of the reaction stoichiometry and of the rate of biomass growth, the generation and production rates of all the species involved in the fermentation reaction can be calculated as a function of the stoichiometric coefficients and of the rate expression used for microbial growth. Table 2.6 summarises the generation

TABLE 2.6 Summary of the Generation and Removal Rates of the Various Substances Involved in Microbial Growth, as a Function of the Microorganisms Growth Rate and Growth Yield. The Notation DX in This Table Refer Only to the Biomass Produced due to Growth, and Does Not Consider Endogenous Metabolism

Microorganisms	Substrate	Growth Yield	Growth Rate	Generation or Removal Rate	
Heterotrophs (oxygen)	$C_wH_xO_yN_z$	$Y_{X/S} = \dfrac{\Delta X}{(-\Delta S)}$	$r_X = \dfrac{\mu_{max}S}{K_{SH}+S} \times X$	Microorganisms	r_X
				Substrate	$-\dfrac{r_X}{Y_{X/S}}$
				Oxygen	$-\dfrac{32\left(w+\dfrac{x}{4}-\dfrac{y}{2}-\dfrac{3}{4}z-5\dfrac{Y_{X/S}MW_{substrate}}{MW_{biomass}}\right)}{Y_{X/S}MW_{substrate}}r_X$
				Ammonia	$-\left(0.12-\dfrac{14z}{Y_{X/S}MW_{substrate}}\right)r_X$
				CO_2	$\left(\dfrac{44w}{Y_{X/S}MW_{substrate}}-1.95z\right)r_X$

(Continued)

TABLE 2.6 (*Continued*)   Summary of the Generation and Removal Rates of the Various Substances Involved in Microbial Growth, as a Function of the Microorganisms Growth Rate and Growth Yield. The Notation DX in This Table Refer Only to the Biomass Produced due to Growth, and Does Not Consider Endogenous Metabolism

Microorganisms	Substrate	Growth Yield	Growth Rate	Generation or Removal Rate	
Heterotrophs (nitrate)	$C_wH_xO_yN_z$	$Y_{X/S} = \dfrac{\Delta X}{(-\Delta S)}$	$r_X = \dfrac{\mu_{max} S}{K_S + S} \times X$	Microorganisms	r_X
				Substrate	$-\dfrac{r_X}{Y_{X/S}}$
				Nitrate	$14\left(\dfrac{4}{5}w + \dfrac{1}{5}x - \dfrac{3}{5}z - \dfrac{2}{5}y - 4\dfrac{Y_{X/S}MW_{substrate}}{MW_{biomass}}\right)\dfrac{r_X}{Y_{X/S}MW_{substrate}}$
				Ammonia	$-\left(0.12 - \dfrac{14z}{Y_{X/S}MW_{substrate}}\right) r_X$
				CO_2	$\left(\dfrac{44w}{Y_{X/S}MW_{substrate}} - 1.95z\right) r_X$
Heterotrophs (aerobic)	COD	$Y_{X/S} = \dfrac{\Delta X}{(-\Delta S_{COD})}$	$r_X = \dfrac{\mu_{max} S}{K_S + S} \times X$	Microorganisms	r_X
				Substrate	$-\dfrac{r_X}{Y_{X/S}}$
				Oxygen	$-\left(\dfrac{1}{Y_{X/SCOD}} - 1.42\right) r_X$
Heterotrophs (anoxic)	COD	$Y_{X/S} = \dfrac{\Delta X}{(-\Delta S_{COD})}$	$r_X = \dfrac{\mu_{max} S}{K_S + S} \times X$	Microorganisms	r_X
				Substrate	$-\dfrac{r_X}{Y_{X/S}}$
				Nitrate	$-\left(\dfrac{1}{Y_{X/S}} - 1.42\right)\dfrac{r_X}{2.86}$

(*Continued*)

TABLE 2.6 (*Continued*) Summary of the Generation and Removal Rates of the Various Substances Involved in Microbial Growth, as a Function of the Microorganisms Growth Rate and Growth Yield. The Notation DX in This Table Refer Only to the Biomass Produced due to Growth, and Does Not Consider Endogenous Metabolism

Microorganisms	Substrate	Growth Yield	Growth Rate	Generation or Removal Rate	
Nitrifiers	NH_3	$Y_{XA/NO3} = \dfrac{\Delta X_A}{(-\Delta NO_3)}$	$r_{XA} = \dfrac{\mu_{maxA} NH_3}{K_{SA} + NH_3} \times X_A$	Microorganisms	r_{XA}
				Ammonia	$-\left(\dfrac{1}{Y_{XA/NO3}} + 0.12\right) r_{XA}$
				Oxygen	$-\left(\dfrac{4.57}{Y_{XA/NO3}} - 1.42\right) r_{XA}$
				Nitrate	$\dfrac{1}{Y_{XA/NO3}} r_{XA}$
				CO_2	$-1.95 r_{XA}$
Fermentative	Glucose	$Y_{X/S} = \dfrac{\Delta X_{GLU}}{(-\Delta S_{GLU})}$	$r_X = \dfrac{\mu_{maxGLU} S_{GLU}}{K_{SGLU} + S_{GLU}} \times X_{GLU}$	Microorganisms	r_X
				Substrate	$-\dfrac{r_X}{Y_{X/S}}$
				Acetic acid	$\left(\dfrac{0.67}{Y_{X/S}} - 0.88\right) r_X$
				Ammonia	$-0.12 r_X$
				CO_2	$\left(\dfrac{0.49}{Y_{X/S}} - 0.65\right) r_X$
				Hydrogen	$\left(\dfrac{0.04}{Y_{X/S}} - 0.059\right) r_X$

(*Continued*)

TABLE 2.6 (*Continued*) Summary of the Generation and Removal Rates of the Various Substances Involved in Microbial Growth, as a Function of the Microorganisms Growth Rate and Growth Yield. The Notation DX in This Table Refer Only to the Biomass Produced due to Growth, and Does Not Consider Endogenous Metabolism

Microorganisms	Substrate	Growth Yield	Growth Rate		Generation or Removal Rate
Methanogens (acetoclastic)	Acetic acid	$Y_{X/S} = \dfrac{\Delta X_{AC}}{(-\Delta S_{AC})}$	$r_X = \dfrac{\mu_{maxAC} S_{AC}}{K_{SAC} + S_{AC}} \times X_{AC}$	Microorganisms	r_X
				Substrate	$-\dfrac{r_X}{Y_{X/S}}$
				Methane	$\left(\dfrac{0.27}{Y_{X/S}} - 0.35\right) r_X$
				Ammonia	$-0.12 r_X$
				CO_2	$\left(\dfrac{0.73}{Y_{X/S}} - 0.97\right) r_X$
Methanogens (hydrogenotrophic)	Hydrogen	$Y_{X/S} = \dfrac{\Delta X_{H2}}{(-\Delta S_{H2})}$	$r_{XH2} = \dfrac{\mu_{maxH2} S_{H2}}{K_{SH2} + S_{H2}} \times X_{H2}$	Microorganisms	r_X
				Substrate	$-\dfrac{r_X}{Y_{X/S}}$
				Methane	$\left(\dfrac{2}{Y_{X/S}} - 0.35\right) r_X$
				Ammonia	$-0.12 r_X$
				CO_2	$-\left(\dfrac{5.5}{Y_{X/S}} + 0.097\right) r_X$

and removal rates of the species generated or removed by various fermentation reactions considered in this book, obtained from the stoichiometry of the respective reactions. Table 2.6 is also based on the applications of the COD balance shown in the next section. Other examples about the generation of rate equations from the overall growth stoichiometry are shown in Examples 2.4 and 2.5.

2.1.3.1 Use of the COD Balance for the Calculation of the Rate Equations

Let us consider the aerobic metabolism of an organic carbon source by heterotrophic microorganisms. We have seen in the previous section that, if we know the molecular formula of the substance, from knowledge of the kinetic parameters of biomass growth and from the growth yield we can calculate the rates of production and consumption of all the species present in the growth stoichiometric equation. However, in many cases the molecular formula of the organic carbon source in wastewater treatment plant is not known, since in reality we have a mixture of many different species. In these cases, the carbon source is expressed as COD, for example $kgCOD/m^3$, and, if the rate of biomass growth is known, the consumption rate of the substrate and of oxygen can still be calculated from the COD balance.

We have seen in Chapter 1 that the COD balance can be expressed as:

$$\begin{pmatrix} \text{Total 'removable'} \\ \text{electrons present in the} \\ \text{removed substrate} \end{pmatrix} = \begin{pmatrix} \text{'removable' electrons present in} \\ \text{the produced microorganisms} \\ + \text{electrons removed by oxygen} \end{pmatrix} \quad (1.8)$$

Equation 1.8 can be written as:

$$\begin{pmatrix} \text{Total COD removed} \\ \text{with the substrate} \end{pmatrix} = \begin{pmatrix} \text{COD of the produced} \\ \text{biomass} + \text{oxygen consumed} \end{pmatrix} \quad (2.82)$$

In symbols:

$$\left(-\Delta S_{COD}\right) = \Delta X_{COD} + \left(-\Delta O_2\right) \quad (1.9)$$

Assuming that biomass is produced by growth on a substrate and is consumed by endogenous metabolism, Equation 1.9 becomes, in terms of rates:

$$-r_{SCOD} = 1.42\left(r_X + r_{end}\right) - r_{O_2} \quad (2.83)$$

In Equation 2.83, we have taken into account that the conversion factor for biomass to COD is 1.42 kg biomass (COD)/kg biomass (see Chapter 1).

If the substrate is expressed as COD, we can define the growth yield as:

$$Y_{X/SCOD} = \frac{kg\,biomass}{kg\,substrate\,(as\,COD)} = \frac{r_X}{-r_{SCOD}} \tag{2.84}$$

So, if r_X is known, the substrate and oxygen consumption rate can be immediately expressed as:

$$r_{SCOD}\left(\frac{kgCOD}{m^3 day}\right) = -\frac{r_X}{Y_{X/SCOD}} \tag{2.85}$$

$$r_{O_2}\left(\frac{kgO_2}{m^3 day}\right) = r_X\left(1.42 - \frac{1}{Y_{X/SCOD}}\right) + 1.42 r_{end} \tag{2.86}$$

Note that both r_{SCOD} and r_{O2} are negative, because both the substrate and oxygen are consumed.

Therefore, if the carbon substrate is expressed as COD, the rate of substrate removal and oxygen consumption can still be expressed as function of the biomass growth rate, by defining the growth yield in terms of substrate COD and using Equations 2.85 and 2.86. However, if the molecular formula of the substrate is not known, we cannot calculate the rate of generation and production of the other species involved in the fermentation reaction, for example carbon dioxide and water.

As seen in Chapter 1, the COD balance does not apply only in the presence of oxygen but it is also valid when nitrate is the electron acceptor or under anaerobic conditions. Under anoxic conditions the COD balance can be written as follows, in terms of rates:

$$-r_{SCOD} = 1.42\left(r_X + r_{end}\right) - r_{N-NO3} \cdot 2.86 \tag{2.87}$$

And introducing the growth yield based on substrate COD defined as above, we obtain:

$$r_{SCOD}\left(\frac{kgCOD}{m^3 day}\right) = -\frac{r_X}{Y_{X/SCOD}} \tag{2.88}$$

$$r_{NO3}\left(\frac{kgN-NO_3}{m^3 day}\right) = \frac{r_X}{2.86}\left(1.42 - \frac{1}{Y_{X/SCOD}}\right) + \frac{1.42 r_{end}}{2.86} \tag{2.89}$$

Example 2.4

Consider the overall growth stoichiometry for the anaerobic fermentation of glycine to acetic acid obtained in Example 2.2. Assuming the rate of microorganisms growth on glycine is r_X (kg microorganisms/m³.day), write the rate equations for glycine, ammonia and hydrogen consumption and for acetic acid, carbon dioxide and water production.

Solution

The overall growth stoichiometry for the anaerobic fermentation of glycine yielding acetic acid and hydrogen has been calculated in Example 2.2 and is reported below:

$$C_2H_5O_2N + (0.66Y_{X/S} - 1)NH_3 + (1 - 2.2Y_{X/S})H_2 \rightarrow$$

$$0.66Y_{X/S}C_5H_7O_2N + (1 - 2.2Y_{X/S})CH_3COOH$$

$$+ 1.1Y_{X/S}CO_2 + 0.88Y_{X/S}H_2O$$

Therefore, using the procedure shown by Equation 2.3, we can calculate the rate of production and consumption of all the species, once the rate of biomass production r_X is known:

$$r_{Glycine}\left(\frac{kg}{m^3 day}\right) = -\frac{1}{0.66Y_{X/S}} r_X \frac{MW_{GLycine}}{MW_{biomass}} = -\frac{1}{Y_{X/S}} r_X$$

(this is an immediate consequence of the definition of $Y_{X/S}$).

$$r_{NH3}\left(\frac{kgN\text{-}NH_3}{m^3 day}\right) = -\frac{(0.66Y_{X/S} - 1)}{0.66Y_{X/S}} r_X \frac{MW_N}{MW_{biomass}} = -\left(0.12 - \frac{0.19}{Y_{X/S}}\right) r_X$$

$$r_{H2}\left(\frac{kg}{m^3 day}\right) = -\frac{(1 - 2.2Y_{X/S})}{0.66Y_{X/S}} r_X \frac{MW_{H2}}{MW_{biomass}} = -\left(\frac{0.0268}{Y_{X/S}} - 0.059\right) r_X$$

$$r_{CH3COOH}\left(\frac{kg}{m^3 day}\right) = \frac{(1 - 2.2Y_{X/S})}{0.66Y_{X/S}} r_X \frac{MW_{CH3COOH}}{MW_{biomass}} = \left(\frac{0.805}{Y_{X/S}} - 1.77\right) r_X$$

$$r_{CO_2}\left(\frac{kg}{m^3 day}\right) = \frac{1.1Y_{X/S}}{0.66Y_{X/S}} r_X \frac{MW_{CO2}}{MW_{biomass}} = 0.65r_X$$

$$r_{H_2O}\left(\frac{kg}{m^3 day}\right) = \frac{0.88Y_{X/S}}{0.66Y_{X/S}} r_X \frac{MW_{H2O}}{MW_{biomass}} = 0.21r_X$$

Example 2.5

Consider the overall growth stoichiometry for the anaerobic fermentation of palmitic acid to acetic acid and hydrogen obtained in Example 2.3. Assuming the rate of microorganisms growth on palmitic acid is r_X (kg microorganisms/m³.day), write the rate equations for palmitic acid, ammonia, water and carbon dioxide consumption and for acetic acid and hydrogen production.

Solution

The procedure is analogous to Example 2.4. The overall growth stoichiometry for the anaerobic fermentation of palmitic acid was obtained in Example 2.3 and is reported below:

$$C_{16}H_{32}O_2 + 2.27Y_{X/S}NH_3 + (14-10.26Y_{X/S})H_2O + 3.45Y_{X/S}CO_2 \rightarrow$$

$$2.27Y_{X/S}C_5H_7O_2N + (8-3.95Y_{X/S})CH_3COOH + (14-6.91Y_{X/S})H_2$$

Therefore, using the procedure shown by Equation 2.3, we can calculate the rate of production and consumption of all the species, once the rate of biomass production r_X is known:

$$r_{palmitic}\left(\frac{kg}{m^3 day}\right) = -\frac{1}{2.27Y_{X/S}}r_X\frac{MW_{palmitic}}{MW_{biomass}} = -\frac{1}{Y_{X/S}}r_X$$

(this is an immediate consequence of the definition of $Y_{X/S}$).

$$r_{NH3}\left(\frac{kgN-NH_3}{m^3 day}\right) = -\frac{2.27Y_{X/S}}{2.27Y_{X/S}}r_X\frac{MW_N}{MW_{biomass}} = -0.12r_X$$

$$r_{H_2O}\left(\frac{kg}{m^3 day}\right) = -\frac{(14-10.26Y_{X/S})}{2.27Y_{X/S}}r_X\frac{MW_{H_2O}}{MW_{biomass}} = -\left(\frac{0.98}{Y_{X/S}}-0.72\right)r_X$$

$$r_{CO_2}\left(\frac{kg}{m^3 day}\right) = \frac{3.45Y_{X/S}}{2.27Y_{X/S}}r_X\frac{MW_{CO_2}}{MW_{biomass}} = -0.59r_X$$

$$r_{CH_3COOH}\left(\frac{kg}{m^3 day}\right) = \frac{(8-3.95Y_{X/S})}{2.27Y_{X/S}}r_X\frac{MW_{CH_3COOH}}{MW_{biomass}} = \left(\frac{1.87}{Y_{X/S}}-0.92\right)r_X$$

$$r_{H_2O}\left(\frac{kg}{m^3 day}\right) = \frac{(14-6.91Y_{X/S})}{2.27Y_{X/S}}r_X\frac{MW_{H_2O}}{MW_{biomass}} = \left(\frac{0.109}{Y_{X/S}}-0.054\right)r_X$$

2.2 MASS TRANSFER

Mass transfer is important in biological wastewater treatment processes. The main reason why mass transfer is important is that in aerobic processes oxygen has to be provided to the system. Oxygen is often provided by generating air bubbles in the system (Figure 2.7) or by mechanical aerators, in which case the surface of the liquid is agitated vigorously to put the liquid in contact with the atmosphere. In the description of the theory in this section we will make the example that aeration is provided via air (or oxygen) bubbles, which is called aeration with diffusers, but the final equation is the same if the aeration is provided with mechanical aerators. Oxygen transfers from the bubbles to the liquid phase where it is used by the microorganisms for their metabolism. Also, another reason why mass transfer is important is that during aeration the carbon dioxide dissolved in the liquid phase transfers to the gas phase and this phenomenon may affect pH, which has to be kept in the right range for the optimum microbial metabolism. This section describes a theory to express the rate of mass transfer of a species from the gas to the liquid phase, or vice versa.

The mass transfer theory used in this book is based on the 'two-film' model, which is schematised in Figure 2.8. In Figure 2.8, oxygen is used as the substance being transferred as an example, but the model can be applied to any substance.

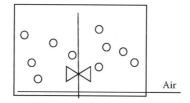

FIGURE 2.7   Conceptual scheme of oxygen transfer in bioreactors.

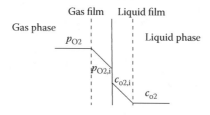

FIGURE 2.8   Scheme of the two-film model (oxygen is used as an example of the substance being transferred).

According to the 'two-film' model, all the resistance to mass transfer is located in two boundary layers (films), located in the gas and in the liquid phases (gas and liquid films). Oxygen (or in general the species being transferred) is present in the bulk gas phase at a partial pressure p_{O2} (typically expressed in atm) and in the bulk liquid phase at a concentration C_{O2} (expressed in mass or mol units, for example mg/L, kg/m³, or mol/L). In Figure 2.8, oxygen concentration in the bulk gas and bulk liquid phases is not in equilibrium, that is p_{O2} and c_{O2} are not in equilibrium with each other. If they were at equilibrium the rate of mass transfer would be zero. The two-film model assumes that at the interface between the liquid and gas phase oxygen concentration on the two sides of the interface is in equilibrium, that is $p_{O2,i}$ and $c_{O2,i}$ are in equilibrium with each other. We assume here that the equilibrium relationship is linear at all concentrations, that is $C_{O_2,i} = k_{eq} \times p_{O_2,i}$.

Since, according to the model, the mass transfer resistance is only present in the gas and liquid films, the rate of mass transfer per unit of transfer area (or mass transfer flux) can be expressed as:

$$J_{O_2}\left(\frac{kg_{O_2}}{m^2 \times day}\right) = k_g \times \left(p_{O_2} - p_{O_2,i}\right) = k_l \times \left(c_{O_2,i} - c_{O_2}\right) \qquad (2.90)$$

where k_g and k_l are the mass transfer coefficients for the gas and liquid phases respectively. The equation above still does not allow for an easy calculation of the mass transfer rate because, even assuming that the mass transfer coefficients are known from the literature or from experiments, the concentrations at the interface, $p_{O2,i}$ and $c_{O2,i}$ are not known. It would be desirable to express the mass transfer rate as a function of the known and easily measurable concentrations p_{O2} and c_{O2}. In order to do this, the concentration C^*_{O2} is introduced (Figure 2.9). C^*_{O2} is the oxygen concentration in the liquid phase that would be in equilibrium with the bulk gas phase, that is $C^*_{O2} = k_{eq} \times p_{O_2}$. While the concentrations p_{O2}, $p_{O2,i}$, $C_{O2,i}$ and

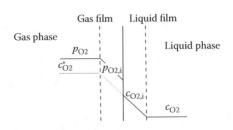

FIGURE 2.9　Two-film model with the introduction of the concentration C^*_{O2}.

C_{O2} exist physically in the system, the concentration C_{O2}^* is fictitious and does not exist physically. However, the advantage of introducing C_{O2}^* is that it can be easily calculated from p_{O2}, if the equilibrium relationship is known. Introducing C_{O2}^* the mass transfer flux can also be expressed as:

$$J_{O2}\left(\frac{\text{kg}_{O2}}{\text{m}^2 \times \text{day}}\right) = k_L \times \left(c_{O2}^* - c_{O2}\right) = k_L \times \left(k_{eq}\, p_{O2} - c_{O2}\right) \qquad (2.91)$$

where k_L is a 'global' mass transfer coefficient, as opposed to the 'local' mass transfer coefficients k_g and k_l introduced earlier. Global coefficient means that it is referred to both phases, while the local coefficients only refer to one phase. The advantage of using Equation 2.88 over Equation 2.87 for expressing the mass transfer rates lies in the fact that in Equation 2.88 the variables p_{O2} and C_{O2} represent values in the bulk values which are readily measurable, while Equation 2.87 includes the interface values $p_{O2,i}$ and $c_{O2,i}$ which are not measurable.

The value of the global coefficient k_L can be related to the values of the local coefficients k_g and k_l. From Equation 2.87 we have:

$$k_g \times \left(p_{O2} - p_{O2,i}\right) = k_l \times \left(k_{eq}\, p_{O2,i} - c_{O2}\right) \qquad (2.92)$$

From Equation 2.92 the value of $p_{O2,i}$ can be expressed as a function of the other variables:

$$p_{O2,i} = \frac{k_g\, p_{O2} + k_l c_{O2}}{k_g + k_{eq} k_l} \qquad (2.93)$$

Introducing this value of $p_{O2,i}$ into $k_g \times \left(p_{O2} - p_{O2,i}\right)$ we obtain:

$$
\begin{aligned}
k_g \times \left(p_{O2} - p_{O2,i}\right) &= k_g \times \left(p_{O2} - \frac{k_g\, p_{O2} + k_l c_{O2}}{k_g + k_{eq} k_l}\right) \\
&= \frac{k_g k_l}{k_g + k_{eq} k_l}\left(k_{eq}\, p_{O2} - c_{O2}\right)
\end{aligned}
\qquad (2.94)
$$

By equalling Equation (2.94) with Equation (2.91), we obtain an expression for k_L as a function of k_g, k_l and k_{eq}:

$$k_L = \frac{k_g k_l}{k_g + k_{eq} k_l} \qquad (2.95)$$

So it is evident that the global coefficient k_l depends on the local coefficients k_g and k_l and on the equilibrium constant k_{eq}. Equation 2.95 shows that if the gas has a high solubility (high value of k_{eq}), then $k_L \approx k_g$, that is the controlling resistance is on the gas side. However, if the gas has a low solubility (low value of k_{eq}), then $k_L \approx k_l$, that is the controlling resistance is on the liquid side. Oxygen is a gas with low solubility; therefore, in this case the controlling resistance is on the liquid side.

So far we have expressed the mass transfer flux J, that is the rate of mass transfer per unit mass transfer area. In the calculations of biological wastewater treatment plants we are interested in the volumetric mass transfer rate, that is mass transfer rate per unit volume of the biological reactor. We call the volumetric mass transfer rate (again using oxygen as an example) $r_{O_2}\left(kg\,O_2/m^3 \times day\right)$. r_{O_2} and J_{O_2} are obviously related as follows:

$$r_{O_2}\left(\frac{kg\;O_2}{m^3 \times day}\right) = J_{O_2}\left(\frac{kg\;O_2}{m^2 \times day}\right)\frac{A}{V} \tag{2.96}$$

where:

A is the total area of the bubbles present in the reactor
V is the liquid volume in the reactor.

We call the ratio $(A/V) = a$, where 'a' obviously represents the specific area of mass transfer, that is the mass transfer area per unit volume. Therefore, the desired volumetric mass transfer rate can be expressed as:

$$r_{O_2}\left(\frac{kg\,O_2}{m^3 \times day}\right) = k_L \times a \times \left(c^*_{O_2} - c_{O_2}\right) = k_L \times a \times \left(k_{eq}\,p_{O_2} - c_{O_2}\right) \tag{2.97}$$

The two parameters k_l and a have a totally different physical meaning. k_l is related to the resistance to mass transfer, while a accounts for the number and size of the bubbles in the reactor. However, it is difficult with simple experimental means to determine the values of the parameters k_l and a independently. In general, the combined values of $k_L \cdot a$ is determined and therefore the term $k_L a$ is usually considered one single parameter, rather than two individual parameters.

Even though it has been derived making the example of aeration from gas bubbles, Equation 2.97 can be applied in the same way also to the case of aeration with mechanical aerators. If aeration is carried out with mechanical aerators, p_{O_2} is very simply the partial pressure of oxygen in the atmosphere, that is approximately 0.21 atm.

It is important to note that Equation 2.97 can be used to describe the mass transfer of any substance, not just oxygen, from the gas to the liquid phase (or vice versa). For example, if the substance being transferred is carbon dioxide, Equation 2.97 can be written as $r_{CO2} \left(kg_{CO_2}/m^3 \times day \right) = k_L a_{CO_2} \times \left(k_{eqCO_2} p_{CO_2} - c_{CO_2} \right)$ and similar equations describe the mass transfer for any other substances.

2.2.1 Correlations for the Mass Transfer Coefficients

We have seen that the rate of mass transfer of a substance (in particular oxygen) from the gas to the liquid phase is proportional to the parameter $k_L a$. The $k_L a$ is usually determined from experimental data and a large number of correlations have been reported in the literature (Appendix A describes a simple method for the experimental measurement of $k_L a$ in the lab). In presenting the correlations for $k_L a$, we need to distinguish between aeration with diffusers (gas bubbles) and mechanical aeration. In both cases the $k_L a$ depends on the physical properties of air, water and oxygen; however, it also depends on the turbulence and fluid dynamics of the system and different correlations have been developed for the two methods of aeration.

For aeration by diffusers the mass transfer coefficient $k_L a$ depends on the turbulence of the system and on the size and number of gas bubbles in the system. Correlations in different forms have been developed in the literature, but in general they often take the form of:

$$k_L a V = k_{diff} \times Q_{gas}^{b_{diff}}$$ (2.98)

where:

V is the volume of the biological reactor

Q_{gas} is the gas flow rate

The coefficients k_{diff} and b_{diff} depend on many factors, for example the type of diffuser, the size of the holes and the geometry of the biological reactor. Therefore, correlations such as the one represented by Equation 2.98 need to be taken carefully and in general are only valid for the system for which they have been determined or for geometrically similar systems. In spite of their limitations, correlations such as Equation 2.98 are useful in understanding the effect of the gas flow rate on the mass transfer coefficients.

For mechanical aerators, the correlation for $k_L a$ includes the power draw of the mechanical agitator, the diameter and the rotational speed of the agitator. It has been found that for a large number of agitators and for relatively high turbulence of the liquid phase, the $k_L a$ for mechanical aerators can be expressed as:

$$k_L a V \cong 2 \times 10^{-6} \frac{P^{1.5}}{D^3 N^{1.5}} \tag{2.99}$$

In Equation 2.99 $k_L a V$ is expressed as m³/s, P is the power draw of the agitator (W), D is its diameter (m), and N is its rotational speed (1/s). Equation 2.99 has been found valid for values of the Froude number $Fr = \left(N^2 D/g \right)$ (where g is the gravitational constant) higher than approximately 0.1. In Equation 2.99, the power draw P is in turn dependent on the agitator diameter and rotational speed, according to the relationship:

$$\frac{P}{\rho N^3 D^5} = k_{mech} \left(\frac{N^2 D}{g} \right)^{b_{mech}} \tag{2.100}$$

In Equation 2.100, in addition to the symbols already defined, ρ represents the density of the liquid phase and k_{mech} and b_{mech} are empirical parameters which are characteristics of the specific type of agitator. Note that the term $(P/\rho N^3 D^5)$ represents the (adimensional) power number for the agitator, often indicated as P_0. Substituting Equation 2.100 into Equation 2.99 and assuming a density of 1000 kg/m³ for the liquid phase and a gravitational constant of 9.8 m/s², we obtain:

$$k_L a V \left(\frac{m^3}{s} \right) = \frac{0.063 k_{mech}^{1.5}}{9.8^{1.5 b_{mech}}} N^{(3+3b_{mech})} D^{(4.5+1.5 b_{mech})} \tag{2.101}$$

Equation 2.101 represents a correlation for $k_L a$ for mechanical aerators as a function of the agitator type (which affects the values of k_{mech} and b_{mech}), rotational speed (N) and diameter (D). As noted previously for aeration with diffusers, correlations such as Equation 2.101 need to be taken carefully and obviously cannot be expected to give accurate predictions, due to the complexity of the mass transfer phenomenon. However, they are useful for an estimation of the mass transfer coefficient that a certain agitator can provide.

It is also important to observe that in the literature many other correlations for mass transfer in biological reactors have been developed, more or less different from Equations 2.98 and 2.99.

2.2.2 Power and Efficiency for Aerators

In mass transfer calculations the most important aspect is the calculation of the required power consumption, because this determines the operating costs of the aeration system. For aeration with diffusers, the power requirement is determined by the power required to compress the air (or the oxygen, in case pure oxygen is used for aeration) from atmospheric pressure to the pressure at the outlet of the diffusers. Since the diffusers are usually placed at the bottom of the reactor, the pressure at the outlet of the diffusers is due to the hydrostatic pressure of the water in the reaction tank, which is proportional to the liquid depth. The equation that gives the compression power is

$$P_{compr}\left(W\right) = \frac{\gamma}{\gamma-1} Q_{air}\, p_{inlet} \left[\left(\frac{p_{outlet}}{p_{inlet}}\right)^{\frac{\gamma-1}{\gamma}} - 1 \right] \qquad (2.102)$$

In Equation (2.102) Q_{air} is the air flow rate in m³/s, p_{inlet} is (usually) the atmospheric pressure and p_{outlet} is the pressure at the outlet of the diffusers, which is proportional to the liquid depth, and γ is a coefficient which for air equals to 1.4. Note that Equation 2.102 gives the power for an ideal compression, and it needs to be divided by the efficiency of the compressor, $\eta_{compressor}$, typically in the range 0.7–0.8. Note that the required compression power is proportional, as expected, to the air flow rate; therefore, it is desirable to achieve the highest possible mass transfer coefficient k_La with the lowest possible air flow rate Q_{air}. Therefore, it is desirable to have diffusers with the most favourable k_La correlation, according to Equation 2.98.

For mechanical aeration, the required power draw by the agitator is given by Equation 2.100, which becomes, assuming a density of 1000 kg/m³:

$$P(W) = \frac{k_{mech}\, 1000\; N^{3+2b_{mech}}\, D^{5+b_{mech}}}{9.8^{b_{mech}}} \qquad (2.103)$$

As expected the power draw is proportional to the agitator speed and diameter, and to the parameters k_{mech} and b_{mech} which are specific for the particular agitator type.

When expressing the mass transfer performance of diffusers or mechanical agitators in biological wastewater treatment, it is common to use the concept of efficiency, which is defined in different ways for diffusers and mechanical aerators. For aeration with diffusers, efficiency is defined as the ratio between the oxygen transferred and the oxygen fed to the system. Typically the value of the efficiency of diffusers is below 20%, meaning that usually over 80% of the oxygen fed to the system is not transferred to the liquid phase and leaves the system with the outlet gas stream. For mechanical aerators the efficiency is usually defined as the ratio between the mass of oxygen transferred to the liquid phase and the power absorbed by the agitator, and is usually expressed with the units of kg of oxygen per kWh. Typical values of the efficiency of mechanical aerators are in the range 0.7–1.2 kg O_2/kWh. Of course, for both aeration with diffusers and mechanical aeration it is desirable to have the highest possible efficiency of mass transfer.

2.3 pH CALCULATION

pH is an important variable in biological wastewater treatment processes, since microorganisms are only active in a certain, often narrow, pH range. The pH in biological reactor depends on the characteristics of the influent, its pH and alkalinity, and is affected by the biological reactions occurring in the system. In this section we will see the fundamental equations for pH calculation in biological process.

pH is defined as:

$$pH = -\log\left[H_3O^+\right] \tag{2.104}$$

From Equation (2.104), it follows immediately that:

$$\left[H_3O^+\right] = 10^{-pH} \tag{2.105}$$

The basis for pH calculation is the charge balance, that is the condition of electroneutrality of the solution:

$$\sum positive\,charges = \sum negative\,charges \tag{2.106}$$

This condition results in one nonlinear equation in one unknown, the concentration of hydrogen ion $\left[H_3O^+\right]$, which can be solved to obtain the pH of the solution.

Example 2.6 pH of a solution of a strong acid

Calculate the pH of a solution of a strong acid, $2 \cdot 10^{-3}$ M HCl.

Solution

The positive charges in this solution will be the H_3O^+ ions, while the negative charges will be the Cl^- and OH^- ions. The charge balance is:

$$\left[H_3O^+ \right] = \left[Cl^- \right] + \left[OH^- \right] \tag{2.107}$$

The concentration of the hydroxyl ion OH^- and H_3O^+ are linked by their equilibrium in water:

$$2H_2O \leftrightarrow H_3O^+ + OH^- \tag{2.108}$$

The equilibrium constant of Equation (2.108) is

$$K_w = \left[H_3O^+ \right]\left[OH^- \right] = 1 \cdot 10^{-14} \tag{2.109}$$

Therefore, we can express OH^- as a function of H_3O^+:

$$\left[OH^- \right] = \frac{K_w}{\left[H_3O^+ \right]} \tag{2.110}$$

And the charge balance becomes

$$\left[H_3O^+ \right] = \left[Cl^- \right] + \frac{K_w}{\left[H_3O^+ \right]} \quad \Rightarrow \quad 10^{-pH} = \left[Cl^- \right] + \frac{K_w}{10^{-pH}} \tag{2.111}$$

Equation 2.111 is a nonlinear equation in the unknown pH. Since $\left[Cl^- \right] = 2 \cdot 10^{-3} \, M$, because strong acids are totally dissociated, solving the equation gives $pH = 2.70$.

Example 2.7 pH of a solution of a weak acid

Calculate the pH of a solution of a weak acid, $2 \cdot 10^{-3} \, M$ acetic acid CH_3COOH.

Solution

In this case the charge balance can be written as:

$$\begin{aligned} \left[H_3O^+ \right] &= \left[CH_3COO^- \right] + \frac{K_w}{\left[H_3O^+ \right]} \Rightarrow 10^{-pH} \\ &= \left[CH_3COO^- \right] + \frac{K_w}{10^{-pH}} \end{aligned} \tag{2.112}$$

In order to solve Equation 2.112, we need to calculate the concentration of the ion CH_3COO^-. We know that $2 \cdot 10^{-3}\,M$ is the total concentration of the acetic acid added (which we will call CH_3COOH_{tot}), which may be present in solution as dissociated ion (CH_3COO^-) and undissociated form (CH_3COOH), so we have:

$$\left[CH_3COOH_{tot}\right] = \left[CH_3COO^-\right] + \left[CH_3COOH\right] \qquad (2.113)$$

The two species CH_3COO^- and CH_3COOH are linked by the equilibrium of acetic acid in water:

$$CH_3COOH + H_2O \leftrightarrow CH_3COO^- + H_3O^+ \qquad (2.114)$$

The equilibrium constant of reaction (Equation 2.114) is:

$$K_{CH3COOH} = \frac{\left[CH_3COO^-\right]\left[H_3O^+\right]}{\left[CH_3COOH\right]} = 1.8 \cdot 10^{-5} \qquad (2.115)$$

Equation 2.115 can be rearranged as:

$$\left[CH_3COOH\right] = \frac{\left[CH_3COO^-\right]\left[H_3O^+\right]}{K_{CH3COOH}} \qquad (2.116)$$

And by substituting in the mass balance for acetic acid, Equation 2.112:

$$\left[CH_3COOH_{tot}\right] = \left[CH_3COO^-\right] + \frac{\left[CH_3COO^-\right]\left[H_3O^+\right]}{K_{CH3COOH}}$$

$$\Rightarrow \left[CH_3COO^-\right] = \frac{\left[CH_3COOH_{tot}\right]}{1 + \dfrac{\left[H_3O^+\right]}{K_{CH3COOH}}} \Rightarrow \qquad (2.117)$$

$$\left[CH_3COO^-\right] = \frac{\left[CH_3COOH_{tot}\right]}{1 + \dfrac{10^{-pH}}{K_{CH3COOH}}}$$

Equation 2.117 can be substituted in the charge balance (2.112):

$$10^{-pH} = \frac{\left[CH_3COOH_{tot}\right]}{1 + \dfrac{10^{-pH}}{K_{CH3COOH}}} + \frac{K_w}{10^{-pH}} \qquad (2.118)$$

Equation 2.118 can now be solved for pH since all the values of the parameters are known. We obtain pH = 3.74.

Note that Equation 2.117 can also be used to calculate the distribution of the dissociated and undissociated forms of acetic acid as a function of the pH of the solution. Indeed:

$$\frac{\left[CH_3COO^-\right]}{\left[CH_3COOH_{tot}\right]} = \frac{1}{1 + \dfrac{10^{-pH}}{K_{CH3COOH}}} \tag{2.119}$$

and

$$\frac{\left[CH_3COOH\right]}{\left[CH_3COOH_{tot}\right]} = \frac{10^{-pH}}{K_{CH3COOH} + 10^{-pH}} \tag{2.120}$$

The plots of Equations 2.119 and 2.120 are shown in Figure 2.10.

2.3.1 pH Buffers

pH buffers are a solution of an acid and its conjugate base and give the benefit that, when they are present, the pH of the solution is more resistant to changes due to addition of acid or base. An example of a pH buffer is the acetate buffer, which is prepared by mixing acetic acid and sodium acetate.

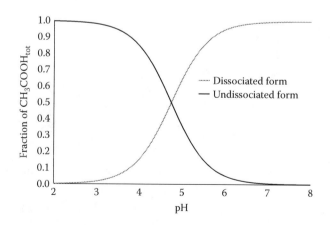

FIGURE 2.10 Example 2.7 Distribution of the dissociated (CH_3COO^-) and undissociated (CH_3COOH) forms of acetic acid as a function of pH.

To understand how the acetate buffer works, we need to consider the acetic acid equilibrium we have written previously:

$$CH_3COOH + H_2O \leftrightarrow CH_3COO^- + H_3O^+ \qquad (2.114)$$

If we have a solution where both CH_3COO^- and CH_3COOH are present, any externally added H_3O^+ ions, due for example to the addition of a strong acid or to biological reactions, will react with the acetate ion to give the undissociated form. Therefore, the added H_3O^+ ions will be at least partially neutralised and the pH of the solution will change much less than if there was no buffer.

Similarly, if a base is added and OH^- ions are added to the system, these ions will react with H_3O^+ giving an increase in pH. However, the undissociated acetic acid will dissociate and generate new H_3O^+ ions, and so the pH will change much less than if there was no buffer present.

We now want to calculate the pH of a buffer solution and its buffering power, that is its ability to resist to pH changes due to the addition of a strong acid or a strong base.

Let us consider we prepare a solution by adding certain concentrations of CH_3COOH and CH_3COONa in water. The pH of this solution can be calculated as usual with the charge balance:

$$\left[H_3O^+\right] + \left[Na^+\right] = \left[CH_3COO^-\right] + \frac{K_w}{\left[H_3O^+\right]} \qquad (2.121)$$

Substituting the expression for $[CH_3COO^-]$ as a function of CH_3COOH_{tot} we obtain:

$$\left[H_3O^+\right] + \left[Na^+\right] = \frac{\left[CH_3COOH_{tot}\right]}{1 + \dfrac{\left[H_3O^+\right]}{K_{CH_3COOH}}} + \frac{K_w}{\left[H_3O^+\right]} \qquad (2.122)$$

Equation 2.122 allows for the calculation of the pH of a solution containing the acetate buffer. Here $[CH_3COOH_{tot}]$ corresponds to the sum of the molar concentrations of the CH_3COOH and CH_3COONa which were used to prepare the solution. So the pH of this solution depends on both the total concentration of acetic acid + sodium acetate added and on the ratio between the added acid and conjugate base.

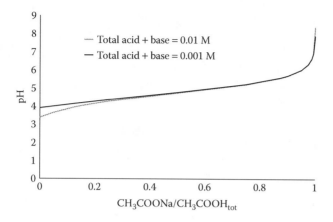

FIGURE 2.11   pH of a solution containing both acetic acid and acetate.

Figure 2.11 shows the pH of a solution containing both acetic acid and sodium acetate as a function of the ratio between the base and the total species added, for two different total concentrations.

The total concentration of acetic acid and acetate determines the buffering power of the solution, that is, which concentration of added acid or base the buffered solution can withstand without changing the pH too much. For example, let us assume that we have a buffer solution of acetic acid and acetate, prepared using equal concentrations of CH_3COOH and CH_3COONa. If we add a base, for example KOH, the pH of the resulting solution can be calculated by solving the equation:

$$\left[H_3O^+\right]+\left[Na^+\right]+\left[K^+\right]=\frac{\left[CH_3COOH_{tot}\right]}{1+\dfrac{\left[H_3O^+\right]}{K_{CH_3COOH}}}+\frac{K_w}{\left[H_3O^+\right]} \qquad (2.123)$$

Where [K+] is the concentration of the added base and, same as before, [Na+] is the concentration of the CH_3COONa added and [CH_3COOHtot] is the total concentration of the buffer. Figure 2.12 shows the results of the calculation for two different values of the total buffer concentration, compared with the case of no buffer, that is when [Na+] = [CH_3COOH_{tot}] = 0.

2.3.2 Equilibrium of Carbonic Acid

WE have seen that the presence of buffers is beneficial in wastewaters because it causes the wastewater to be more resistant to pH changes.

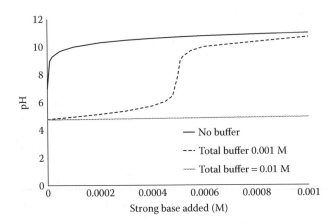

FIGURE 2.12 Effect of buffer concentration on the capacity of a solution to resist pH changes due to the addition of a strong base (acetate buffer prepared with $[CH_3COONa] = [CH_3COOH]$).

One typical buffer present in many wastewaters is carbonic acid, present in the various forms, undissociated (H_2CO_3), as bicarbonate (HCO_3^-) or carbonate (CO_3^{2-}). It is important to consider the equilibrium of carbon acid in water because it can act as a buffer against pH changes.

The equilibrium reactions for carbonic acid in water are written below. Equilibrium between carbonic acid and dissolved carbon dioxide:

$$H_2CO_3 \leftrightarrow CO_2 + H_2O \qquad K_{CO2} = \frac{[CO_2]}{[H_2CO_3]} = 588 \qquad (2.124)$$

Equilibrium between carbonic acid and bicarbonate:

$$H_2CO_3 + H_2O \leftrightarrow HCO_3^- + H_3O^+$$

$$K_{H2CO3} = \frac{[HCO_3^-][H_3O^+]}{[H_2CO_3]} = 2.5 \cdot 10^{-4} \qquad (2.125)$$

Equilibrium between bicarbonate and carbonate:

$$HCO_3^- + H_2O \leftrightarrow CO_3^{2-} + H_3O^+$$

$$K_{HCO3} = \frac{[CO_3^{2-}][H_3O^+]}{[HCO_3^+]} = 4.7 \cdot 10^{-11} \qquad (2.126)$$

The reactions (2.124–2.126) show that the system carbonic acid/bicarbonate/carbonate can act as a buffer against pH changes since any generation of H_3O^+ ions due to the biological reactions occurring in the system will be at least partially counterbalanced by these equilibria.

The equilibrium in Equations 2.124 through 2.126 allow for the calculation of the distribution of the various forms of carbonic acid as a function of pH. Indeed, having defined $[H_2CO_{3tot}]$ as the total concentration of carbonic acid in its various possible forms, we have:

$$[H_2CO_{3tot}] = [CO_2] + [H_2CO_3] + [HCO_3^-] + [CO_3^{2-}] \quad (2.127)$$

By introducing Equation 2.127 in Equations 2.124 through 2.126, we obtain, after rearrangements:

$$\frac{[CO_2]}{[H_2CO_{3tot}]} = \frac{K_{CO_2}}{1 + K_{CO_2} + \dfrac{K_{H_2CO_3}}{10^{-pH}} + \dfrac{K_{HCO_3}K_{H_2CO_3}}{10^{-2pH}}} \quad (2.128)$$

$$\frac{[H_2CO_3]}{[H_2CO_{3tot}]} = \frac{1}{1 + K_{CO_2} + \dfrac{K_{H_2CO_3}}{10^{-pH}} + \dfrac{K_{HCO_3}K_{H_2CO_3}}{10^{-2pH}}} \quad (2.129)$$

$$\frac{[HCO_3^-]}{[H_2CO_{3tot}]} = \frac{K_{H_2CO_3}}{10^{-pH}} \frac{1}{1 + K_{CO_2} + \dfrac{K_{H_2CO_3}}{10^{-pH}} + \dfrac{K_{HCO_3}K_{H_2CO_3}}{10^{-2pH}}} \quad (2.130)$$

$$\frac{[CO_3^{2-}]}{[H_2CO_{3tot}]} = \frac{K_{HCO_3}K_{H_2CO_3}}{10^{-2pH}} \frac{1}{1 + K_{CO_2} + \dfrac{K_{H_2CO_3}}{10^{-pH}} + \dfrac{K_{HCO_3}K_{H_2CO_3}}{10^{-2pH}}} \quad (2.131)$$

Equations 2.128 through 2.18 give the distribution of the various forms of carbonic acid, CO_2, H_2CO_3, HCO_3^- and CO_3^{2-} as a function of the pH of the solution. Figure 2.13 shows the distribution of these species.

2.3.3 Alkalinity

Alkalinity is defined as the capacity of a solution to neutralise acids. It is measured by adding a strong acid to the solution and recording the pH of the solution as a function of the concentration of acid added. The end point of the alkalinity measurement is typically taken as pH = 4.5, because it is

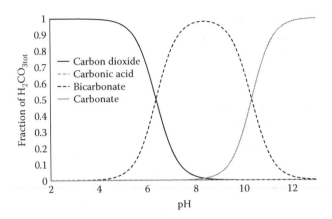

FIGURE 2.13   Distribution of carbonic acid between free dissolved CO_2, H_2CO_3, HCO_3^- and CO_3^{-2}.

assumed that at this pH all the basic species have been converted to their acid counterparts.

Often it can be assumed that the alkalinity of a wastewater is entirely due to the dissolved carbonate and bicarbonate ions. In this case it can be shown that by measuring the alkalinity and the initial pH of the solution, the solution is totally characterised in terms of its concentration of total carbonic acid and of strong acid and bases. Indeed, for a generic solution where the only weak acids and bases present are due to the equilibrium of carbonic acid, and where the concentrations of strong acid and bases are $[\Sigma An]$ and $[\Sigma Cat]$ the charge balance can be written as:

$$\left[H_3O^+\right]+\left[\Sigma Cat\right]=\left[HCO_3^-\right]+2\left[CO_3^{2-}\right]+\left[\Sigma An\right]+\left[OH^-\right] \quad (2.132)$$

which, taking into account the equilibrium of carbonic acid, can be re-written as:

$$10^{-pH}+\left[\Sigma Cat-\Sigma An\right]$$

$$=\frac{\left[H_2CO_{3,tot}\right]}{1+K_{CO2}+\dfrac{K_{H_2CO_3}}{10^{-pH}}+\dfrac{K_{HCO_3}K_{H_2CO_3}}{10^{-2pH}}} \quad (2.133)$$

$$\left(\frac{K_{H_2CO_3}}{10^{-pH}}+2\frac{K_{HCO_3}K_{H_2CO_3}}{10^{-2pH}}\right)+\frac{K_w}{10^{-pH}}$$

Equation 2.133 is the charge balance for the solution at the initial pH. If we measure the alkalinity of this solution and we terminate the measurement when pH has reached the value of 4.5, at the final conditions all the HCO_3^- and CO_3^{2-} will have been converted to H_2CO_3 (which in turn will be in equilibrium with the dissolved CO_2) and so will no longer appear in the charge balance. Since the alkalinity is defined as the concentration of acid that is required to bring the solution from the initial pH to the final pH, at the final pH conditions (pH = 4.5) the charge balance for the solution will be written as:

$$[\Sigma Cat - \Sigma An] = [Alk_{mol}] + \frac{K_w}{10^{-4.5}} - 10^{-4.5} \qquad (2.134)$$

Therefore, from the value of the alkalinity the net concentration [ΣCat − ΣAn] that is the net concentration of strong acids and bases in the wastewater can be calculated. Once this is known, from the charge balance written at the initial pH the term [H_2CO_{3tot}], that is the total carbonic acid concentration in the initial solution, can be calculated by re-arranging Equations 2.133:

$$[H_2CO_{3tot}] = \frac{10^{-pHinit} - \dfrac{K_W}{10^{-pHinit}} + [\Sigma Cat - \Sigma An]}{\dfrac{K_{H_2CO_3}}{10^{-pHinit}} + 2\dfrac{K_{HCO_3}K_{H_2CO_3}}{10^{-2pHinit}}} \qquad (2.135)$$

$$\left(1 + K_{CO_2} + \frac{K_{H_2CO_3}}{10^{-pHinit}} + \frac{K_{HCO_3}K_{H_2CO_3}}{10^{-2pHinit}}\right)$$

Knowledge of the terms [Σ Cat − Σ An] and [H_2CO_{3tot}] for the wastewater under consideration is important, because it will allow to carry out the charge balance for the wastewater in the biological processes, and in turn to calculate the pH of the biological reactor under different process conditions. This will be shown in the next chapters.

It is important to note that in Equation 2.134 the alkalinity has to be expressed in mol/L, while usually it is conventionally expressed as mgCaCO$_3$/L. The equivalence between these two different units of the alkalinity is shown in Example 2.8.

So far we have assumed that the alkalinity is only due to the carbonate the bicarbonate ions. In reality many other weak bases can contribute to the alkalinity, and one important example is ammonia. Ammonia is an important example because it is one of the parameters to be removed in wastewaters, and if it is present in significant concentrations affect the wastewater alkalinity and the relationship between alkalinity, [ΣCat − ΣAn] and [H_2CO_{3tot}]. Ammonia dissolved in water is subject to the chemical equilibrium:

$$NH_3 + H_2O \leftrightarrow NH_4^+ + OH^- \quad K_{eqNH3} = \frac{\left[NH_4^+\right]\left[OH^-\right]}{\left[NH_3\right]} = 1.8 \cdot 10^{-5} \quad (2.136)$$

Similarly as we have done with other species we define the total ammonia concentration as:

$$\left[NH_{3tot}\right] = \left[NH_3\right] + \left[NH_4^+\right] \quad (2.137)$$

The pH of a solution containing carbonic acid and ammonia, as well as strong acid and bases is given by a simple modification of Equation 2.133:

$$10^{-pH} + \left[\Sigma Cat - \Sigma An\right] + \left[NH_4^+\right] =$$

$$\frac{\dfrac{\left[H_2CO_{3,tot}\right]}{1 + K_{CO_2} + \dfrac{K_{H2CO3}}{10^{-pH}} + \dfrac{K_{HCO3} K_{H2CO3}}{10^{-2pH}}}}{\left(\dfrac{K_{H2CO3}}{10^{-pH}} + 2\dfrac{K_{HCO3} K_{H2CO3}}{10^{-2pH}}\right) + \dfrac{K_w}{10^{-pH}}} \quad (2.138)$$

Equation 2.138 can be re-written considering the equilibrium of ammonia in water, Equation 2.136, and Equation 2.137:

$$10^{-pH} + \left[\Sigma Cat - \Sigma An\right] + \frac{K_{eqNH3} NH_{3tot}}{K_{eqNH3} + \dfrac{K_w}{10^{-pH}}} =$$

$$\frac{\dfrac{\left[H_2CO_{3,tot}\right]}{1 + K_{CO_2} + \dfrac{K_{H2CO3}}{10^{-pH}} + \dfrac{K_{HCO3} K_{H2CO3}}{10^{-2pH}}}}{\left(\dfrac{K_{H2CO3}}{10^{-pH}} + 2\dfrac{K_{HCO3} K_{H2CO3}}{10^{-2pH}}\right) + \dfrac{K_w}{10^{-pH}}} \quad (2.139)$$

Equation 2.139 allows for the calculation of the pH of a solution of ammonia and carbonic acid in water. In terms of alkalinity, we can relate the alkalinity of this solution to its total concentration of carbonic acid, total concentration of ammonia and concentration of strong acid and bases with the same approach we used previously. If we measure the alkalinity of this solution by titration with a strong acid, when we reach the final pH 4.5 ammonia will all be present as NH_4^+ ($[NH_4^+] = [NH_{3tot}]$) and there will be no carbonates or bicarbonates present, so the charge balance will be:

$$[\Sigma Cat - \Sigma An] = [Alk_{mol}] + \frac{K_w}{10^{-4.5}} - 10^{-4.5} - [NH_{3tot}] \qquad (2.140)$$

Assuming the total concentration of ammonia is known, Equation 2.140 allows the calculation of $[\Sigma Cat - \Sigma An]$ if the alkalinity of the solution is known by experimental measurement. From the alkalinity and the initial pH we can calculate the total concentration of carbonic acid:

$$[H_2CO_{3tot}] = \frac{\left(10^{-pHinit} - \dfrac{K_W}{10^{-pHinit}} + [\Sigma Cat - \Sigma An] \right) + \dfrac{K_{eqNH_3} NH_{3tot}}{K_{eqNH_3} + \dfrac{K_W}{10^{-pHinit}}}}{\dfrac{K_{H_2CO_3}}{10^{-pHinit}} + 2\dfrac{K_{HCO_3} K_{H_2CO_3}}{10^{-2pHinit}}} \Bigg/ \left(1 + K_{CO_2} + \frac{K_{H_2CO_3}}{10^{-pHinit}} + \frac{K_{HCO_3} K_{H_2CO_3}}{10^{-2pHinit}} \right) \qquad (2.141)$$

Equations 2.140 and 2.141 are analogous to Equations 2.134 and 2.135 (which were derived under the assumption that carbonic acid was the only contributor to alkalinity in addition to strong acids and bases) and allow for the calculation of the total net concentrations of acids and bases and of the total concentration of carbonic acid, if we know the initial pH of the wastewater, its total ammonia concentration and its alkalinity.

Example 2.8

Show the equivalence between $[Alk_{mol}]$ and $[Alk_{CaCO3}]$ by calculating the alkalinity of a solution 100 mg $CaCO_3$/L.

Solution

Since the MW of $CaCO_3$ is 100 g/mol, 100 mg $CaCO_3$/L correspond to $1 \cdot 10^{-3}$ mol/L. Since Ca^{2+} is a bivalent cation, for this solution we have:

$$\left[\Sigma Cat - \Sigma An\right] = 2 \cdot 10^{-3} M$$

and therefore:

$$\left[Alk_{mol}\right] = 2 \cdot 10^{-3} - \frac{K_w}{10^{-4.5}} + 10^{-4.5} = 2.03 \cdot 10^{-3} M$$

Since, obviously: $\left[Alk_{CaCO3}\right] = 100 \dfrac{mgCaCO_3}{L}$

we have that:

$$\left[Alk_{mol}\right] = \left[Alk_{CaCO3}\right] \cdot 2.03 \cdot 10^{-5} \frac{mol}{mgCaCO_3}$$

In this book we will use the factor $2.0 \cdot 10^{-5}$ to convert the alkalinity expressed as mgCaCO$_3$/L to mol/L.

Example 2.9

Calculate the net concentration of strong acids and bases and the total concentration of carbonic acid for a solution for which the alkalinity curve in Figure 2.14 has been obtained.

Solution

From Figure 2.14 it can be seen that the initial pH of the solution is 9. Also, from close inspection of the curve it can be estimated that the concentration of strong acid required to bring the pH down to 4.5 is approximately 0.0103 M. Therefore, the terms [$\Sigma Cat - \Sigma An$] and [H_2CO_{3tot}] can be calculated as follows:

$$\left[\Sigma Cat - \Sigma An\right] = 0.0103 + \frac{K_w}{10^{-4.5}} - 10^{-4.5} \cong 0.0103 M$$

$$\left[H_2CO_{3tot}\right] = \frac{10^{-9} - \dfrac{K_w}{10^{-9}} + 0.0103}{\dfrac{K_{H_2CO_3}}{10^{-9}} + 2\dfrac{K_{HCO_3}K_{H_2CO_3}}{10^{-18}}}$$

$$\left(1 + K_{CO2} + \frac{K_{H_2CO_3}}{10^{-9}} + \frac{K_{HCO_3}K_{H_2CO_3}}{10^{-18}}\right) = 0.0099 M$$

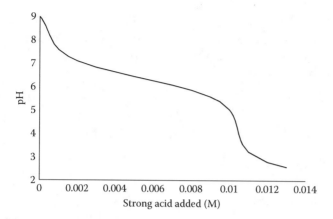

FIGURE 2.14   pH vs strong acid added for Example 2.9.

Example 2.10

A solution has a concentration of $[\Sigma Cat - \Sigma An] = 1.0 \cdot 10^{-3} M$ and $[H_2CO_{3tot}] = 1.1 \cdot 10^{-3} M$.

1. Calculate the pH and alkalinity of the solution;
2. Assume that a strong base is added to the solution so that its pH is increased to 9.6. Calculate the concentration of the base added and the new alkalinity of the solution;
3. Show that, using the pH and alkalinity values of case b), the total concentration of carbonic acid, which has remained unchanged in this problem, can be calculated correctly.

Solution

1. The pH of the solution is given by the charge balance under the initial conditions:

$$10^{-pH} + 1.0 \cdot 10^{-3} = \frac{1.1 \cdot 10^{-3}}{1 + K_{CO_2} + \dfrac{K_{H_2CO_3}}{10^{-pH}} + \dfrac{K_{HCO_3} K_{H_2CO_3}}{10^{-2pH}}}$$

$$\left(\frac{K_{H_2CO_3}}{10^{-pH}} + 2 \frac{K_{HCO_3} K_{H_2CO_3}}{10^{-2pH}} \right) + \frac{K_w}{10^{-pH}}$$

This equation, solved for pH, gives pH = 7.37. The alkalinity of the solution is given by:

$$[Alk] = 1.0 \cdot 10^{-3} + \frac{K_w}{10^{-4.5}} - 10^{-4.5} = 1.03 \cdot 10^{-3} M = 51 \text{ mgCaCO}_3/L$$

2. Under the new conditions of pH 9.6, the new net concentration of cations and anions $[\Sigma Cat - \Sigma An]$ can be calculated by the charge balance, which this time can be written as:

$$10^{-9.6} + [\Sigma Cat - \Sigma An] = \cfrac{1.1 \cdot 10^{-3}}{1 + K_{CO_2} + \cfrac{K_{H_2CO_3}}{10^{-9.6}} + \cfrac{K_{HCO_3} K_{H_2CO_3}}{10^{-2 \cdot 9.6}}}$$

$$\left(\frac{K_{H_2CO_3}}{10^{-9.6}} + 2 \frac{K_{HCO_3} K_{H_2CO_3}}{10^{-2 \cdot 9.6}} \right) + \frac{K_w}{10^{-9.6}}$$

Solving for $[\Sigma Cat - \Sigma An]$ we obtain: $[\Sigma Cat - \Sigma An] = 1.31 \cdot 10^{-3}$ M. Therefore, the concentration of strong base that has been added is equal to: $1.31 \cdot 10^{-3} M - 1.0 \cdot 10^{-3} M = 3.1 \cdot 10^{-4} M$. The new alkalinity of the solution is:

$$[Alk] = 1.31 \cdot 10^{-3} + \frac{K_w}{10^{-4.5}} - 10^{-4.5} = 1.34 \cdot 10^{-3} M = 67\, mgCaCO_3/L$$

3. The concentration of total carbonic acid has not changed due to the addition of the strong base and it can be calculated either from the initial values of pH and alkalinity or from the values of pH and alkalinity of case (b). For the initial case we have pH = 7.37 and $[\Sigma Cat - \Sigma An] = 1.0 \cdot 10^{-3} M$ so we calculate:

$$[H_2CO_{3tot}] = \cfrac{10^{-7.37} - \cfrac{K_w}{10^{-7.37}} + 1.0 \times 10^{-3}}{\cfrac{K_{H_2CO_3}}{10^{-7.37}} + 2 \cfrac{K_{HCO_3} K_{H_2CO_3}}{10^{-2 \times 7.37}}}$$

$$\left(1 + K_{CO_2} + \frac{K_{H_2CO_3}}{10^{-7.37}} + \frac{K_{HCO_3} K_{H_2CO_3}}{10^{-2 \times 7.37}} \right) = 1.1 \times 10^{-3} M$$

For case (b), we have pH = 9.60 and so:

$$[H_2CO_{3tot}] = \cfrac{10^{-9.6} - \cfrac{K_w}{10^{-9.6}} + 1.0 \cdot 10^{-3}}{\cfrac{K_{H_2CO_3}}{10^{-9.6}} + 2 \cfrac{K_{HCO_3} K_{H_2CO_3}}{10^{-2 \cdot 9.6}}}$$

$$\left(1 + K_{CO_2} + \frac{K_{H_2CO_3}}{10^{-9.6}} + \frac{K_{HCO_3} K_{H_2CO_3}}{10^{-2 \cdot 9.6}} \right) = 1.1 \cdot 10^{-3} M$$

The two methods give the same value of the total concentration of carbonic acid, as expected.

2.3.4 Acidic Wastewaters

If a wastewater has an acidic pH, that is a pH below approximately 5, its alkalinity is close to 0. Its acidic pH is determined by the presence of strong or weak acids. We can follow a procedure similar to the one used in the previous section to calculate the term [ΣCat – ΣAn] and the total concentration of acids.

The increase in pH for a solution of a weak acid in correspondence with the addition of a strong base is reported in Figure 2.15. The pH increases slowly at first, then it shows a sharp increase. The sharp increase in pH corresponds to the fact that the weak acid becomes totally dissociated, that is totally present as anion, and therefore its buffer capacity disappears. In real wastewaters the situation is more complex since a mixture of various weak acids may be present. Therefore, the increase in pH is often less sharp than in Figure 2.15.

Even though it is an ideal situation, a model with one single weak acid allows to describe a relatively simple procedure for the characterisation of a wastewater in terms of its buffer capacity. Let HA be the weak acid in the wastewater. In general, in the initial solution HA will be present in part as dissociated species H_3O^+ and A^- and in part as undissociated species HA. We can estimate the total concentration of weak acid, [HA_{tot}], its equilibrium constant K_{HA} and the concentration of total ions, [ΣCat – ΣAn] with

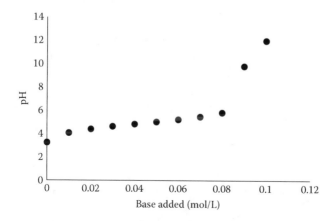

FIGURE 2.15   Increase in pH with addition of base for a real wastewater.

the following procedure, which is essentially analogous to the procedure used for an alkaline solution.

Under the initial conditions (no base added), the pH of the solution is expressed as:

$$\left[H_3O^+\right]+\left[\Sigma Cat\right]=\left[A^-\right]+\left[\Sigma An\right]+\left[OH^-\right] \tag{2.142}$$

that is

$$\left[10^{-pHinit}\right]+\left[\Sigma Cat-\Sigma An\right]=\frac{\left[HA_{tot}\right]}{1+\dfrac{\left[10^{-pHinit}\right]}{K_{HA}}}+\left[\frac{K_W}{10^{-pHinit}}\right] \tag{2.143}$$

In this equation we have three unknowns, $[\Sigma Cat - \Sigma An]$, $[HA_{tot}]$ and K_{HA}. K_{HA} can be estimated from the titration curve, by considering that when the slope of the curve increases this means that most of the HA species is present as dissociated form A^-. In general it can be assumed that when the slope changes about 90% of HA is present as A^-. This information can be used to calculate K_{HA}. Indeed:

$$\left[A^-\right]=\frac{\left[HA_{tot}\right]}{1+\dfrac{\left[10^{-pH}\right]}{K_{HA}}} \tag{2.144}$$

If at the pH where the slope changes $\left[A^-\right]=0.90\left[HA_{tot}\right]$, K_{HA} is given by:

$$K_{HA}=9\cdot10^{-pHslopechange} \tag{2.145}$$

Once K_{HA} is known, the values of $[\Sigma Cat - \Sigma An]$ and $[HA_{tot}]$ can be calculated by applying the charge balance under the initial conditions and under the conditions where the slope changes. The two charge balances are:

$$\left[10^{-pHinit}\right]+\left[\Sigma Cat-\Sigma An\right]=\frac{\left[HA_{tot}\right]}{1+\dfrac{\left[10^{-pHinit}\right]}{K_{HA}}}+\left[\frac{K_W}{10^{-pHinit}}\right] \tag{2.146}$$

$$\left[10^{-pHslopechange}\right]+\left[\Sigma Cat-\Sigma An\right]+\left[Na^+_{slopechange}\right]$$

$$=0.9\left[HA_{tot}\right]+\frac{K_W}{10^{-pHslopechange}} \tag{2.147}$$

In Equation 2.147 $[Na^+{}_{slope\,change}]$ represents the concentration of strong base added up to the point where the slope changes rapidly. Combining Equations 2.146 and 2.147, we solve for the unknowns $[\Sigma Cat - \Sigma An]$ and $[HAtot]$:

$$[HA_{tot}] = \frac{\left(10^{-pH_{init}} - 10^{-pH_{slope\,change}} - \left[Na^+{}_{slope\,change}\right] - K_W\left(\dfrac{1}{10^{-pH_{init}}} - \dfrac{1}{10^{-pH_{slope\,change}}}\right)\right)}{1 + \dfrac{\left[10^{-pH_{init}}\right]}{K_{HA}} - 0.9} \tag{2.148}$$

$$[\Sigma Cat - \Sigma An] = \frac{[HA_{tot}]}{1 + \dfrac{\left[10^{-pH_{fnit}}\right]}{K_{HA}}} + \frac{K_W}{10^{-pH_{init}}} - \left[10^{-pH_{init}}\right] \tag{2.149}$$

Taking Figure 2.15 as an example, the value of $pH_{slope\,change}$ is equal to approximately 5.8 and the amount of base added up to that point is 0.08 mol/L. This means that:

$$K_{HA} = 1.5 \cdot 10^{-5}$$

$$\left[Na^+{}_{slope\,change}\right] = 0.08M$$

From these values, we calculate from Equations 2.148 and 2.149:

$$[HA_{tot}] = 0.09M$$

$$[\Sigma Cat - \Sigma An] = 1.82 \cdot 10^{-3}M$$

With the calculated values of HA_{tot}, $[\Sigma Cat - \Sigma An]$ and K_{HA} we can simulate the behaviour of this wastewater to any addition of base, for example we can calculate the pH as a function of the amount of base added and double-check the agreement with the experimental data. Figure 2.16 shows the comparison of the calculated curve with the curve used to calculate the parameters.

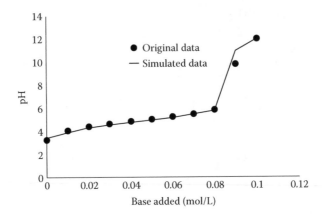

FIGURE 2.16 Comparison of the original data with the data simulated with the procedure described in this section.

2.4 SETTLING

In many cases[*], especially for aerobic processes such as activated sludge, the biological reactor is followed by a settling tank, which has the role of both clarifying the liquid effluent and generating a concentrated sludge stream to be recycled to the reactor. In this section we will cover a simplified theory of settling, which will enable us to calculate the area required for settling the sludge coming from the biological reactor. Usually the concentration of microorganisms coming out of biological processes is high enough that we can apply the theory of hindered settling. Hindered settling occurs when settling of the solids is affected by other solids and, as a consequence, the higher the solids concentration the lower the settling velocity.

Figure 2.17 shows the typical profiles of solids during batch sedimentation in a cylinder. At the start of the sedimentation the solids concentration is uniform. Then solids start to settle and we can easily identify two zones in the cylinder: the top zone, where (ideally) only liquid is present with no solids and the lower zone with solids at the same concentration as at the start of the test. A sediment is also present at the bottom, where the solids deposit, even though it is not always easy to visually distinguish the

---

[*] This section has been adapted from the lectures of Dr. Marcus Campbell Bannermann and Dr. Anirhudda Majumder for the course EX4530 Separation Processes 2, undergraduate programmes in Chemical Engineering, University of Aberdeen.

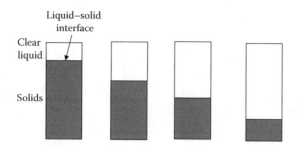

FIGURE 2.17   Solids profiles during batch sedimentation in a cylinder.

sediment from the zone at uniform solids concentration. Over time, during the sedimentation process, the interface between the top and the lower zone moves downwards, due to settling, which occurs at constant speed, and the volume of the sediment at the bottom increases. The critical point is reached when the interface between the suspension and sediment meets the suspension/supernatant interface. Further settling usually occurs after this, but the bulk of the sedimentation is complete.

To obtain a quantitative measurement of the settling velocity as a function of the solids concentration, simple batch tests in cylinders can be carried out. In these tests the volume corresponding to the interface between the supernatant and the zone at uniform solids concentration can be measured as a function of time. Typically it is observed that this volume decreases linearly with time, up to when the critical point is reached, after which the decrease in volume is much slower. Tests can be repeated at different concentrations of solids, and it is observed that the settling velocity decreases as the solids concentration increases. Figure 2.18 reports

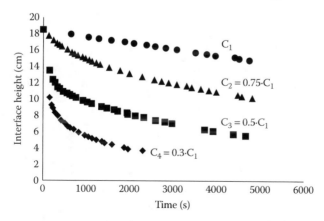

FIGURE 2.18   Interface height vs time in a batch settling tests in a cylinder.

a typical plot obtained from batch settling data with solids at different concentrations.

From each of the curves in Figure 2.18 the settling velocity, u_c, can be calculated as the initial slope of the curve and a correlation between settling velocity and solids concentration, C, can be found. This correlation is often expressed by a power function like:

$$u_C = \alpha C^{-\beta} \tag{2.150}$$

The data from Figure 2.18 are plotted in Figure 2.19, where they are also correlated using Equation 2.150.

Once an expression for the settling velocity as a function of the solids concentration is obtained, the area required for settling of a suspension at a given flow rate and concentration can be calculated with the approach described below. A scheme of the settling tank is shown in Figure 2.20. Let Q_{in} be the influent flow rate (e.g. m³/day) and C_{in} be the influent solids concentration (e.g. kg/m³). We assume that all the solids settle and there are no solids losses with the supernatant. This means that all the solids that enter the settling tank move downwards in the thickening zone, that is at any height in this zone the mass flow of solids moving downwards is:

$$\psi_T \left(\frac{kg}{day} \right) = Q_{in} \times C_{in} \tag{2.151}$$

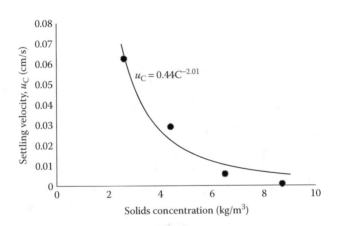

FIGURE 2.19   Interface height vs time in a batch settling tests in a cylinder for the settling curves in Figure 2.18.

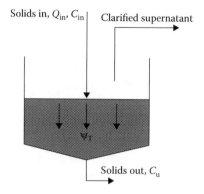

FIGURE 2.20 Scheme of a settling tank. (Courtesy of Dr. Marcus Campbell Bannerman.)

For a particular height, h, the solid particles are at a concentration $C(h)$ and have a settling velocity $u_c(h)$. There is also an additional velocity of the sediment, u_r, due to the withdrawal of solids in the underflow. Therefore, at any height of the settling tank the mass flow rate (typically in kg/day) of the solids directed downwards is:

$$\left(u_C(h) + u_r \right) AC(h) \tag{2.152}$$

where A is the cross-sectional area of the clarifier.

In order for the settling tank to be able to settle all the solids in the feed, in any section of the tank the downward mass flow needs to be all the solids to move downwards, this mass flow must be greater or equal than the total mass flow, that is:

$$\left(u_C(h) + u_r \right) AC(h) \geq \psi_T \tag{2.153}$$

At the bottom of the tank we have:

$$\psi_T = \left(u_U + u_r \right) AC_U \tag{2.154}$$

where u_U is the settling velocity at the underflow concentration C_U.

We can Equation 2.154 to eliminate u_r from Equation 2.153 and we obtain:

$$\psi_T \leq A \frac{u_C - u_U}{(1/C) - (1/C_U)} \tag{2.155}$$

The minimum area required in order to settle all the suspended solids that enter the clarifier is therefore:

$$A_{min} = Q_{in} C_{in} \left[\frac{(1/C)-(1/C_U)}{u_C - u_U} \right]_{max} \tag{2.156}$$

Therefore, in order to calculate the minimum area required to settle a suspension with flow rate Q_{in} and concentration C_{in}, we need to generate a plot of $\left[(1/C)-(1/C_U)/u_C - u_U\right]$ vs C, in the range of solids concentrations present in the settling tank, that is in the range from C_{in} to C_u. The maximum value of this function will correspond to the minimum area required for settling. It is evident that the higher the influent solids concentration, the larger the area required for settling, and this has important consequences in the design of biological processes coupled with settling tanks. Of course, the actual area has to be larger than the minimum area calculated with this procedure.

It is important to observe that the procedure discussed in this section only refers to the thickening zone of the settling tank. The design of the clarification zone is not presented here; however, the requirements for a proper clarification also need to be considered in the design of the settling tank.

Example 2.11: Calculation of the minimum area for a settling tank

Calculate the minimum area of a settling tank required to settle the suspended solids in a stream with a flow rate of 10,000 m^3/day and with a suspended solids concentration of 5 kg/m^3. The desired underflow concentration is 10 kg/m^3. Use the settling curve from Figure 2.19.

Solution

The minimum required area is obtained from Equation 2.156. We need to calculate the maximum value of the function

$$\left[\frac{(1/C)-(1/C_U)}{u_C - u_U} \right]$$

in the concentration range 5–10 kg/m^3. In this case we have from Figure 2.19:

$$u_C \left(\frac{cm}{s} \right) = 0.44 C^{-2.01}$$

Also:

$$C_U = 10 \, kg/m^3$$

$$u_U = 0.44 \times 10^{-2.01} = 0.0043 \, cm/s$$

The plot of $\left[(1/C) - (1/C_U)/u_C - u_U \right]$ vs C is reported in Figure 2.21. From this Figure it is evident that the maximum occurs for a concentration close to the underflow value. The maximum of this function is 11.04 m²·s/kg. Therefore, the minimum required area is:

$$A_{min} = \frac{10,000 \, (m^3/day)}{86,400 \, (s/day)} 5 \frac{kg}{m^3} \times 11.04 \frac{m^2 s}{kg} = 6.4 \, m^2$$

This minimum area corresponds to a diameter of 2.85 m.

2.4.1 Filamentous Bulking

Clearly it is desirable, in biological wastewater treatment processes that include a settling stage, to have microorganisms with good settling properties. The main measure of the settling properties, is, as we have seen, the

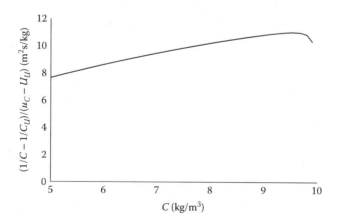

FIGURE 2.21 Example 2.11 plot to calculate the minimum required area for the settling tank.

settling velocity. Among the various factors that can prevent microorganisms from settling fast, one of the most important is filamentous bulking, which can be a particularly important phenomenon for activated sludge processes. Filamentous bulking is due to the proliferation of microorganisms that grow as filaments, instead than as flocs as in well settling activated sludge processes. When filaments prevail, they tend to form bridges between flocs, lowering the settling velocity and causing poor compaction during settling. The result is poor settling, with possibly loss of microorganisms with the clarified effluent. Many theories have been developed to understand the causes of filamentous bulking in activated sludge processes. One of the main theories is called 'kinetic selection'. According to this theory filaments are kinetically favoured when the substrate concentration is low because they can easily protrude from the flocs and reach for low concentrations of substrate. However, according to this theory, floc-formers (the 'normal' microorganisms in activated sludge process) are favoured at high substrate concentration because they have an intrinsic higher growth rate. Using the Monod kinetic model (Section 2.1.2.1) the growth kinetics of both filaments and floc-formers can be described by the Monod model, Equations 2.55 and 2.56, but the two populations have different kinetic parameters μ_{max} and K_S. In particular, according to this theory, floc-formers have higher values of both μ_{max} and K_s, which makes them favoured at high substrate concentration and disadvantaged at lower substrate concentration. The specific growth rates according to the Monod model and to the kinetic selection theory for floc-formers and filaments are shown in Figure 2.22.

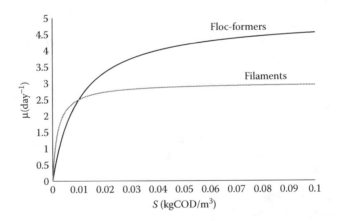

FIGURE 2.22 Monod growth curves for floc-formers and filaments according to the kinetic theory (numerical values are arbitrary are for example only).

In addition to the settling rate, the settling properties of microorganisms are also often evaluated by another parameter, called the sludge volume index (SVI). The SVI (ml/g) represents the volume occupied by a certain mass of microorganisms after settling for a fixed period of time (typically 30 min). The SVI test is typically done in graduated cylinders where a suspension of microorganisms is allowed to settle, and the volume occupied by the settled biomass is measured at the end. Well settling sludge has a SVI of less than 100 ml/g, while bulking sludge can have SVI higher than 150 ml/g.

2.5 HEAT GENERATION AND HEAT TRANSFER

2.5.1 Heat Generation

Biological reactions are chemical reactions and, like any chemical reactions, they generate (exothermic reactions) or absorb (endothermic reactions) heat. The heat generated or absorbed by the reaction is the difference between the enthalpy of the reactants and products. For a generic chemical reaction:

$$\alpha A + \beta B \rightarrow \gamma C + \delta D \tag{2.157}$$

The heat generated or absorbed by the reaction is called the enthalpy of reaction, which is defined as:

$$\Delta H_r \left(\frac{J}{mol} \right) = \gamma H_C + \delta H_D - \alpha H_A - \beta H_B \tag{2.158}$$

In Equation 2.158 H_A, H_B and others (J/mol or kJ/kmol) are the enthalpies of the species involved in the reaction. The reaction is exothermic if $\Delta H_r < 0$ and it is endothermic if $\Delta H_r > 0$.

The enthalpy of the reaction depends on the temperature and pressure at which the reaction is carried out. The enthalpy of reaction calculated at 25°C and 1 atm is called standard enthalpy of reaction, and is indicated with ΔH_r^0.

The enthalpies of the species which are involved in the reaction can be calculated by assuming a reference state. The reference state is usually taken as 25°C and 1 atm and for each species it has to be specified whether the substance is in the solid, liquid or gaseous (or vapour) state. The enthalpy of the reaction which produces a substance from its elements is called enthalpy of formation (ΔH_f) of the substance. The enthalpy of formation is usually given at 25°C and 1 atm and has therefore the meaning of standard enthalpy of formation. By convention, the enthalpies of the elements at 25°C and 1 atm are assumed to be equal to 0.

For example the standard enthalpy of formation of water is, by definition, the enthalpy at 25°C of the reaction:

$$H_2 + 0.5O_2 \rightarrow H_2O \qquad (2.159)$$

$$\Delta H_r^0 = \Delta H_{fH2O}^0 = H_{H2O}(25°C) - H_{H2}(25°C) - 0.5H_{O2}(25°C)$$
$$= H_{H2O}(25°C) \qquad (2.160)$$

The enthalpy of species A at the generic temperature T can be calculated as follows:

$$H_A(T)\left(\frac{J}{mol}\right) = \Delta H_{fA}(25°C) + \left[\lambda_{phase\,change}(25°C)\right] + \int_{25}^{T} c_{PA}dt \qquad (2.161)$$

where $\Delta H_{fA}(25°C)$ is the enthalpy of formation of species A from the elements at 25°C, $\lambda_{phase\,change}(25°C)$ is the enthalpy of phase change (e.g. vaporisation and fusion), if present, and c_{PA} is the specific heat (or heat capacity) of species A. The enthalpy of phase change is only required if the physical state of the species at the temperature T is different from the physical state of the species in the reference state at which the enthalpy of formation $\Delta H_{fA}(25°C)$ has been calculated.

If the specific heat of species A can be considered constant between 25°C and the temperature T, we have:

$$H_A(T)\left(\frac{J}{mol}\right) = \Delta H_{fA}(25°C) + \left[\lambda_{phase\,change}(25°C)\right] + c_{PA}(T-25) \qquad (2.162)$$

Example 2.12 shows the calculation of the heat of reaction for some fermentation reactions typical of biological wastewater treatment processes. The example shows that fermentation reactions are usually exothermic and that the heat of reaction of aerobic or anoxic reactions is typically much larger than for anaerobic reactions.

Example 2.12

Compare the enthalpy of reaction for biomass growth on glucose under the three conditions:

1. Aerobic growth;
2. Anoxic growth (using nitrate as electron acceptor);
3. Anaerobic growth, with production of acetic acid.

Assume in all cases a growth yield $Y_{X/S} = 0.2$ kg biomass/kg glucose. Assume that the reaction temperature is 25°C, that oxygen and ammonia are present in the liquid phase and that all the produced carbon dioxide is in the gas phase.

Solution

1. For aerobic growth on glucose the growth stoichiometry is (Section 2.1):

$$C_6H_{12}O_6 + (6-7.96Y_{X/S})O_2 + 1.59Y_{X/S}NH_3 \rightarrow$$
$$1.59Y_{X/S}C_5H_7O_2N + (6-7.96Y_{X/S})CO_2$$
$$+ (6-3.18Y_{X/S})H_2O$$

which, for $Y_{X/S} = 0.2$ kg biomass/kg glucose becomes:

$$C_6H_{12}O_6 + 4.41O_2 + 0.32NH_3 \rightarrow$$
$$0.32C_5H_7O_2N + 4.41CO_2 + 5.36H_2O$$

Therefore, the required ΔH_r can be calculated from:

$$\Delta H_r^0 \left(\frac{J}{mol\,glucose} \right) = 0.32H_{biomass}(25°C, solid) + 4.41H_{CO2}(25°C, gas)$$

$$+ 5.36H_{H2O}(25°C, liquid) - H_{glucose}(25°C, dissolved)$$
$$- 4.41H_{O2}(25°C, dissolved) - 0.32H_{NH3}(25°C, dissolved)$$

For biomass the reference state is a solid therefore its enthalpy at 25°C coincides with the standard enthalpy of formation:

$$H_{biomass}(25°C, solid) = -7.7 \cdot 10^5 \, J/mol$$
$$\Rightarrow 0.32H_{biomass}(25°C, solid)$$
$$= -2.46 \cdot 10^5 \, J/mol\,glucose$$

Similarly for carbon dioxide the required enthalpy coincides with its standard enthalpy of formation:

$$H_{CO2}(25°C, gas) = -3.9 \cdot 10^5 \, J/mol \Rightarrow 4.41H_{CO2}(25°C, gas)$$
$$= -1.72 \cdot 10^6 \, J/mol\,glucose$$

Similarly for water:

$$H_{H2O}(25°C, liquid) = -2.9 \times 10^5 J/mol \implies$$

$$5.36 H_{H2O}(25°C, liquid) = -1.55 \cdot 10^6 J/mol\,glucose$$

For glucose the standard enthalpy of formation is given as a solid, but in biological reactions glucose is dissolved in the liquid phase:

$$H_{glucose}(25°C, dissolved) = \Delta H^0_{fglucose} + \lambda_{dissolution}(25°C)$$

$$= -1.28 \cdot 10^6 J/mol + 1.1 \cdot 10^4 J/mol$$

$$= -1.27 \times 10^6 J/mol \implies -H_{glucose}(25°C, dissolved)$$

$$= 1.27 \times 10^6 J/mol$$

Similarly, for oxygen we need to add the enthalpy of dissolution to the standard enthalpy of formation (which is 0, because oxygen is an element):

$$H_{O2}(25°C, dissolved) = \Delta H^0_{foxygen} + \lambda_{dissolution}(25°C)$$

$$= 0 - 1.47 \cdot 10^4 J/mol \implies -4.41 H_{O2}(25°C, dissolved)$$

$$= 6.48 \times 10^4 J/mol\,glucose$$

For ammonia the standard enthalpy of formation is given in the liquid phase, therefore we have:

$$H_{NH3}(25°C, dissolved) = \Delta H^0_{fNH3} = -8.1 \cdot 10^4 J/mol \implies$$

$$-0.32 H_{NH3}(25°C, dissolved) = \Delta H^0_{fNH3} = 2.59 \times 10^4 J/mol\,glucose$$

In conclusion, the required enthalpy of reaction is given by adding up algebraically all these enthalpy terms:

$$\Delta H^0_r \left(\frac{J}{mol\,glucose} \right) = -2.2 \times 10^6 \frac{J}{mol}$$

2. For anoxic growth on glucose, with $Y_{X/S} = 0.2$ kg biomass/kg glucose the stoichiometry is (from Section 2.1):

$$C_6H_{12}O_6 + 3.52 HNO_3 + 0.32 NH_3 \rightarrow$$
$$0.32 C_5H_7O_2N + 4.41 CO_2 + 7.12 H_2O + 1.76 N_2$$

And the enthalpy of reaction is:

$$\Delta H_r^0 \left(\frac{J}{mol\,glucose} \right) = 0.32 H_{biomass}(25°C, solid)$$
$$+4.41 H_{CO2}(25°C, gas) + 7.13 H_{H2O}(25°C, liquid)$$
$$+1.76 H_{N2}(25°C, gas) - H_{glucose}(25°C, dissolved)$$
$$-3.52 H_{NO3^-}(25°C, dissolved) - 0.32 H_{NH3}(25°C, dissolved)$$

The contributions of the enthalpies of biomass, carbon dioxide and ammonia are the same as calculated in part (1). For water we have:

$$H_{H2O}(25°C, liquid) = -2.9\ 10^5 J/mol \implies$$

$$7.13 H_{H2O}(25°C, liquid) = -2.07 \cdot 10^6\,J/mol\,glucose$$

For nitrogen:

$$H_{N2}(25°C, gas) = 0 \implies 1.76 H_{N2}(25°C, gas) = 0$$

For nitrate:

$$H_{NO3^-}(25°C, dissolved) = -2.07 \cdot 10^5 J/mol \implies$$

$$-3.52 H_{NO3^-}(25°C, dissolved) = 7.3 \times 10^5 J/mol$$

In conclusion, the required enthalpy of reaction is given by:

$$\Delta H_r^0 \left(\frac{J}{mol\,glucose} \right) = -2.0 \cdot 10^6\,\frac{J}{mol}$$

3. For anaerobic growth on glucose with production of acetic acid, the reaction stoichiometry is, again from Section 2.1:

$$C_6H_{12}O_6 + 1.59 Y_{X/S} NH_3 \rightarrow$$

$$1.59 Y_{X/S} C_5H_7O_2N + (2 - 2.65 Y_{X/S}) CH_3COOH$$

$$+ (4 - 5.3 Y_{X/S}) H_2 + (2 - 2.65 Y_{X/S}) CO_2 + (7.42 Y_{X/S} - 2) H_2O$$

which, for $Y_{X/S} = 0.2$ kg biomass/kg glucose becomes:

$$C_6H_{12}O_6 + 0.32NH_3 + 0.52H_2O \rightarrow$$

$$0.32C_5H_7O_2N + 1.47CH_3COOH + 2.94H_2 + 1.47CO_2$$

The enthalpy of the reaction is given by:

$$\Delta H_r \left(\frac{J}{mol\,glucose} \right) = 0.32H_{biomass}\left(25°C, solid\right)$$

$$+1.47H_{AC}\left(25°C, liquid\right) + 2.94H_{H2}\left(25°C, gas\right)$$

$$+1.47H_{CO2}\left(25°C, gas\right) - H_{glucose}\left(25°C, dissolved\right)$$

$$-0.32H_{NH3}\left(25°C, dissolved\right)$$

$$-0.52H_{H2O}\left(25°C, liquid\right)$$

The contribution of biomass, glucose and ammonia are the same calculated in part (1), the remaining contributions are as follows.
Acetic acid:

$$H_{AC}\left(25°C, dissolved\right) = \Delta H_{fAC}^0 + \lambda_{dissolutionAC}\left(25°C\right)$$

$$= -4.8 \cdot 10^5 \, J/mol - 1.5 \cdot 10^3 \, J/mol$$

$$= -4.8 \cdot 10^5 \, J/mol \Rightarrow 1.47H_{AC}\left(25°C, dissolved\right)$$

$$= -7.1 \cdot 10^5 \, J/mol\,glucose$$

Hydrogen:

$$H_{H2}\left(25°C, gas\right) = 0 \Rightarrow 2.94H_{H2}\left(25°C, gas\right) = 0$$

Carbon dioxide:

$$H_{CO2}\left(25°C, gas\right) = -3.9 \cdot 10^5 \, J/mol \Rightarrow 1.47H_{CO2}\left(25°C, gas\right)$$

$$= -5.7 \cdot 10^5 \, J/mol\,glucose$$

Water:

$$H_{H2O}\left(25°C, liquid\right) = -2.9 \cdot 10^5 \, J/mol \Rightarrow$$

$$-0.52H_{H2O}\left(25°C, liquid\right) = 1.5 \cdot 10^5 \, J/mol\,glucose$$

Therefore, the required reaction enthalpy for the anaerobic conversion of glucose to acetic acid is:

$$\Delta H_r^0 \left(\frac{J}{mol\,glucose} \right) = -8 \cdot 10^4 \, \frac{J}{mol}$$

The heat of reaction under anaerobic conditions is much lower than under aerobic or anoxic conditions, which, however, are very similar.

2.5.2 Heat Transfer

The rate of heat transfer $\dot{Q}$ between a fluid at temperature T_1 and another fluid at temperature T_2 separated by a wall of conductive material of thickness x and area A can in general be expressed by the following equation:

$$\dot{Q}\left(\frac{J}{s}\right) = UA(T_1 - T_2) \tag{2.163}$$

where:

U (typically expressed in W/m².°C) is the overall heat transfer coefficient
A is the area available for heat transfer
T_1 and T_2 are the temperatures of the two fluids which exchange heat

With reference to a flat solid surface which separated the hot and cold fluid (Figure 2.23), an expression for the overall heat transfer coefficient U can be derived as described below.

The heat flux across the heat transfer area A is given by:

$$\frac{\dot{Q}}{A}\left(\frac{J}{m^2 s}\right) = h_1 (T_1 - T_{w1}) \tag{2.164}$$

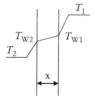

FIGURE 2.23   Scheme of heat transfer between two fluid separated by a solid surface.

$$\frac{\dot{Q}}{A}\left(\frac{J}{m^2s}\right) = \frac{k}{x}\left(T_{w1} - T_{w2}\right) \tag{2.165}$$

$$\frac{\dot{Q}}{A}\left(\frac{J}{m^2s}\right) = h_2\left(T_{w2} - T_2\right) \tag{2.166}$$

In Equations 2.164 through 2.166 h_1 and h_2 are the heat transfer coefficients for the two fluids (they have the same units as U, typically W/m².°C) and k is the conductivity of the solid material (typically given in W/m².°C).

Assuming the heat transfer coefficients and the bulk temperatures T_1 and T_2 are known, by making equal Equations 2.164, 2.165 and 2.166 we obtain a system of two equations in the two unknowns T_{w1} and T_{w2} which can be therefore expressed as a function of the other variables. For example for T_{w2} we obtain:

$$T_{w2} = \frac{kh_1T_1 + h_2\left(h_1x + k\right)T_2}{h_2\left(h_1x + k\right) + kh_1} \tag{2.167}$$

Combing Equation 2.167 with Equation 2.16:

$$\frac{\dot{Q}}{A}\left(\frac{J}{m^2s}\right) = h_2\left(T_{w2} - T_2\right) = \frac{h_2h_1k\left(T_1 - T_2\right)}{h_2\left(h_1x + k\right) + kh_1} \tag{2.168}$$

Comparing Equation 2.168 with Equation 2.164 we obtain:

$$U = \frac{h_2h_1k}{h_2\left(h_1x + k\right) + kh_1} \tag{2.169}$$

or:

$$\frac{1}{U} = \frac{1}{h_1} + \frac{1}{h_2} + \frac{x}{k} \tag{2.170}$$

Equations 2.169 or 2.170 express the overall heat transfer coefficient U as a function of the local heat transfer coefficients h_1 and h_2 and of the properties of the wall, k and x.

If the heat transfer area is not plan but cylindrical, with external radius r_1 and internal radius r_2, the expression for U modifies as follows:

$$\frac{1}{U_1} = \frac{1}{h_1} + \frac{r_1}{r_2h_2} + \frac{r_1\ln\left(r_1/r_2\right)}{k} \tag{2.171}$$

or

$$\frac{1}{U_2} = \frac{1}{h_2} + \frac{r_2}{r_1 h_1} + \frac{r_2 \ln(r_1/r_2)}{k} \tag{2.172}$$

In these equations U_1 refers to the external heat transfer area and U_2 to the internal one. Therefore, the heat transfer rate can be expressed with the following equations:

$$\dot{Q}\left(\frac{J}{s}\right) = U_1 A_1 (T_1 - T_2) \tag{2.173}$$

$$\dot{Q}\left(\frac{J}{s}\right) = U_2 A_2 (T_1 - T_2) \tag{2.174}$$

In Equations 2.173 and 2.174 A_1 and A_2 are the external and internal heat transfer areas respectively. In many cases the surface across which heat transfer occurs is curve and not flat, so Equations 2.173 through 2.174 should apply. However, in many cases biological reactors are large vessels and the curvature of the wall is small and can be neglected. This is shown in the Example 2.13.

Example 2.13

Compare the values of the overall heat transfer coefficient referred to the internal and external area of the vessel wall for an agitated vessel having an internal diameter of 1 m and a wall thickness of 1 cm. Assume h_1 and h_2 are both equal to 1000 W/m².°C and the conductivity of the wall material is 16 W/m.°C.

Solution

For the external area we use Equation 2.171:

$$\frac{1}{U_1} = \frac{1}{1000} + \frac{1.01}{1 \cdot 1000} + \frac{1.01 \ln(1.01/1)}{16} \Rightarrow U_1 = 379.06 \ \ W/m^2.°C$$

For the internal area we use Equation 2.172:

$$\frac{1}{U_2} = \frac{1}{1000} + \frac{1}{1.01 \cdot 1000} + \frac{1 \ln(1.01/1)}{16} \Rightarrow U_2 = 379.95 \ \ W/m^2.°C$$

Assuming the vessel heat transfer area is flat (internal and external area coincide) we use Equation 2.170:

$$\frac{1}{U} = \frac{1}{1000} + \frac{1}{1000} + \frac{0.01}{16} \implies U = 380.95\,\text{W/m}^2.{}^{\circ}\text{C}$$

The values of U calculated with the three equations are all essentially the same; therefore, it is usually reasonable, for agitated vessels, to assume the value of U obtained with the formula for flat heat transfer surfaces.

We have seen that essentially the rate of heat transfer between two fluids separated by a solid surface depends on the temperature difference between the two fluids, on the area available for heat transfer and on the overall heat transfer coefficient. The overall heat transfer coefficient in turn depends on the heat transfer coefficients for the two fluids and on the conductivity and thickness of the solid wall between the fluids. Typically the main resistance to heat transfer lies in the two fluids, while the solid wall is usually a conductive material (high value of the conductivity k), and therefore its resistance to heat transfer is often small or negligible.

An important difference between the heat transfer resistances of the two fluids, $1/h_1$ and $1/h_2$, and the resistance of the solid wall between the fluids, (x/k), is that the latter can be immediately calculated from a physical property of the wall material (its thermal conductivity) and from its thickness, while the former depend not only on the physical properties of the but also on the turbulence and fluid dynamics of the fluids. For agitated vessels, the heat transfer coefficient for the fluid inside the vessel is often given by correlations of the form:

$$\text{Nu} = \frac{h_2 D_{\text{vessel}}}{k_{\text{fluid}}} = a_1 \left(\frac{ND_{\text{ag}}^2 \rho_{\text{fluid}}}{\mu_{\text{fluid}}} \right)^{a_2} \left(\frac{c_{P\text{fluid}}\, \mu_{\text{fluid}}}{k_{\text{fluid}}} \right)^{\frac{1}{3}} \left(\frac{\mu_{\text{fluid}}}{\mu_{\text{Wfluid}}} \right)^{a_3} \quad (2.175)$$

where:
   Nu is the Nusselt number
   D_{ag} is the diameter of the agitator
   N is the agitator speed (revs/s)

The parameters a_1, a_2, a_3 mainly depends on the type of agitator used and on the turbulence of the system, indicated by the Reynolds

TABLE 2.7   Values of the Parameters for Correlation (2.175)

Agitator	A_1	A_2	A_3	Range of Reynolds Numbers
Paddle	0.36	2/3	0.21	300–300,000
Disc, flat blade turbine	0.54	2/3	0.14	40–30,000
Anchor	0.36	2/3	0.18	300–40,000

number $Re = \left(ND_{ag}^2 \rho_{fluid} / \mu_{fluid} \right)$. The parameters for Equation 2.175 are reported in Table 2.7 for some common agitators.

Equation 2.171 shows that the heat transfer coefficient for the fluid inside a stirred vessel depends on the type, size and speed of the agitator and on the physical properties of the fluid.

If heat transfer occurs via a jacket, the heat transfer side on the jacket side, h_1, depends on the velocity of the fluid inside the jacket and on the physical properties of the fluid in the jacket. Correlations for the jacket-side heat transfer coefficient have a similar form as Equation 2.175, for example they often have the form:

$$Nu_{jacket} = \frac{h_1 D_{jacketl}}{k_{fluid}} = b_1 Re^{b_2} Pr^{\frac{1}{3}} \left(\frac{\mu_{fluid}}{\mu_{Wfluid}} \right)^{b_3} \qquad (2.176)$$

In Equation 2.176, D_{jacket} is a characteristic dimension of the jacket and the Reynolds number is defined on the basis of the velocity of the fluid in the jacket and of its characteristic dimension. The parameters b_1, b_2, b_3 depend on the jacket type. Essentially the heat transfer coefficient for the fluid in the jacket is proportional to its velocity, because higher velocity corresponds to higher turbulence.

Even though correlations such as (Equation 2.175) and (Equation 2.176) have been developed for mechanically agitated vessels and they are not always applicable to biological wastewater treatment processes, they are important in understanding the effect of process parameters on the heat transfer rate in these processes.

Example 2.14

Calculate the overall heat transfer coefficient for a jacketed vessel where the jacket side heat transfer coefficient is $h_2 = 1000 \ W/m^2.°C$. The vessel wall is made of stainless steel ($k = 16 \ W/m.°C$) and has a thickness of 1 cm. The agitator is a paddle with 0.5 m diameter and the vessel diameter is 1.5 m and it rotates at 40 rpm. Assume the physical properties of the fluid inside the vessel are the ones of water.

Solution

Applying Equation 2.170 the overall heat transfer coefficient is given by:

$$\frac{1}{U} = \frac{1}{1000} + \frac{1}{h_2} + \frac{0.01}{16} = 1.625 \cdot 10^{-3} + \frac{1}{h_2}$$

We need to determine the heat transfer coefficient inside the vessel, which is given by Equation 2.175. We have:

$$h_2 = \frac{0.6(W/m°C)}{1.8m} 0.36 \left(\frac{(40/60) \cdot 0.5^2 \cdot 1000}{1 \cdot 10^{-3}} \right)^{2/3} \left(\frac{4186 \cdot 1 \cdot 10^{-3}}{0.6} \right)^{1/3}$$

$$= 697 \frac{W}{m^2 °C}$$

We obtain:

$$U = 327 \frac{W}{m^2 . °C}$$

2.6 REMOVAL OF XENOBIOTICS IN BIOLOGICAL PROCESSES

In general in this book we will assume that the only removal mechanism for organic substrates is biodegradation. The only exceptions are the so-called xenobiotics, that is man-made chemicals which can be present, usually at low or very low concentration, in the influent of wastewater treatment plants. Based on the research evidence, the following mechanisms can be important for the removal of xenobiotics in biological wastewater treatment processes: biodegradation, stripping and adsorption. The relative importance of the various mechanisms depends on the nature of the xenobiotic and on the operating parameters of the process. In this section, we will present the kinetic models used for the various removal mechanisms, while the mass balance and calculation of the effluent concentration of the xenobiotics in the activated sludge process will be shown in Chapter 4.

2.6.1 Biodegradation

The easiest way to model biodegradation of xenobiotics is to model it exactly as the biodegradation of any organic substrate. Therefore, we will assume that there is a class of microorganisms, called X_{XOC}, which use xenobiotics as only carbon source. The growth rate of X_{XOC} can be written in the usual way:

$$r_{XXOC} \left(\frac{kg\,biomass}{m^3 day} \right) = \frac{\mu_{maxXOC} S_{XOC}}{K_{SXOC} + S_{XOC}} X_{XOC} \qquad (2.177)$$

In Equation 2.177, S_{XOC} is the concentration of the xenobiotic in the liquid phase and μ_{maxXOC} and K_{SXOC} are the kinetic parameters for the growth of the xenobiotic-degrading microorganisms. Similarly as for the heterotrophic microorganisms, we assume that xenobiotic-degrading microorganisms undergo endogenous metabolism:

$$r_{endXOC} \left(\frac{kg\,biomass}{m^3 day} \right) = -b_{XOC} X_{XOC} \qquad (2.178)$$

Once the rate of microorganisms' growth on the xenobiotic is known, the rate of removal of the xenobiotic due to biodegradation can be expressed using the growth yield $Y_{X/SXOC}$, analogously to what we have done for readily biodegradable substrates:

$$r_{bio} \left(\frac{kg\,xenobiotic}{m^3 day} \right) = -\frac{r_{XXOC}}{Y_{X/SXOC}} \qquad (2.179)$$

2.6.2 Adsorption

Adsorption is the process by which soluble substances adhere to the external surface of microorganisms and are therefore removed from solution by a physical mechanism, rather than by biodegradation. For most organic species in biological wastewater treatment plants adsorption on biomass surface is not an important mechanism for their removal and can be neglected. However, adsorption can be important for hydrophobic substances of low solubility such as many xenobiotics. In biological processes adsorption is usually described by a simplified linear relationship between the concentration of a substance in the liquid phase and the concentration of the substance in the solid phase, that is on the external surface of the microorganisms. This equilibrium relationship can be written as:

$$S_{XOC,biom} \left(\frac{kg\,XOC}{kg\,biomass} \right) = K_P \times S_{XOC} \qquad (2.180)$$

In Equation 2.180, $S_{XOC,biom}$ is the concentration of the species on the biomass surface and S_{XOC} is its concentration in the liquid phase. K_p is an empirical constant to be determined from experimental data.

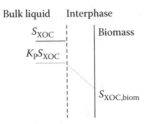

FIGURE 2.24   Scheme of the adsorption of a xenobiotic on the biomass surface.

It is important to observe that xenobiotics are adsorbed on all the biomass, not only on the xenobiotic-degrading microorganisms. In reality usually the concentration of xenobiotics in wastewaters is only a minor fraction of the overall biodegradable COD; therefore, the concentration of xenobiotic-degrading microorganisms will be neglected in the expression of the rate of adsorption.

The rate of adsorption is usually expressed as:

$$r_{ads}\left(\frac{kg}{m^3 day}\right) = k_{ads}\left(K_P S_{XOC} - S_{XOC,biom}\right)X \qquad (2.181)$$

In Equation 2.181, the term $K_P S_{XOC}$ represents the concentration of the xenobiotic on the biomass surface that would be in equilibrium with the actual concentration S_{XOC} in the liquid phase, while $S_{XOC,biom}$ represents the actual concentration of the xenobiotic on the biomass (Figure 2.24). r_{ads} is taken as positive in the adsorption direction, that is when the substance moves from the liquid to the biomass phase.

2.6.3 Stripping

Xenobiotics can transfer from the liquid phase to the gas phase. The rate of this process can be described by the same model used for oxygen transfer from the gas to the liquid phase, even though in this case the direction of transfer will be reversed (from the liquid to the gas phase). Therefore, the rate of transfer of a xenobiotic is given by:

$$r_{strip}\left(\frac{kg}{m^3 day}\right) = k_L a_{XOC}\left(k_{eqXOC} p_{XOC} - S_{XOC}\right) \qquad (2.182)$$

In Equation (2.182), $k_L a_{XOC}$ is the mass transfer coefficient for the xenobiotic, p_{XOC} is the partial pressure of the xenobiotic in the gas phase and

S_{XOC} is the concentration of the xenobiotic in the liquid phase. k_{eqXOC} is the equilibrium constant for the xenobiotic between the gas and the liquid phase, defined (in the same way as the equilibrium constants for the partition of oxygen and other gases defined earlier in the mass transfer section) as: $k_{eqXOC} = \left(S_{XOC} / p_{XOC} \right)_{eq}$. Therefore, according to the mass transfer theory described earlier in this chapter, the term $k_{eqXOC} p_{XOC}$ represents the hypothetical concentration of xenobiotic in the liquid phase that would be in equilibrium with its partial pressure in the gas phase. Usually the partial pressure of the xenobiotic in the gas phase is very small, especially if aeration is carried out with mechanical aerators, when p_{XOC} represents the partial pressure of the xenobiotic in the atmosphere (which is virtually equal to 0). If aeration is carried out with gas bubbles, p_{XOC} is the partial pressure of the xenobiotic in the bubbles, which is higher than in the atmosphere, but probably still very small in most cases. Therefore, we will assume that $p_{XOC} \approx 0$ in all cases. With this assumption Equation 2.182 becomes:

$$r_{strip} \left(\frac{kg}{m^3 day} \right) = -k_L a_{XOC} S_{XOC} \qquad (2.183)$$

Equation 2.183 shows that the rate of stripping of the xenobiotic is proportional to its concentration in the liquid phase. The minus sign indicates that the xenobiotic transfer from the liquid to the gas phase. According to the mass transfer theory described earlier, the mass transfer coefficient of the xenobiotic, $k_L a_{XOC}$ depends both on its physical properties and on the fluid dynamics and turbulence of the system.

2.7 KEY POINTS

- Growth of microorganisms on organic or inorganic substrates (fermentation) can be described by chemical reactions, analogously to non-fermentation chemical reactions. The stoichiometry of microorganisms' growth can be obtained by applying elemental balances, using information from biochemistry on the substrates consumed and on the products produced and introducing the parameter 'growth yield' ($Y_{X/S}$), which needs to be measured experimentally;

- The rate of microorganisms' growth is proportional to the microorganisms and to the substrate concentration. The dependency on the substrate concentration is often expressed by the Monod kinetic model;

- Once the stoichiometry for microorganisms' growth and the rate equation are both known, the rate of production and consumptions of all the species consumed and produced by the fermentation reaction can be obtained from simple stoichiometric calculations;

- Mass transfer can be very important in biological wastewater treatment processes, for example for the supply of oxygen in aerobic processes. The rate of mass transfer of a substance from the gas to the liquid phase (or vice versa) is proportional to the mass transfer coefficient and to the driving force, which is the difference between the hypothetical concentration in the liquid phase that would be in equilibrium with the partial pressure in the gas phase and the actual concentration in the liquid phase. The mass transfer coefficient is in turn dependent on the physical properties of the system and of the species which transfers and on the fluid dynamics of the system;

- Settling can also be very important in biological wastewater treatment, when the process includes a solid-liquid separation (e.g. the activated sludge process with biomass recycle). The settling velocity of microorganisms usually decreases with increasing concentration. From knowledge of the settling velocity, of the mass of solids to be settled and of the inlet and outlet (underflow) concentrations, the minimum area required for settling can be calculated. This minimum area only refers to the thickening zone, and not to the clarifying zone;

- pH needs to be maintained in the appropriate range in biological processes. The pH of a process can be calculated from the charge balance. In order to perform the charge balance in biological wastewater treatment processes, the total net concentration of acid and bases and the total carbonic acid content of the inlet wastewater need to be determined. Under certain simplifying assumptions, these two variables can be determined if the alkalinity and the initial pH of the wastewaters are known;

- Fermentation reactions are exothermic and generate heat. The heat generated in a fermentation reaction can be calculated, like for any chemical reaction, as the difference between the enthalpies of the products and of the reactants. The rate of heat transfer can be expressed as a function of a heat transfer coefficient and of the driving force, which is the temperature difference between the hot and cold fluid;

- Man-made chemicals, such as pharmaceuticals and others (generally called xenobiotics), can be removed in biological treatment processes with different mechanisms, that is biodegradation, adsorption on the microorganisms and stripping.

Questions and Problems

2.1 a. Write the overall growth stoichiometry for the aerobic metabolism on the substrate glutamic acid ($C_5H_9O_4N$), as a function of the growth yield $Y_{X/S}$ (kg biomass/kg substrate). On the basis of the obtained stoichiometry, determine the maximum possible value for $Y_{X/S}$ on glutamic acid;

b. As a function of the obtained overall growth stoichiometry, calculate the rate of production/consumption of ammonia and of carbon dioxide when the growth yield $Y_{X/S}$ is equal to 0.2 kg biomass/kg substrate and the rate of microorganisms' growth on glutamic acid is 2 kg biomass/m³.day.

2.2 a. Consider the hydrolysis of a slowly biodegradable substrate made of proteins. Assume, for simplicity, that the proteins are all made of the monomer alanine ($C_3H_7O_2N$). The empirical formula of the protein is therefore $(C_3H_5ON)_n$. Calculate the rate of alanine production by hydrolysis of the protein, assuming the rate of protein hydrolysis is 100 g/m³.day;

b. Consider the hydrolysis of a slowly biodegradable substrate made of fats. Assume, for simplicity, that the fat is all made of triglycerides of stearic acid ($C_{18}H_{36}O_2$). The empirical formula of the fat is therefore $C_{57}H_{110}O_6$. Calculate the rate of production of glycerol and stearic acid when the rate of hydrolysis of the fat is 100 g/m³.day.

2.3 Assuming that the specific growth rate of certain microorganisms on a given substrate S is expressed by the equation:

$$\mu = \mu_{max} \frac{S}{K_S + S} \frac{O_2}{K_{O2} + O_2}$$

And that $S \gg K_S$, which oxygen concentration needs to be maintained in the liquid phase in order to have a rate that is at least 90%

of the maximum rate in the absence of oxygen limitation? Assume $K_{O2} = 1$ mg/L.

2.4 Calculate the pH of a solution 0.1 M in sodium bicarbonate and 0.1 M in acetic acid.

CHAPTER 3

Mass Balances, Energy Balances and Parameter Estimation

I N THIS CHAPTER, WE will present the general methodology to develop mass and enthalpy balances for biological wastewater treatment processes. We will also show how to obtain the model parameters from experimental data. A general assumption that we will make in this chapter and throughout this book is that the reactors will always be considered perfectly mixed.

3.1 MASS BALANCES

Mass balances have the general form:

$$\text{Accumulation} = \text{Input} - \text{Output} + \text{Generation} - \text{Consumption} \quad (3.1)$$

This general equation takes different forms depending on whether we are considering a batch or a continuous reactor.

3.1.1 Mass Balances in Batch Reactors

In a batch reactor, we have no input and output terms, so the general mass balance can be written simply as:

$$\text{Accumulation} = \text{Generation} - \text{Consumption} \quad (3.2)$$

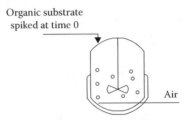

FIGURE 3.1 Batch reactor where microorganisms consume an organic substrate.

For example, consider a batch reactor where an organic substrate is spiked at time 0 and is consumed by microorganisms under aerobic conditions (Figure 3.1).

Let us suppose, for example, that the substrate is glucose. We have seen in Chapter 2 that the stoichiometry of microbial growth on glucose can be written as follows:

$$C_6H_{12}O_6 + \left(6 - 7.96Y_{X/S}\right)O_2 + 1.59Y_{X/S}NH_3 \rightarrow$$

$$1.59Y_{X/S}C_5H_7O_2N + \left(6 - 7.96Y_{X/S}\right)CO_2 + \left(6 - 3.18Y_{X/S}\right)H_2O$$

The most important species for which we want to write mass balances are biomass, substrate (i.e. glucose), nitrogen and oxygen. First of all, we have to agree on the units we will use for the various species in the mass balances. Here we decide to express the concentrations of all the species in kg/m³. We assume Monod kinetics (Chapter 2) for biomass growth, i.e.:

$$r_X = \frac{\mu_{max} S}{K_S + S} X$$

The use of this rate equation for biomass growth means that we are assuming that all the other substrates are not rate limiting and in excess. In case other substrates may be limiting (a frequent case is oxygen, if the aeration rate is not enough to maintain oxygen concentration to non-limiting values), the approach described in Section 2.1.2.1 needs to be used (e.g. Equation 2.55, if oxygen concentration is also rate limiting).

For endogenous metabolism, we assume a first-order dependence on biomass concentration (Chapter 2):

$$r_{end} = -bX$$

Biomass balance:

Biomass accumulated = biomass generated − biomass consumed

Biomass is generated by growth and consumed by endogenous metabolism, so

$$\frac{dX}{dt}\left(\frac{\text{kg biomass}}{\text{m}^3 \cdot \text{day}}\right) = r_X + r_{\text{end}} = \left(\frac{\mu_{\text{max}} S}{K_S + S} - b\right) X \qquad (3.3)$$

Substrate balance:

Substrate accumulated = substrate generated − substrate consumed

There is no substrate generation in this process and substrate is consumed by biomass growth:

$$\frac{dS}{dt}\left(\frac{\text{kg substrate}}{\text{m}^3 \cdot \text{day}}\right) = r_S = -\frac{r_X}{Y_{X/S}} = -\frac{\mu_{\text{max}} S}{K_S + S} \frac{X}{Y_{X/S}} \qquad (3.4)$$

Where in this case, the units of the growth yield, $Y_{X/S}$, will be kg biomass/kg substrate (glucose).

Ammonia balance:

Ammonia accumulated = ammonia generated − ammonia removed

Ammonia is generated by endogenous metabolism and is removed by biomass growth:

$$\frac{d(NH_3)}{dt}\left(\frac{\text{kg N}}{\text{m}^3 \cdot \text{day}}\right) = -0.12 \cdot r_{\text{end}} - 0.12 \cdot r_X = -\left(\frac{\mu_{\text{max}} S}{K_S + S} - b\right) X \cdot 0.12 \quad (3.5)$$

Here we have assumed that 12% of the microorganisms' dry weight is composed of nitrogen (this follows from the empirical formula of the microorganisms $C_5H_7O_2N$).

Oxygen balance:

Here we are interested in the oxygen balance in the liquid phase.

Oxygen accumulated = Oxygen in + oxygen generated − oxygen removed

Oxygen enters the liquid phase due to transfer from the gas phase, and it is removed by biomass growth and endogenous metabolism. There is no oxygen generated.

$$\frac{dC_{O2}}{dt}\left(\frac{\text{kg oxygen}}{\text{m}^3 \cdot \text{day}}\right) = -\frac{\mu_{\text{max}} S}{K_S + S} \cdot \left(\frac{1.07}{Y_{X/S}} - 1.42\right) X - 1.42 b X$$

$$+ k_L a\left(C_{O2}^* - C_{O2}\right) \qquad (3.6)$$

In this equation, the term $(\mu_{max} S/K_S + S) \cdot [(1.07/Y_{X/S}) - 1.42]X + 1.42bX =$ OUR represents the oxygen uptake rate (OUR) by the microorganisms per unit volume of the reactor (in Equation 3.6, it is taken with the negative sign because it causes the dissolved oxygen concentration to decrease), while the term $k_L a(C^*_{O2} - C_{O2})$ represents the rate of oxygen transfer from the gas to the liquid phase.

Assuming that the parameters μ_{max}, K_S, b and $Y_{X/S}$ are known, the mass balances above allow for the calculation of the biomass, substrate, ammonia and oxygen profiles over time.

Often in wastewater treatment processes, the feed is not composed of a single carbon source, and in this case, the substrate can be characterised only by its chemical oxygen demand (COD). Therefore, in this case, the units for the substrate will be kg COD/m³ (instead of kg substrate/m³) and the units for the growth yield $Y_{X/S}$ will have to be kg biomass/kg COD. Also, the substrate might not be (entirely) readily biodegradable, but (part of it) might be slowly biodegradable. The slowly biodegradable substrate (X_S) needs to be hydrolysed before being metabolised and here we will assume the rate equation for hydrolysis seen in Chapter 2. The mass balances for a batch reactor where the carbon source is measured as COD are shown below, where the mass balances for the slowly biodegradable substrate X_S are also shown:

$$\frac{dX_S}{dt}\left(\frac{\text{kgCOD}}{\text{m}^3 \cdot \text{day}}\right) = r_{hydr} = -k_h \frac{X_S/X}{K_X + X_S/X} X \tag{3.7}$$

$$\frac{dS}{dt}\left(\frac{\text{kg COD}}{\text{m}^3 \cdot \text{day}}\right) = r_S - r_{hydr} = k_h \frac{X_S/X}{K_X + X_S/X} X - \frac{\mu_{max} S}{K_S + S} \frac{X}{Y_{X/S}} \tag{3.8}$$

$$\frac{d(NH_3)}{dt}\left(\frac{\text{kg N}}{\text{m}^3 \cdot \text{day}}\right) = -\left(\frac{\mu_{max} S}{K_S + S} - b\right) X \cdot 0.12 \tag{3.9}$$

$$\frac{dX}{dt}\left(\frac{\text{kg biomass}}{\text{m}^3 \cdot \text{day}}\right) = r_X + r_{end} = \left(\frac{\mu_{max} S}{K_S + S} - b\right) X \tag{3.10}$$

$$\frac{dC_{O2}}{dt}\left(\frac{\text{kg oxygen}}{\text{m}^3 \cdot \text{day}}\right) = -\frac{\mu_{max} S}{K_S + S} \cdot X \cdot \left(\frac{1}{Y_{X/S}} - 1.42\right) - 1.42bX$$

$$+ k_L a\left(C^*_{O2} - C_{O2}\right) \tag{3.11}$$

Where the OUR by the microorganisms is:

$$OUR\left(\frac{kg\,oxygen}{m^3 \cdot day}\right) = \frac{\mu_{max}\,S}{K_S + S} \cdot X \cdot \left(\frac{1}{Y_{X/S}} - 1.42\right) + 1.42bX \qquad (3.12)$$

If nitrate is used as electron acceptor instead of oxygen (anoxic conditions), the nitrate balance in batch tests is the following:

$$\frac{dNO_3}{dt}\left(\frac{kg\,N-NO_3}{m^3 \cdot day}\right) = \frac{-\dfrac{\mu_{max}\,S}{K_S + S} \cdot X \cdot \left(\dfrac{1}{Y_{X/S}} - 1.42\right) - 1.42bX}{2.86} \qquad (3.13)$$

Note that the nitrate consumption rate is equal to the oxygen consumption rate (assuming the same values of the parameters) divided by the factor 2.86, as explained in Chapter 2.

For another example, let us consider anaerobic digestion carried out in a batch reactor. For simplicity, we assume that glucose is the substrate. We assume that glucose is converted into acetic acid and hydrogen by fermentative microorganisms. Acetic acid is then converted into methane by acetoclastic methanogens and hydrogen is converted into methane by hydrogenotrophic methanogens. A scheme of a batch reactor for anaerobic digestion is shown in Figure 3.2.

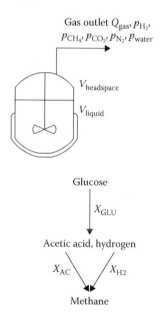

FIGURE 3.2 Scheme of a batch anaerobic digester and of the model of the anaerobic digestion of glucose used here.

Mass balances for batch anaerobic digestion are more complicated than for aerobic systems because in anaerobic processes, we are also interested in calculating the composition of the gas phase, which in general is not of interest for aerobic systems. Also, anaerobic digestion inevitably requires multiple species of microorganisms and mass balances need to be written for each of them.

We will assume that glucose and acetic acid are present only in the liquid phase, while hydrogen, methane and carbon dioxide are present in both phases.

The mass balances for the various components in a batch reactor are written below.

Fermentative microorganisms (X_{GLU}) convert glucose into acetic acid and hydrogen. These microorganisms grow on glucose and are removed by endogenous metabolism.

$$\frac{dX_{GLU}}{dt}\left(\frac{\text{kg microorganisms}}{\text{m}^3.\text{day}}\right) = r_{XGlu} + r_{endGlu}$$

$$= \left(\mu_{\max GLU}\frac{GLU}{K_{SGLU} + GLU} - b_{GLU}\right)X_{GLU} \tag{3.14}$$

Acetoclastic methanogens (X_{Ac}). These microorganisms grow on acetic acid and are removed by endogenous metabolism.

$$\frac{dX_{Ac}}{dt}\left(\frac{\text{kg microorganisms}}{\text{m}^3.\text{day}}\right) = r_{XAc} + r_{endAc}$$

$$= \left(\mu_{\max AC}\frac{AC}{K_{SAC} + AC} - b_{AC}\right)X_{AC} \tag{3.15}$$

Hydrogenotrophic methanogens (X_{H2}). These microorganisms grow on acetic acid and are removed by endogenous metabolism.

$$\frac{dX_{H2}}{dt}\left(\frac{\text{kg microorganisms}}{\text{m}^3.\text{day}}\right) = r_{XH2} + r_{endH2}$$

$$= \left(\mu_{\max H2}\frac{H_2}{K_{SH2} + H_2} - b_{H2}\right)X_{H2} \tag{3.16}$$

Inert biomass. These are inactive microorganisms produced by the endogenous metabolism of X_{GLU}, X_{Ac} and X_{H2}.

$$\frac{dX_{inert}}{dt}\left(\frac{\text{kg microorganisms}}{\text{m}^3.\text{day}}\right) = -r_{endGLU} - r_{endAc} - r_{endH2}$$

$$= b_{GLU}X_{GLU} + b_{Ac}X_{Ac} + b_{H2}X_{H2}$$

(3.17)

Glucose (GLU). Glucose is removed by fermentative microorganisms.

$$\frac{dGLU}{dt} = \frac{-r_{XGlu}}{Y_{X/SGlu}}$$

(3.18)

Acetic acid. Acetic acid is produced by fermentative microorganisms that grow on glucose and is removed by acetoclastic methanogens.

$$\frac{dAc}{dt} = \left(\frac{0.67}{Y_{X/SGlu}} - 0.88\right)r_{XGlu} - \frac{r_{XAc}}{Y_{X/SAc}}$$

(3.19)

Hydrogen in the liquid phase. Hydrogen in the liquid phase is produced by fermentative microorganisms, is removed by hydrogenotrophic microorganisms and transfers to or from the gas phase.

$$\frac{dH_2}{dt} = \left(\frac{0.044}{Y_{X/SGlu}} - 0.058\right)r_{XGlu} - \frac{r_{XH2}}{Y_{X/SH2}} + k_La\left(k_{eqh2} \cdot p_{H2} - H_2\right)$$

(3.20)

Methane in the liquid phase. Methane in the liquid phase is produced by acetoclastic and hydrogenotrophic microorganisms, and transfers to and from the gas phase.

$$\frac{dCH_4}{dt} = \left(\frac{0.267}{Y_{X/SAc}} - 0.354\right)r_{XAc} + \left(\frac{2}{Y_{X/SH2}} - 0.352\right)r_{XH2}$$

$$+ k_La\left(k_{eqCH4} \cdot p_{CH4} - CH_4\right)$$

(3.21)

Carbon dioxide in the liquid phase. Carbon dioxide in the liquid phase is produced by fermentative microorganisms and acetoclastic methanogens, is removed by hydrogenotrophic methanogens, and transfers to and from the gas phase.

$$\frac{dCO_2}{dt} = \begin{bmatrix} \left(\dfrac{0.49}{Y_{X/SGlu}} - 0.65\right) r_{XGlu} + \left(\dfrac{0.73}{Y_{X/SAc}} - 0.97\right) r_{XAc} \\ -\left(\dfrac{5.5}{Y_{X/SH2}} + 0.097\right) r_{XH2} + k_L a \left(k_{eqCO2} \cdot p_{CO2} - CO_2\right) \end{bmatrix} \cdot$$

$$\left(\frac{K_{CO2}}{1 + K_{CO2} + \dfrac{K_{H2CO3}}{10^{-pH}} + \dfrac{K_{HCO3} K_{H2CO3}}{10^{-2pH}}}\right) \qquad (3.22)$$

In Equation 3.22, the term

$$\left(\frac{K_{CO2}}{1 + K_{CO2} + \dfrac{K_{H2CO3}}{10^{-pH}} + \dfrac{K_{HCO3} K_{H2CO3}}{10^{-2pH}}}\right)$$

accounts for the fact that only a fraction of the generated carbon dioxide is present as such in the liquid phase, and part of it is present as other forms of carbonic acid (undissociated carbonic acid, bicarbonate, carbonate), which we have assumed to be in equilibrium (the equilibrium is function of pH as seen in Chapter 2).

Hydrogen in the head space. It transfers to and from the liquid phase, and is removed by the outlet flow of the gas phase.

$$\frac{dp_{H2}}{dt} = \frac{-k_L a \left(k_{eqh2} \cdot p_{H2} - H_2\right)}{\rho_{H2}} \frac{V_{liquid}}{V_{head\,space}} p_{tot} - \frac{Q_{gas} \, p_{H2}}{V_{head\,space}} \qquad (3.23)$$

Methane in the head space. It transfers to and from the liquid phase, and is removed by the outlet flow of the gas phase.

$$\frac{dp_{CH4}}{dt} = \frac{-k_L a \left(k_{eqCH4} \cdot p_{CH4} - CH_4\right)}{\rho_{CH4}} \frac{V_{liquid}}{V_{head\,space}} p_{tot} - \frac{Q_{gas} \, p_{CH4}}{V_{head\,space}} \qquad (3.24)$$

Carbon dioxide in the head space. It transfers to and from the liquid phase, and is removed by the outlet flow of the gas phase.

$$\frac{dp_{CO2}}{dt} = \frac{-k_L a \left(k_{eqCO2} \cdot p_{CO2} - CO_2\right)}{\rho_{CO2}} \frac{V_{liquid}}{V_{head\,space}} p_{tot} - \frac{Q_{gas} \, p_{CO2}}{V_{head\,space}} \qquad (3.25)$$

Gas flow rate from the head space. The gas flow rate from the head space is equal to the sum of the gases leaving the headspace.

$$\frac{Q_{gas}}{V_{liquid}} = \frac{\left(-\dfrac{k_La\left(k_{eqCO2} \cdot p_{CO2} - CO_2\right)}{p_{CO2}} - \dfrac{k_La\left(k_{eqH2} \cdot p_{H2} - H_2\right)}{p_{H2}} - \dfrac{k_La\left(k_{eqCH4} \cdot p_{CH4} - CH_4\right)}{p_{CH4}} \right)}{1 - \dfrac{p_{swat}}{p_{tot}}}$$

(3.26)

Assuming the kinetic parameters and physical properties of the system are known, Equations 3.14 through 3.26 can be solved to give the time profiles of all the variables in a batch anaerobic reactor.

Example 3.1

A batch reactor is fed with wastewater at 500 mgCOD/l. The initial concentration of the microorganisms is 100 mg/l. Assume the substrate in the wastewater is all readily biodegradable. Calculate the initial rate of the following processes:

- Substrate removal;
- Net micro-organism production;
- Oxygen consumption by the microorganisms

Kinetic parameters:

$$\mu_{max} = 4 \ day^{-1}$$

$$b = 0.1 \ day^{-1}$$

$$k_{hydr} = 3 \ kg \ COD/kg \ biomass.day$$

$$K_X = 0.2 \ kg \ COD/kg \ biomass$$

$$K_S = 0.004 \ kg \ COD/m^3$$

$$Y_{X/S} = 0.2 \ kg \ biomass/kg \ COD$$

Solution

The initial rates can be calculated using Equations 3.8, 3.10 and 3.11, the last without including the k_La term, since the rate of oxygen consumption by the microorganisms is required and not the overall rate of decrease in the oxygen concentration (which depends also on the rate of oxygen supply, i.e. on the $k_L a$).

$$\frac{dS}{dt}\left(\frac{\text{kg COD}}{\text{m}^3 \cdot \text{day}}\right) = -\frac{4\,\text{day}^{-1} \cdot 0.500\dfrac{\text{kg COD}}{\text{m}^3}}{0.004\dfrac{\text{kg COD}}{\text{m}^3} + 0.500\dfrac{\text{kg COD}}{\text{m}^3}}\,0.2\frac{0.1\,\text{kg biomass}}{\dfrac{\text{kg biomass}}{\text{kg COD}}}$$

$$= -1.98\frac{\text{kg COD}}{\text{m}^3 \cdot \text{day}}$$

$$\frac{dX}{dt}\left(\frac{\text{kg biomass}}{\text{m}^3 \cdot \text{day}}\right)$$

$$= \left(\frac{4\,\text{day}^{-1} \cdot 0.500\dfrac{\text{kg COD}}{\text{m}^3}}{0.004\dfrac{\text{kg COD}}{\text{m}^3} + 0.500\dfrac{\text{kg COD}}{\text{m}^3}} - 0.1\,\text{day}^{-1}\right) \cdot 0.1\frac{\text{kg biomass}}{\text{day}} \quad (3.10)$$

$$= 0.39\frac{\text{kg biomass}}{\text{m}^3 \cdot \text{day}}$$

$$r_{\text{O2biomass}}\left(\frac{\text{kg oxygen}}{\text{m}^3 \cdot \text{day}}\right)$$

$$= -\frac{4\,\text{day}^{-1} \cdot 0.500\dfrac{\text{kg COD}}{\text{m}^3}}{0.004\dfrac{\text{kg COD}}{\text{m}^3} + 0.500\dfrac{\text{kg COD}}{\text{m}^3}}$$

$$\cdot 0.1\frac{\text{kg biomass}}{\text{day}} \cdot \left(\frac{1}{0.2\dfrac{\text{kg biomass}}{\text{kg COD}}} - 1.42\right) - 1.42 \cdot 0.1 \cdot 0.1$$

$$= -1.43\frac{\text{kg oxygen}}{\text{m}^3 \cdot \text{day}}$$

Note that the rate of oxygen consumption could also have been calculated immediately using the COD balance from the rate of substrate removal and biomass formation (converted into COD):

$$r_{O2biomass}\left(\frac{kg\, oxygen}{m^3\cdot day}\right)$$

$$= -\left(1.98\frac{kg\, COD}{m^3\cdot day} - 0.39\frac{kg\, biomass}{m^3\cdot day}1.42\frac{kg\, COD}{kg\, biomass}\right)$$

$$= -1.43\frac{kg\, oxygen}{m^3\cdot day}$$

Example 3.2

A batch reactor is inoculated with only nitrifying microorganisms at a concentration of 20 mg/l. The initial concentration of ammonia is 10 mg $N\text{-}NH_3$/l. Calculate the initial rates of:

- Ammonia removal;
- Nitrate production;
- Net micro-organism production;
- Oxygen consumption

$$\mu_{max\, A} = 0.8\ day^{-1}$$

$$b_A = 0.05\ day^{-1}$$

$$K_{SA} = 0.001\ kg\ N/m^3$$

$$Y_{XA/NO3} = 0.15\ kg\, biomass/kg\, N - NO_3$$

Solution

The calculation follows immediately from mass balances and from the stoichiometry and kinetics for nitrifying microorganisms seen in Chapter 2.

The growth rate of nitrifying microorganisms is given by:

$$r_{XA} = \frac{\mu_{maxA} NH_3}{K_{SA} + NH_3} \cdot X_A$$

$$= \frac{0.8 \, day^{-1} \, 0.01 \dfrac{kg \, N - NH_3}{m^3}}{0.001 \dfrac{kg \, N - NH_3}{m^3} + 0.01 \dfrac{kg \, N - NH_3}{m^3}} 0.02 \dfrac{kg \, biomass}{m^3}$$

$$= 0.015 \dfrac{kg \, biomass}{m^3 day}$$

And the net rate of microorganisms production is:

$$\frac{dX_A}{dt} \left(\frac{kg \, biomass}{m^3 \cdot day} \right)$$

$$= r_{XA} - b_A X_A = 0.015 \dfrac{kg \, biomass}{m^3 day} - 0.05 \, day^{-1} \cdot 0.02 \dfrac{kg \, biomass}{m^3}$$

$$= 0.014 \dfrac{kg \, biomass}{m^3 day}$$

$$\frac{dNH_3}{dt} \left(\frac{kg \, N - NH_3}{m^3 \cdot day} \right)$$

$$= -\left(\frac{1}{Y_{XA/NO3}} + 0.12 \right) r_{XA}$$

$$= -\left(\frac{1 \dfrac{kg \, N - NH_3}{kg \, N - NO_3}}{0.15 \dfrac{kg \, biomass}{kg \, N - NO_3}} + 0.12 \dfrac{kg \, N - NH_3}{kg \, biomass} \right) 0.015 \dfrac{kg \, biomass}{m^3 day}$$

$$= -0.102 \dfrac{kg \, N - NH_3}{m^3 day}$$

$$\frac{dNO_3}{dt}\left(\frac{kg\,N-NO_3}{m^3 \cdot day}\right) = \frac{1}{Y_{XA/NO_3}}r_{XA}$$

$$= \frac{1}{0.15\dfrac{kg\,biomass}{kg\,N-NO_3}}0.015\frac{kg\,biomass}{m^3 day}$$

$$= 0.100\frac{kg\,N-NO_3}{m^3 day}$$

$$r_{O2}\left(\frac{kg\,oxygen}{m^3 \cdot day}\right)$$

$$= -\left(\frac{4.54}{Y_{XA/NO3}}-0.04\right)r_{XA}$$

$$= -\left(\frac{4.54\dfrac{kg\,oxygen}{kg\,N-NO_3}}{0.15\dfrac{kg\,biomass}{kg\,N-NO_3}}-0.04\frac{kg\,oxygen}{biomass}\right)0.015\frac{kg\,biomass}{m^3 day}$$

$$= -0.45\frac{kg\,oxygen}{m^3 day}$$

Example 3.3

Calculate the time profiles of substrate, ammonia, biomass and oxygen concentration for a biological reactor fed with glucose as the only carbon source. Assume an initial concentration of glucose equal to 1 g/l and an initial biomass concentration equal to 0.1 g/l. Assume oxygen is transferred with a $k_L a$ equal to 100 day^{-1} and the saturation concentration of oxygen in water is 9 mg/l.

Kinetic parameters:

$$\mu_{max} = 6\,day^{-1}$$

$$b = 0.2\,day^{-1}$$

$$Y_{X/S} = 0.3\,kg\,biomass/kg\,substrate$$

Solution

The solution comes from the integration of Equations 3.3 through 3.6 and is reported in Figure 3.3. Substrate concentration drops at an increasing rate as biomass concentration increases. After the substrate is removed completely, biomass concentration starts to decrease slowly due to endogenous metabolism. Ammonia concentration decreases during biomass growth and increases very slowly during endogenous metabolism because ammonia is released due to biomass decay. The OUR increases during substrate removal because biomass grows and removes the substrate at an increasing rate, and

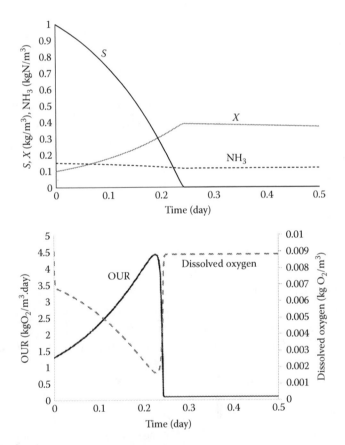

FIGURE 3.3   Example 3.3. Profiles of substrate, biomass, ammonia, OUR and dissolved oxygen during an aerobic batch test.

then, after the substrate is removed completely from the medium, the OUR falls sharply and it is only due to endogenous metabolism. Correspondingly with the OUR profile, the dissolved oxygen concentration decreases when the substrate is present and increases rapidly when the substrate is removed completely.

3.1.2 Mass Balances in Continuous Reactors

Assume the biological process is carried out in a continuous reactor, such as the one shown in Figure 3.4.

Substrate and ammonia are continuously fed to the reactor at a concentration S_0 and $NH3_0$, and the feed flow rate is Q. In the reactor, substrate and ammonia are consumed and biomass is produced. The outlet stream has the same flow rate of the feed, Q, and the same composition of the reactor, due to the assumption of perfect mixing. The mass balances have the general form:

$$\text{Accumulation} = \text{Input} - \text{Output} + \text{Generation} - \text{Consumption}$$

After the initial startup phase, continuous reactors reach steady state (the accumulation terms becomes equal to 0), and the steady-state mass balance can be written as:

$$\text{Input} + \text{Generation} = \text{Output} + \text{Consumption}$$

Assuming glucose as a substrate, we have the following mass balances.
Biomass:
There is no biomass in the feed (no input term), biomass is generated by growth and consumed by endogenous metabolism, and is present in the output stream.

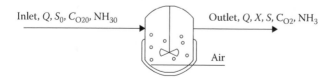

Inlet, Q, S_0, C_{O20}, $NH3_0$ Outlet, Q, X, S, C_{O2}, NH_3 Air

FIGURE 3.4 Scheme of a continuous reactor (no microorganisms in the feed) where microorganisms remove the substrate under aerobic conditions.

$$\frac{d(XV)}{dt}\left(\frac{\text{kg biomass}}{\text{day}}\right) = r_X V + r_{end} V - QX$$

$$= \left(\frac{\mu_{max} S}{K_S + S} - b\right) XV - QX \tag{3.27}$$

And at steady state, where $d(XV)/dt = 0$:

$$\left(\frac{\mu_{max} S}{K_S + S} - b\right) XV = QX \tag{3.28}$$

Substrate:

Substrate is present in the feed and in the output stream, is not generated and is removed by biomass growth.

$$\frac{d(SV)}{dt}\left(\frac{\text{kg substrate}}{\text{day}}\right) = QS_0 + r_S V - QS = QS_0 - \frac{\mu_{max} S}{K_S + S}\frac{XV}{Y_{X/S}} - QS \tag{3.29}$$

And at steady state:

$$QS_0 = \frac{\mu_{max} S}{K_S + S}\frac{XV}{Y_{X/S}} + QS \tag{3.30}$$

Ammonia:

Ammonia is present in the feed and in the output stream, is generated by endogenous metabolism and removed by biomass growth.

$$\frac{d(NH_3 V)}{dt}\left(\frac{\text{kg N}}{\text{day}}\right) = QNH_{30} - QNH_3 - 0.12 \cdot r_{end} V - 0.12 \cdot r_X V$$

$$= QNH_{30} - QNH_3 - \left(\frac{\mu_{max} S}{K_S + S} - b\right) XV \cdot 0.12 \tag{3.31}$$

And at steady state:

$$QNH_{30} = QNH_3 + \left(\frac{\mu_{max} S}{K_S + S} - b\right) XV \cdot 0.12 \tag{3.32}$$

Oxygen:

Oxygen is present in the feed, is transferred from the gas phase (input), is not generated, is consumed by the biomass and is present in the output stream.

$$\frac{d(O_2 V)}{dt}\left(\frac{kg\,oxygen}{day}\right) = QC_{O20} - \frac{\mu_{max} S}{K_S + S} \cdot \left(\frac{1}{Y_{X/S}} - 1.42\right)XV$$

$$-1.42bXV + k_L a\left(C_{O2}^* - C_{O2}\right)V - QC_{O2} \tag{3.33}$$

And at steady state:

$$QC_{O20} + k_L a\left(C_{O2}^* - C_{O2}\right)V = \frac{\mu_{max} S}{K_S + S} \cdot \left(\frac{1}{Y_{X/S}} - 1.42\right)XV$$

$$+1.42bXV + QC_{O2} \tag{3.34}$$

Assuming the kinetic parameters, the volume of the reactor and the flow rate Q are known, Equations 3.28, 3.30, 3.32 and 3.34 can be solved simultaneously to calculate the steady state of the biological reactor, i.e. the values of the substrate, biomass, ammonia and oxygen concentration at steady state.

3.2 ENTHALPY BALANCES

The general form of enthalpy balances corresponds with the general form of mass balances and can be written as:

$$\text{Enthalpy accumulated} = \text{Enthalpy in} - \text{Enthalpy out}$$
$$+ \text{Enthalpy added to the system} - \text{Enthalpy removed} \tag{3.35}$$
$$\text{from the system}$$

3.2.1 Enthalpy Balances for Batch Systems

In a purely batch system, we do not have in and out terms, and therefore, the enthalpy balance can be written as:

$$\text{Enthalpy accumulated} = \text{Enthalpy added to the}$$
$$\text{system} - \text{Enthalpy removed from the system} \tag{3.36}$$

However, in biological reactions, we never have a completely batch process because even when we do not have any liquid inlet and outlet streams, we may have an inlet gas stream and we always have an outlet gas stream.

For example, let us consider an aerobic adiabatic batch process, where a carbon source is being removed by microorganisms. The reactor is filled with a liquid phase that contains the substrate, the nitrogen source and the inoculum of microorganisms. Oxygen is supplied by sparging the reactor with the chosen gas, air or pure oxygen. In this case, even though the process can be considered batch as far as the liquid phase is concerned, we do have an inlet stream, i.e. the gas phase that is used to provide oxygen, and an outlet stream, i.e. the gas phase that contains the carbon dioxide generated by the reaction. Since the process is adiabatic, there is no enthalpy added or removed from the system. In the following, we will ignore, for simplicity, the contribution to the enthalpy balance of the inlet gas stream (the contribution of dissolved oxygen, however, will be included in the enthalpy of the liquid phase). However, it is important to include in the enthalpy balance the contribution of the outlet gas stream because it includes carbon dioxide, which is a product of the reaction. In this example, we will ignore any water vapour that may be present in the outlet gas stream (otherwise, its enthalpy should also be taken into account).

With these assumptions, the generic form of the enthalpy balance is:

$$\text{Enthalpy accumulated} = -\text{Enthalpy out} \qquad (3.37)$$

Since biological reactions always take place in the water phase and the liquid phase can always be considered a dilute solution with its physical properties equal to the properties of water, the total enthalpy of the liquid phase can be written as $V\rho H$ where H (J/kg) is the enthalpy of the liquid phase per unit mass and the density of the liquid phase can be considered constant. The volume of the liquid phase can also be considered constant because there are no inlet or outlet liquid streams and the change in volume due to the biological reactions and to the dissolution of oxygen can be ignored. Therefore, the enthalpy accumulated, i.e. the rate of enthalpy change, is given by $V\rho(dH/dt)$. The enthalpy leaving the system is due to the carbon dioxide generated by the reaction that leaves the system with the gas phase. For simplicity, here we ignore the solubility of carbon dioxide and assume that all the carbon dioxide generated by the reaction evolves as a gas. Therefore, the term 'enthalpy out' can be written as $r_{CO2} V H_{CO2}$, where r_{CO2} (kg/m³.day) is the rate of carbon dioxide generation per unit volume of the biological reactor.

Therefore, with all these assumptions, the enthalpy balance for the biological process is:

$$V\rho \frac{dH}{dt} = -r_{CO2} V H_{CO2} \quad \Rightarrow \quad \rho \frac{dH}{dt} = -r_{CO2} H_{CO2} \qquad (3.38)$$

H is the enthalpy of the liquid phase per unit mass and can be written as:

$$H = \frac{M_S H_S + M_{O2} H_{O2} + M_{NH3} H_{NH3} + M_X H_X + M_{H2O} H_{H2O}}{M_{tot}} \quad (3.39)$$

Where M is the mass (kg) of the various substances, the subscript S stands for substrate, O_2 for oxygen, NH_3 for ammonia, X for biomass and H_2O for water. The mass of the various substances can be expressed as a function of their respective concentrations, S, C_{O2}, C_{NH3}, X, C_{H2O} (kg/m³):

$$M_S = S \cdot V; \quad M_{O2} = C_{O2} \cdot V; \quad M_{NH3} = C_{NH3} \cdot V;$$

$$M_X = X \cdot V; \quad M_{H2O} = C_{H2O} V; \quad M_{tot} = \rho V \quad (3.40)$$

Therefore, the derivative of the specific enthalpy H can be expressed as:

$$\frac{dH}{dt}\left(\frac{J}{kg.day}\right) = \frac{1}{V\rho}\begin{pmatrix} H_S V \dfrac{dS}{dt} + H_{O2} V \dfrac{dC_{O2}}{dt} + H_{NH3} V \dfrac{dC_{NH3}}{dt} \\[2mm] + H_X V \dfrac{dX}{dt} + H_{H2O} V \dfrac{dC_{H2O}}{dt} + S V \dfrac{dH_S}{dt} \\[2mm] + C_{O2} V \dfrac{dH_{O2}}{dt} + C_{NH3} V \dfrac{dH_{NH3}}{dt} + X V \dfrac{dH_X}{dt} \\[2mm] + C_{H2O} V \dfrac{dH_{H2O}}{dt} \end{pmatrix}$$

$$\quad (3.41)$$

$$= \frac{1}{\rho}\begin{pmatrix} H_S \dfrac{dS}{dt} + H_{O2} \dfrac{dC_{O2}}{dt} + H_{NH3} \dfrac{dC_{NH3}}{dt} \\[2mm] + H_X \dfrac{dX}{dt} + H_{H2O} \dfrac{dC_{H2O}}{dt} + S \dfrac{dH_S}{dt} \\[2mm] + C_{O2} \dfrac{dH_{O2}}{dt} + C_{NH3} \dfrac{dH_{NH3}}{dt} \\[2mm] + X \dfrac{dH_X}{dt} + C_{H2O} \dfrac{dH_{H2O}}{dt} \end{pmatrix}$$

From Chapter 2, assuming that the specific heat is independent on temperature (in the temperature range considered here), the derivative of the specific enthalpy for a generic species A is:

$$\frac{dH_A}{dt} = c_{PA}\frac{dT}{dt} \tag{3.42}$$

And therefore the various terms dH_S/dt, dH_{O2}/dt and so on can be written as $dH_S/dt = c_{PS}(dT/dt)$, $dH_{O2}/dt = c_{PO2liq}(dT/dt)$, etc.

Therefore, the derivative of the specific enthalpy of the liquid phase can be rewritten as:

$$\frac{dH}{dt}\left(\frac{J}{kg.day}\right) = \frac{1}{\rho}\begin{pmatrix} H_S\dfrac{dS}{dt} + H_{O2}\dfrac{dC_{O2}}{dt} + H_{NH3}\dfrac{dC_{NH3}}{dt} + H_X\dfrac{dX}{dt} \\[2mm] + H_{H2O}\dfrac{dC_{H2O}}{dt} + Sc_{PS}\dfrac{dT}{dt} + C_{O2}c_{PO2}\dfrac{dT}{dt} \\[2mm] + C_{NH3}c_{PNH3}\dfrac{dT}{dt} + Xc_{PX}\dfrac{dT}{dt} + C_{H2O}c_{PH2O}\dfrac{dT}{dt} \end{pmatrix} \tag{3.43}$$

$$= \frac{1}{\rho}\begin{pmatrix} H_S\dfrac{dS}{dt} + H_{O2}\dfrac{dC_{O2}}{dt} + H_{NH3}\dfrac{dC_{NH3}}{dt} \\[2mm] + H_X\dfrac{dX}{dt} + H_{H2O}\dfrac{dC_{H2O}}{dt} \\[2mm] + \dfrac{dT}{dt}\left(Sc_{PS} + C_{O2}c_{PO2} + C_{NH3}c_{PNH3} + Xc_{PX} + C_{H2O}c_{PH2O}\right) \end{pmatrix}.$$

This expression can be simplified if we assume that since we are in a dilute solution and water is by far the main component in it:

$$Sc_{PS} + C_{O2}c_{PO2} + C_{NH3}c_{PNH3} + Xc_{PX} + C_{H2O}c_{PH2O} = \rho \cdot c_{PH2O} \tag{3.44}$$

And so we obtain:

$$\frac{dH}{dt}\left(\frac{J}{kg.day}\right) = \frac{1}{\rho}\begin{pmatrix} H_S\dfrac{dS}{dt} + H_{O2}\dfrac{dC_{O2}}{dt} + H_{NH3}\dfrac{dC_{NH3}}{dt} \\[2mm] + H_X\dfrac{dX}{dt} + H_{H2O}\dfrac{dC_{H2O}}{dt} + \rho c_{pH2O}\dfrac{dT}{dt} \end{pmatrix} \tag{3.45}$$

And, therefore, the enthalpy balance for the adiabatic batch system can be rewritten as:

$$\rho c_{PH2O}\frac{dT}{dt} = -H_S\frac{dS}{dt} - H_{O2}\frac{dC_{O2}}{dt} - H_{NH3}\frac{dC_{NH3}}{dt}$$

$$-H_X\frac{dX}{dt} - H_{H2O}\frac{dC_{H2O}}{dt} - r_{CO2}H_{CO2} \tag{3.46}$$

In order to calculate the temperature change in the system, we need to be able to calculate the various terms dS/dt, dC_{O2}/dt, etc. These terms depend on the rate and stoichiometry of the biological process. If, for example, the substrate is glucose, the enthalpy balance can be written as:

$$\rho c_p \frac{dT}{dt} = -H_S \frac{-r_X}{Y_{X/S}}$$

$$-H_{O2}\left(-r_X \cdot \left(\frac{1.07}{Y_{X/S}} - 1.42\right)X - 1.42bX + k_L a\left(C_{O2}^* - C_{O2}\right)\right)$$

$$-H_{NH3}\left(-0.12 \cdot r_{end} - 0.12 \cdot r_X\right) - H_X\left(r_X + r_{end}\right)$$

$$-H_{H2O}\left(\left(\frac{0.6}{Y_{X/S}} - 0.28\right)r_X - 0.32 r_{end}\right)$$

$$-H_{CO2}\left(\left(\frac{1.47}{Y_{X/S}} - 1.95\right)r_X - 1.95 r_{end}\right)$$

(3.47)

From Equation 3.47, the change in the reactor temperature as the batch reaction proceeds can be calculated from the rate of biomass growth and oxygen transfer, and from the stoichiometry of the growth reaction and of the endogenous metabolism.

If the batch reactor is not adiabatic, then we need to include the term for the heat transfer to or from the external environment, i.e.:

Enthalpy accumulated = Enthalpy added to the system
– Enthalpy removed from the system   (3.48)

Let us assume that enthalpy is added or removed from the reactor through a jacket, where a heating or cooling fluid flows. The enthalpy added to (or removed from) the system per unit time depends on the overall heat transfer coefficient, as described in Chapter 2:

$$\dot{Q} = UA\Delta T$$

where ΔT is the temperature difference between the fluid in the reactor and the fluid in the jacket (here we are assuming that the jacket is perfectly mixed, i.e. that the temperature in the jacket is uniform and equal to the outlet temperature of the jacket fluid).

Therefore, referring to the batch process we have considered so far, with the addition of heat addition/removal via the jacket, the enthalpy balance can be written as:

$$V\rho \frac{dH}{dt} = -r_{CO2}VH_{CO2} - UA\Delta T \implies \rho \frac{dH}{dt} = -r_{CO2}H_{CO2} - \frac{UA(T-T_j)}{V} \quad (3.49)$$

which becomes, with the assumptions made above:

$$\rho c_{PH2O} \frac{dT}{dt} = -H_S \frac{dS}{dt} - H_{O2} \frac{dC_{O2}}{dt} - H_{NH3} \frac{dC_{NH3}}{dt}$$

$$-H_X \frac{dX}{dt} - H_{H2O} \frac{dC_{H2O}}{dt} - r_{CO2}H_{CO2} - \frac{UA(T-T_j)}{V} \quad (3.50)$$

If we assume, as we have done previously, that the substrate is glucose, the temperature profile in the batch reactor is given by the equation below:

$$\rho c_p \frac{dT}{dt} = -H_S \frac{-r_X}{Y_{X/S}}$$

$$-H_{O2}\left(-r_X \cdot \left(\frac{1.07}{Y_{X/S}} - 1.42\right)X - 1.42bX + k_La\left(C_{O2}^* - C_{O2}\right)\right)$$

$$-H_{NH3}\left(-0.12 \cdot r_{end} - 0.12 \cdot r_X\right) - H_X\left(r_X + r_{end}\right) \quad (3.51)$$

$$-H_{H2O}\left(\left(\frac{0.6}{Y_{X/S}} - 0.28\right)r_X - 0.32r_{end}\right)$$

$$-H_{CO2}\left(\left(\frac{1.47}{Y_{X/S}} - 195\right)r_X - 1.95r_{end}\right) - \frac{UA(T-T_j)}{V}$$

Assuming that the heat transfer coefficient U and the geometry of the reactor (A and V) are known, this equation can be integrated to calculate the temperature profile in the reactor $T(t)$ if the temperature in the jacket T_j is known and constant. In practice, the temperature in the jacket can be maintained constant by using a control loop that adjusts the flow rate of the fluid in the jacket. However, in the general case, the temperature in the jacket can be calculated by means of an enthalpy balance for the fluid in the jacket.

For the fluid in the jacket, we need the general enthalpy balance for a continuous flow system, i.e.:

$$\text{Enthalpy accumulated} = \text{Enthalpy in} - \text{Enthalpy out} \\ + \text{Enthalpy added to the system} - \text{Enthalpy removed} \\ \text{from the system} \tag{3.35}$$

For the fluid in the jacket, Equation 3.35 can be written as

$$V_J \rho_J \frac{dH_J}{dt} = w_J c_{pJ} \left(T_{JIN} - 25 \right) - w_J c_{pJ} \left(T_J - 25 \right) + UA \left(T - T_J \right) \tag{3.52}$$

Equation 3.52 becomes, assuming that the specific heat for the fluid in the jacket is independent of temperature in the considered temperature range:

$$\frac{dT_J}{dt} = \frac{w_J c_{pJ} \left(T_{JIN} - T_J \right) + UA \left(T - T_J \right)}{c_{pJ} V_J \rho_J} \tag{3.53}$$

This equation gives the temperature profile for the fluid in the jacket and can be integrated together with the enthalpy balance for the fluid in the reactor. In summary, if we have a jacketed batch reactor, the temperature profile in the reactor and in the jacket can be obtained by writing the enthalpy balances both in the reactor and in the jacket. They result in a system of two differential equations in the two unknowns T and T_J, Equations 3.51 and 3.53, which can be solved with the appropriate initial conditions for the two variables.

Example 3.4

a) Calculate the temperature profile for the batch reaction in Example 3.3, assuming adiabatic conditions;

b) Calculate the temperature profile for the same batch reaction, but assuming an initial glucose concentration of 10 g/l. A higher $k_L a$ is required in this case, so assume a $k_L a$ value equal to 6000 day^{-1};

c) For the case with an initial substrate concentration equal to 10 g/l, assume that the reactor is cooled with a jacket with inlet water at a temperature of 10°C. Assume that the reactor has a cylindrical shape, volume of 10 m^3 and a diameter of 2 m. Assume that the jacket has a void space, where water flows, of 3 cm. Assume that the overall heat transfer coefficient is 100 W/m^2/K. Calculate the flow rate of the cooling fluid that is required to limit the temperature increase of the reactor to 10°C.

Solution

The temperature profiles for cases (a) and (b) are obtained from the integration of the enthalpy balance, Equation 3.47. The only difference between cases (a) and (b) is the initial substrate concentration.

In the enthalpy balances, r_X and r_{end} are given by the usual equations:

$$r_X = \frac{\mu_{max} S}{K_S + S} X \quad r_{end} = -bX$$

with the parameter values given in Example 3.3.

The first step is to calculate the profiles of biomass, substrate and oxygen with time, and then the corresponding values of r_X and r_{end} at each time step. Since we are assuming that the kinetic parameters are independent of temperature, the profiles are the same as those obtained in Example 3.3.

The next step is to calculate the enthalpy terms H_S, H_{O2}, and so on. The enthalpies of the various species are expressed by the general formula, in which we need to pay attention to express all the terms as J/kg, instead of J/mol, because the rates are expressed as kg and not mol (note that for the dissolved species and for biomass the specific heat has been taken equal to the value for water):

$$H_A(T)\left(\frac{J}{kg}\right) = \Delta H_{fA}(25°C) + \left[\lambda_{phase\,change}(25°C)\right] + c_{PA}(T - 25)$$

For glucose:

$$H_S(T)\left(\frac{J}{kg}\right) = \Delta H_{fGLU}(25°C) + \lambda_{dissolutionGLU}(25°C) + c_{PH2O}(T - 25)$$

$$= -7.1 \cdot 10^6 + 6.1 \cdot 10^4 + 4186(T - 25)$$

For oxygen:

$$H_{O2}(T)\left(\frac{J}{kg}\right) = \Delta H_{fO2}(25°C) + \lambda_{dissolutionO2}(25°C) + c_{PH2O}(T - 25)$$

$$= 0 - 4.6 \cdot 10^5 + 4186(T - 25)$$

For ammonia:

$$H_{O2}(T)\left(\frac{J}{kg}\right) = \Delta H_{fNH3}(25°C) + c_{PH2O}(T - 25)$$

$$= -4.8 \cdot 10^6 + 4186(T - 25)$$

For biomass:

$$H_X(T)\left(\frac{J}{kg}\right) = \Delta H_{fX}(25°C) + c_{PH2O}(T-25)$$

$$= -6.8 \cdot 10^6 + 4186(T-25)$$

For water:

$$H_{H2O}(T)\left(\frac{J}{kg}\right) = \Delta H_{fH2O}(25°C) + c_{PH2O}(T-25)$$

$$= -1.6 \cdot 10^7 + 4186(T-25)$$

For carbon dioxide:

$$H_{CO2}(T)\left(\frac{J}{kg}\right) = \Delta H_{fCO2}(25°C) + c_{PCO2gas}(T-25)$$

$$= -8.9 \cdot 10^6 + 910(T-25)$$

After defining the various enthalpy terms, the problem can be solved by numerical integration of Equation 3.47, which can be done easily in Microsoft Excel as described in the Appendices (of course, any other mathematical software can also be used to solve this equation). The obtained profiles are reported in Figure 3.5.

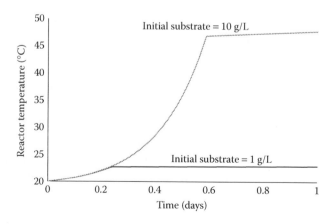

FIGURE 3.5   Example 3.4. Calculated temperature profiles for a batch reactor during the removal of the substrate.

As expected, the temperature increases during the biological process because biological reactions are exothermic and the temperature increase is much larger for the higher initial substrate concentration. The temperature profiles show a sharp change in slope when the substrate is removed completely. This is due to the fact that the rate of endogenous metabolism is much lower than the growth rate and, therefore, the rate of heat generation is much lower and the temperature increases at a much lower, almost insignificant, rate. The final temperature is below 25°C when the initial substrate concentration is 1 g/l, but it is higher than 45°C when the initial substrate is 10 g/l. In the latter case, cooling is probably required in order to maintain the reactor at the desired temperature, and this is shown in part c).

c) If heat is removed via a jacket, the temperature profile in the reactor is given by Equations 3.51 and 3.53, which need to be solved simultaneously. In order to solve the equations, we need to calculate the area of the jacketed area A and the volume of the jacket V_J. We know the volume of the reactor and its diameter, from which we calculate the height:

$$V = \frac{\pi D^2}{4} H \quad \Rightarrow \quad H = \frac{4V}{\pi D^2} = 3.2\,\text{m}$$

Therefore, the jacketed area A is, assuming that all the surface of the reactor is jacketed:

$$A = \frac{\pi D^2}{4} + \pi D H = 23.1\,\text{m}^2$$

And the volume of the jacket can be reasonably approximated by multiplying the jacketed area by the jacket gap:

$$V_J = A \cdot 0.03 = 0.69\,\text{m}^3$$

With these values, Equations 3.51 and 3.53 can be solved for different values of the flow rate of the cooling fluid in the jacket, obtaining the profiles for the reactor temperature shown in Figure 3.6. A flow rate of the cooling fluid of at least 80,000 kg/day is required in order to keep the maximum temperature increase in the reactor to within 10°C.

Note that with a fixed value of the flow rate of the cooling fluid, the reactor temperature will inevitably be variable, because the rate of heat generation changes with time. Therefore, after the substrate is removed entirely, the reactor temperature will start decreasing

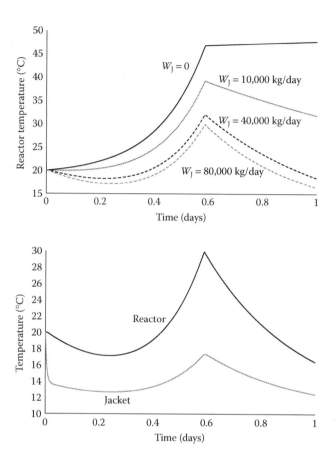

FIGURE 3.6   Example 3.4. Top: profiles of the temperature in a batch biological reactor for various flow rates of the cooling fluid. Bottom: temperature profiles in the reactor and in the jacket for a flow rate of the jacket fluid of 80,000 kg/hr.

because the rate of heat removal will be higher than the rate of heat generation. If we wanted to keep the reactor temperature really constant, we should use a control loop that regulates the cooling fluid flow rate as a function of the reactor temperature. The control loop would increase the flow rate of the cooling fluid when the reactor temperature tends to increase, i.e. when the substrate is removed rapidly, and would decrease it when the rate of heat generation is lower, for example, after the substrate has been removed completely and the only process occurring is endogenous metabolism.

Figure 3.6 also compares the temperature profile in the reactor and in the jacket when the cooling fluid flow rate is 80,000 kg/day.

3.2.2 Enthalpy Balances for Continuous Systems

In a continuous process at steady state, i.e. with no accumulation of enthalpy, the generic enthalpy balance can be written as

$$\text{Enthalpy in} + \text{Enthalpy added to the system}$$
$$= \text{Enthalpy out} - \text{Enthalpy removed from the system} \tag{3.54}$$

If the reactor is adiabatic, there is no heat transfer to or from the external environment, and therefore, the generic form of the heat balance is

$$\text{Enthalpy in} = \text{Enthalpy out} \tag{3.37}$$

Let us consider a continuous adiabatic biological reactor where a substrate S is removed by microorganisms. The feed contains the substrate and ammonia in water, while the effluent of the reactor will contain the residual substrate and ammonia, the biomass, dissolved oxygen and carbon dioxide. Air or pure oxygen is supplied to the reactor to maintain the oxygen concentration in the reactor to the desired value. Similar to what we have done for the enthalpy balances on the batch reactor, we ignore the enthalpy content of the inlet gas, and we assume that all the produced carbon dioxide evolves from the reactor as a gas.

With these assumptions, we have:

$$\text{Enthalpy in} = Q\left(S_0 H_S(T_{feed}) + NH_{30} H_{NH3}(T_{feed}) + H_2 O H_{H2O}(T_{feed})\right) \tag{3.55}$$

and

$$\text{Enthalpy out} = Q\left(S H_S(T) + NH_3 H_{NH3}(T) + X H_X(T)\right.$$
$$\left. + H_2 O H_{H2O}(T) + C_{O2} H_{O2}(T)\right) + r_{CO2} H_{CO2}(T)V \tag{3.56}$$

Therefore, the enthalpy balance for this system is:

$$Q\left(S_0 H_S(T_{feed}) + NH_{30} H_{NH3}(T_{feed}) + H_2 O H_{H2O}(T_{feed})\right)$$
$$= Q\left(S H_S(T) + NH_3 H_{NH3}(T) + X H_X(T)\right. \tag{3.57}$$
$$\left. + H_2 O H_{H2O}(T) + C_{O2} H_{O2}(T)\right) + r_{CO2} H_{CO2}(T)V$$

If the concentration of all the variables at steady state is known, Equation 3.57 can be solved to calculate the temperature in a continuous adiabatic biological reactor.

If the reactor is not adiabatic, but heat is exchanged, for example, using a cooling fluid in the jacket, the general form of the enthalpy balance has to be used and the enthalpy balances becomes:

$$
\begin{aligned}
Q\big(S_0 H_S(T_{feed}) &+ NH_{30} H_{NH3}(T_{feed}) + H_2 OH_{H2O}(T_{feed})\big) \\
&= Q\big(S_0 H_S(T) + NH_3 H_{NH3}(T) + XH_X(T) + H_2 OH_{H2O}(T) \\
&\quad + C_{O2} H_{O2}(T)\big) + r_{CO2} H_{CO2}(T)V + UA\big(T - T_j\big)
\end{aligned}
\tag{3.58}
$$

The jacket temperature T_j can be calculated through an enthalpy balance on the jacket fluid:

$$
\text{Enthalpy in} + \text{Enthalpy added to the system} = \text{Enthalpy out} \tag{3.59}
$$

Equation 3.59 corresponds to

$$
w_J c_{pJ}\big(T_{JIN} - T_{REF}\big) + UA\big(T - T_j\big) = w_J c_{pJ}\big(T_j - T_{REF}\big) \tag{3.60}
$$

which becomes

$$
w_J c_{pJ}\big(T_j - T_{JIN}\big) = UA\big(T - T_j\big) \tag{3.61}
$$

So, assuming the heat transfer area A, the jacket flow rate W_j, the inlet temperature of the cooling fluid T_j and the heat transfer coefficient U are known, the steady-state temperatures of the reactor and of the jacket can be calculated by solving the system of the two Equations 3.59 and 3.61 with the two unknowns T and T_j.

Example 3.5

Consider a continuous biological reactor where microorganisms grow aerobically on glucose as the only carbon source. Assume an inlet concentration of glucose of 10 g/l and a mass transfer coefficient for oxygen equal to 6000 day^{-1}. Assume an inlet temperature of the feed of 20°C. In the enthalpy balances, ignore the contribution of dissolved ammonia.

a) Calculate the concentrations of glucose, oxygen and biomass in the reactor and the reactor temperature as function of the residence time, assuming adiabatic operation;

b) Calculate the reactor temperature as a function of the heat transfer coefficient U for a residence time of 0.2 days and assuming the reactor is jacketed, with a flow rate of the cooling fluid of 50,000 kg/day and with inlet temperature of the jacket fluid of 5°C. Assume that the reactor has a volume of 10 m³ and the geometrical dimensions of Example 3.4;

c) Repeat the calculations of part (b) assuming that the reactor volume is 100 m³, with the geometrical dimensions in the same ratio as in Example 3.4. Assume the jacket fluid has a flow rate of 500,000 kg/day.

Solution

a) The first step is to calculate the steady-state concentrations of glucose, dissolved oxygen and biomass as a function of the residence time. This can be done by solving the mass balances for glucose, biomass and oxygen as described in Section 3.1. The results are shown in Figure 3.7. Glucose concentration is very close to zero for all the values of the residence time, above a certain minimum value for which there is no removal (and there is washout of the microorganisms). Biomass concentration is zero if the residence time is below the minimum value; it rapidly increases when the minimum residence time is reached and then it slowly decreases, due to endogenous metabolism, as the residence time increases further. The

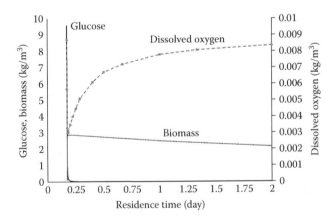

FIGURE 3.7 Example 3.5. Steady-state concentrations of substrate, biomass and dissolved oxygen in a continuous biological reactor as a function of the residence time.

steady-state concentration of dissolved oxygen increases as the residence time increases above the minimum value because the overall rate of oxygen transfer is proportional to $k_L aV$ and V increases as the residence time increases (while $k_L a$ is assumed to remain constant for any value of the reactor volume).

The temperature profile as a function of the residence time can be calculated by solving the steady-state enthalpy balance, Equation 3.57, using the values of the concentration just calculated and the values of the specific enthalpies as a function of the temperature reported in Example 3.4. The obtained temperature profile is shown in Figure 3.8. There is a significant increase in the reactor temperature, which reaches a value higher than 45°C. For all the values of the residence time above the minimum required, the temperature of the reactor is affected only slightly by the residence time. This is due to the fact that when the substrate is removed completely, the heat is generated only by the endogenous metabolism, and this phenomenon has a slow rate.

b) If the reactor is jacketed, the temperatures in the reactor and in the jacket are given by the simultaneous solutions of Equations 3.59 and 3.61. The results are shown in Figure 3.9, as a function of the overall heat transfer coefficient. Clearly, by increasing the heat transfer coefficient, the temperature in the reactor decreases because more heat is removed from the system.

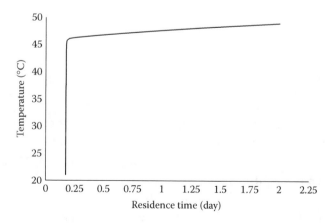

FIGURE 3.8 Example 3.5. Steady-state temperature profile in a continuous adiabatic biological reactor.

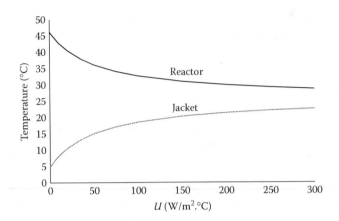

FIGURE 3.9 Example 3.5. Steady-state temperature profile in a continuous jacketed biological reactor as a function of the overall heat transfer coefficient.

c) If the reactor volume is 100 m³, we need to calculate the diameter and height of the reactor taking into account that their ratio is the same as that in the previous case. In that case, the ratio between height and diameter was

$$\frac{H}{D} = \frac{3.2}{2} = 1.6$$

Therefore:

$$V = \frac{\pi D^2}{4} H = \frac{1.6 \pi D^3}{4} \Rightarrow D = 4.3\,\text{m}, \quad H = 6.8\,\text{m}$$

This gives a heat transfer area equal to:

$$A = \frac{\pi D^2}{4} + \pi D H = 106.6\,\text{m}^2$$

The enthalpy balances for the reactor and the jacket can now be solved with the new values of A, V, and W_J. The plot of the temperature in the reactor as a function of the heat transfer coefficient U is shown in Figure 3.10. For a given value of U, the temperature in the reactor is higher for the larger reactor. This is because the A/V ratio is lower for the larger reactor, and therefore, the heat removal rate per unit volume of reactor decreases as the size increases. This shows that in maximising heat transfer, smaller reactors are better than larger ones.

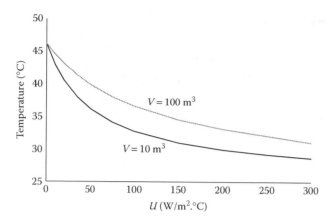

FIGURE 3.10 Example 3.5. Steady-state temperature profile in a continuous jacketed biological reactor as a function of the overall heat transfer coefficient and of the reactor size.

3.3 PARAMETER ESTIMATION

In order to simulate and design a biological wastewater treatment plant, the values of the kinetic parameters and of the growth yield have to be known. They are usually determined from experimental data.

Parameter fitting to the experimental data can be done using two approaches: linear and nonlinear regression. Linear regression requires linearisation of the model equations, while nonlinear regression requires direct comparison of the nonlinear equation with the experimental data. Regardless of the use of liner or nonlinear regression, the general procedure to determine the model parameters that best fit the experimental data can be summarised as follows:

- Generate a set of experimental data. The experimental data can be obtained in batch or continuous experiments;

- Write the mass balances for the relevant species in the (batch or continuous) experiments. In writing the mass balances, the model parameters, which are so far unknown, will appear;

- Compare the profiles generated by the model with the experimental data. The profiles generated by the model will be dependent on the values of the model parameters. The optimum set of model parameters is the one that gives profiles which are as close to the experimental data as possible;

Experiments aimed at determining the kinetic parameters of biomass growth on a substrate can be done in many different ways and there is no set rule on which way is preferable. In the following, a few typical experiments that can be used to determine the model parameters are described. The data shown in this section are used only to explain the procedures and to show which experimental data can be expected in the various tests, and therefore, they do not represent real experiments.

3.3.1 Estimation of the Endogenous Metabolism Coefficient b by Batch Tests

Let us assume we want to measure the coefficient b for heterotrophic microorganisms. The easiest procedure is to take the microorganisms out of the biological reactor and carry out a batch test measuring the OUR in the absence of any external carbon source. In this case, the only contribution to the OUR is endogenous metabolism and the OUR is expressed as:

$$OUR\left(\frac{kgO_2}{m^3 day}\right) = 1.42bX \qquad (3.62)$$

So one very simple approach to measure b is to measure the OUR of the biomass sample in the absence of external substrate and then calculate b from:

$$b\left(day^{-1}\right) = \frac{OUR}{1.42X} \qquad (3.63)$$

This procedure requires only one determination of OUR and the measurement of the biomass concentration X. This procedure is very easy to implement, but the main limitation is that the measurement of X is often not straightforward. Even though X can be approximated in many cases by the volatile suspended solids (VSS), it is important to observe that VSS might give an overestimation of X, since not all the VSS are necessarily composed of active biomass.

A more rigorous method to measure b is to measure the OUR profile over time, starting from a condition where no external substrate is present. This method does not require any information on the initial biomass concentration. The OUR profile during this extended-time experiment is still given by Equation 3.62, but in this case, the biomass concentration during the test decreases according to the equation:

$$\frac{dX}{dt} = -bX \qquad (3.64)$$

which becomes:

$$X = X_0 e^{-bt} \tag{3.65}$$

And by substituting it in Equation 3.62, we obtain:

$$OUR = 1.42 b X_0 e^{-bt} \tag{3.66}$$

If we want to use the linear-regression procedure, we have to linearise Equation 3.66. Since the initial OUR, OUR_0, is given by:

$$OUR_0 = 1.42 b X_0 \tag{3.67}$$

we have:

$$OUR = OUR_0 e^{-bt} \tag{3.68}$$

and:

$$\ln(OUR) = \ln(OUR_0) - b \cdot t \tag{3.69}$$

Therefore, according to Equation 3.69, if we plot $\ln(OUR)$ versus time, we should obtain a straight line with a negative slope, from which we obtain b. Figure 3.11 shows a typical OUR plot and the linearisation procedure to obtain b. The value of b obtained in this case is 0.2019 day^{-1}.

Alternatively, Equation 3.66 can be used in a nonlinear-regression procedure. According to this procedure, the OUR is calculated as a function of time for given values of the parameters b and X_0. The calculated OUR values are compared with the experimental data and the optimum values of the parameters b and X_0 are those which minimise the difference between the model and the experimental data. Figure 3.12 shows the plot obtained with the nonlinear-regression procedure. The nonlinear procedure gives an optimum value of b equal to 0.203 day^{-1}, which is almost identical to the b value calculated with the linear-regression procedure. However, in general, the values obtained with the linear and nonlinear regression will be different, even though usually not largely.

3.3.2 Estimation of Kinetic Parameters on a Readily Biodegradable Substrate by Batch Tests

In this section, two typical procedures to measure the kinetic parameters for biomass growth on a readily biodegradable substrate are shown. The procedures are quite similar and are both based on the measurement of the dissolved oxygen concentration.

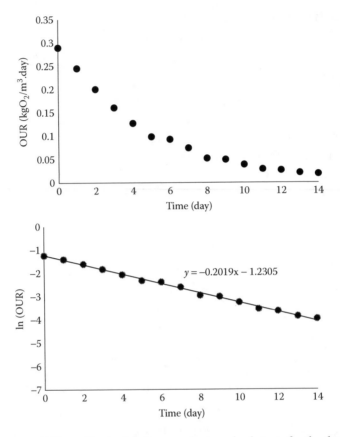

FIGURE 3.11   OUR profiles in the absence of external substrate for the determination of parameter b. Top: OUR data. Bottom: linearization of ln(OUR) to calculate b.

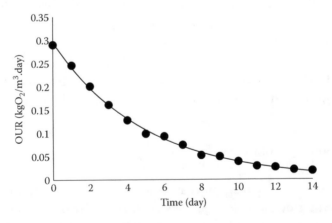

FIGURE 3.12   Fitting of the OUR profile using the nonlinear regression for the calculation of the parameter b. OUR data are the same as in Figure 3.11.

3.3.2.1 Intermittent Aeration Procedure

The experimental procedure can be summarised as follows:

1. Biomass acclimated to the substrate is used, for example, the biomass may be taken from a continuous reactor where the same substrate was used in the feed;

2. Biomass is initially aerated, using diffusers, in the absence of any carbon source;

3. The readily biodegradable substrate is spiked in the reactor at a known concentration. Aeration with diffusers continues until the end of the test.

During the whole length of the test, the OUR is measured at regular intervals, both before and after the addition of the substrate, until the substrate is removed completely from the liquid phase. In order to measure the OUR, aeration is stopped at regular intervals. When aeration is stopped, the oxygen concentration decreases and the slope of the decrease is measured. During this phase, when the oxygen concentration is decreasing, there is typically still some mass transfer of oxygen from the atmosphere to the liquid phase. Therefore, even when the aeration with diffusers is interrupted, the $k_L a$ for the oxygen transfer from the gas phase to the liquid phase will not be zero, although it will be much lower than when the aeration is on with the diffusers. Therefore, the rate of oxygen consumption by the microorganisms, OUR, when aeration with diffusers is off, is given by:

$$OUR = slope + k_L a \cdot \left(C_{O2}^* - C_{O2} \right) \tag{3.70}$$

Where 'slope' is the slope of the oxygen concentration versus time curve when aeration with diffusers is off, $k_L a$ is the mass transfer coefficient for oxygen when aeration with diffusers is off, C_{O2}^* is the saturation concentration of oxygen and C_{O2} is the concentration of oxygen in the liquid phase when aeration is off. Obviously C_{O2} is decreasing when aeration is off, but a good approximation is to use a constant value equal to the average value of the oxygen concentration during the period when aeration is off. In a good experimental setup, the term $k_L a \cdot (C_{O2}^* - C_{O2})$ is much lower than the 'slope' term, and may be neglected in some cases.

When enough points have been collected to calculate the slope, and therefore, the OUR, aeration with diffusers is started again, so that the oxygen concentration comes back up. This procedure of setting the aeration on and off is repeated at regular intervals during the test so that the evolution of the OUR versus time can be obtained.

A typical profile obtained in these tests is shown in Figure 3.13. Before the addition of the external substrate, the OUR is very low because it is only due to the endogenous metabolism. Immediately after the addition of the external substrate, the OUR shows a sharp increase and then increases it further due to growth on the substrate. When the substrate is removed completely, the OUR shows a sharp drop because the only metabolism is again endogenous metabolism.

In order to obtain the model parameters from the experimental data, we need to write the equations that describe the OUR evolution versus time as a function of the model parameters. This equation is:

$$\text{OUR}\left(\frac{\text{kg oxygen}}{\text{m}^3 \cdot \text{day}}\right) = \frac{\mu_{max} S}{K_S + S} \cdot X \cdot \left(\frac{1}{Y_{X/S}} - 1.42\right) + 1.42bX \qquad (3.12)$$

In order to determine the values of the parameters μ_{max}, K_S, b, $Y_{X/S}$, Equation 3.12 needs to be compared with the experimental data, trying to find a set of parameters that minimises the difference between the model and the data. This can be done using either linear or nonlinear regression. An important observation is that the value of the parameter K_S for readily biodegradable substrates is usually very difficult to determine with these tests and indeed with any types of tests. The reason for this is that K_S for

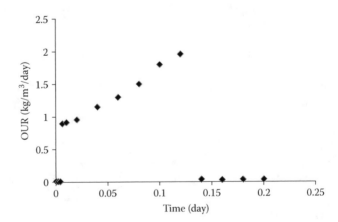

FIGURE 3.13 Typical OUR profile in a batch test with readily biodegradable substrate.

readily biodegradable substrates is usually very low and is usually $\ll S$ during most of the respirometric tests. Therefore, K_S can be usually taken from the literature, for example, a reasonable value can be assumed as 0.004 kg COD/m³.

The linear-regression procedure is presented first. A limitation of the linear procedure is that it only allows for the determination of the parameter $\mu_{max} - b$, from which the value of μ_{max} can be calculated if the value of b is known. Assuming that the substrate is in excess, i.e. $S \gg K_S$, Equation 3.12 can be rewritten as:

$$OUR\left(\frac{kg\,oxygen}{m^3 \cdot day}\right) = \mu_{max} \cdot X \cdot \left(\frac{1}{Y_{X/S}} - 1.42\right) + 1.42bX$$

$$= \left[\mu_{max}\left(\frac{1}{Y_{X/S}} - 1.42\right) + 1.42b\right]X \tag{3.71}$$

The biomass concentration during the test can be expressed as:

$$\frac{dX}{dt} = \left(\mu_{max} - b\right)X \quad \Rightarrow \quad X = X_0 e^{(\mu_{max} - b)t} \tag{3.72}$$

where X_0 is the biomass concentration at the time the external substrate is added, which, for the linear-regression procedure, is considered equal to time 0. Combining Equations 3.71 and 3.72, we obtain:

$$OUR = \left[\mu_{max}\left(\frac{1}{Y_{X/S}} - 1.42\right) + 1.42b\right]X_0 e^{(\mu_{max} - b)t} \tag{3.73}$$

At the time of the addition of the substrate, the value of OUR is OUR_0, given by:

$$OUR_0 = \left[\mu_{max}\left(\frac{1}{Y_{X/S}} - 1.42\right) + 1.42b\right]X_0 \tag{3.74}$$

Therefore, we have:

$$\frac{OUR}{OUR_0} = e^{(\mu_{max} - b)t} \quad \Rightarrow \quad \ln(OUR) = \ln(OUR_0) + \left(\mu_{max} - b\right)\cdot t \tag{3.75}$$

Therefore, a plot of ln(OUR) versus time should give a straight line with a slope equal to $\mu_{max} - b$. The linearisation of the experimental data of Figure 3.13 is shown in Figure 3.14.

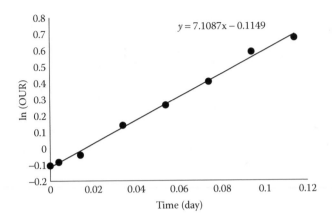

FIGURE 3.14   Linearised plot for the estimation of $\mu_{max} - b$.

From the slope of the regression line, we obtain:

$$\mu_{max} - b = 7.1 \, day^{-1}$$

The linear procedure is simple and only requires measurement of the OUR after the external substrate addition. It does not require any information on the endogenous metabolism, or on the initial biomass or substrate concentration. However, the main limitation of the linear-regression procedure is that it only allows determination of the difference $\mu_{max} - b$.

If all the parameters that describe microbial growth need to be determined from an experiment such as the one described above, then the nonlinear-regression procedure needs to be used. The nonlinear procedure uses all the data from the experiment and allows the simultaneous estimation of the parameters μ_{max}, b, and Y. However, the nonlinear procedure requires knowledge of the initial values of the substrate and the biomass concentration. If not known, the initial biomass concentration can be treated as an additional parameter and estimated on the experimental data, but this would add some additional uncertainty to the fitting procedure.

In using the nonlinear procedure, the OUR simulated by the model (Equation 3.12), needs to be calculated over the whole length of the test, from before the endogenous phase to the end of the test. Obviously during the test, the substrate and biomass concentration, which determine the OUR values, change. Therefore, the values of the variables S and X in Equation 3.12 need to be calculated as a function of time during the OUR test. The variables S and X during the experiment can be calculated by

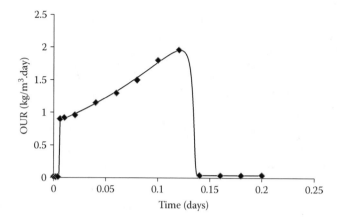

FIGURE 3.15   Fitting of the experimental data in Figure 3.6 using the nonlinear-regression procedure.

integrating the respective mass balance Equations 3.8 and 3.10 (note that in Equation 3.8, the term due to the slowly biodegradable substrate will be ignored in this case). Integration of differential equations is shown in Appendix B.

The results of the fitting with the nonlinear procedure are shown in Figure 3.15. In the fitting, the following values of the initial substrate and biomass concentrations were used: $S_0 = 0.3$ kgCOD/m^3 and $X_0 = 0.05$ kg biomass/m^3. The obtained values of the parameters are reported below:

$$\mu_{max} = 8.1 \text{ day}^{-1}$$

$$b = 0.2 \text{ day}^{-1}$$

$$Y_{X/S} = 0.28 \text{ kg biom/kgCOD}$$

It is worth noting that the values of the difference $\mu_{max} - b$ are similar but not the same as the value obtained on the same data with the linear procedure.

3.3.2.2 Continuous Aeration Procedure

Similarly to the intermittent aeration procedure, this procedure is based on the measurement of dissolved oxygen and requires a biomass that is already acclimated to the readily biodegradable substrate(s). The micro-organisms are put in the reactor in the absence of the external substrate and aerated continuously, either by means of diffusers or by mechanical aeration. The concentration of dissolved oxygen is measured continuously for the whole length of the test. We assume that $k_L a$ and the concentration of oxygen at saturation are both known. The initial biomass concentration is also known (e.g. from a measurement of VSS). Initially, the oxygen concentration will be slightly lower than the saturation value because the microorganisms are consuming oxygen due to endogenous metabolism. After having recorded the initial concentration of oxygen for a short period of time (typically a few minutes), the external substrate is spiked at a known concentration. After the spiking of the external substrate, the oxygen concentration starts decreasing because OUR by the microorganisms increases. When the substrate is removed entirely, the oxygen concentration will rise again because of the decrease in the microbial OUR. Figure 3.16 shows a typical oxygen profile obtained from this procedure.

The model parameters can be calculated from the dissolved oxygen profile using nonlinear regression. The equations that describe the microorganisms' substrate and oxygen profiles during this test are Equations 3.8 (ignoring the slowly biodegradable substrate) and (3.10 and 3.11). The equations need to be integrated to calculate the profiles of X, S and O_2. These profiles, and in particular the O_2 profile, depend on the values of the parameters b, μ_{max}, K_S and $Y_{X/S}$. The optimum values of the parameters are the ones that make the O_2 curve as close to the experimental data as possible. As discussed for the intermittent aeration procedure, the value of K_S for readily biodegradable substrates is usually very low and very difficult

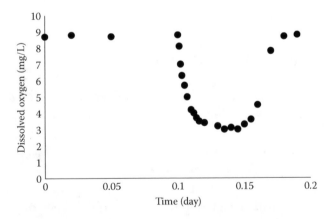

FIGURE 3.16 Example of the dissolved oxygen profile that can be obtained with the continuous aeration procedure.

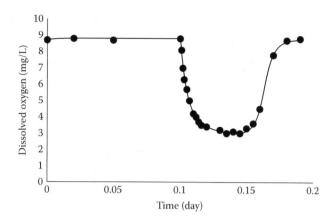

FIGURE 3.17   Example of model fitting to the experimental data of Figure 3.9.

to determine, and therefore, it can often be assumed from the literature, for example, $K_S = 0.004 \ kgCOD/m^3$.

Figure 3.17 shows an example of fitting the model to the dissolved oxygen profile of Figure 3.9. In this case, the results of the fitting were $\mu_{max} = 5.08 \ day^{-1}$, $Y_{X/S} = 0.30$ kg biomass/kg COD and $b = 0.23 \ day^{-1}$. K_S was fixed at 0.004 $kgCOD/m^3$ and the initial biomass concentration was 0.1 kg biomass/m^3.

3.3.3 Estimation of Kinetic Parameters with a Real Wastewater by Batch Tests

In case the substrate is made of a real wastewater, in general, it can be assumed that the substrate will be composed of both readily and slowly biodegradable substrates. In this case, therefore, the kinetic model will have to include, in addition to growth and endogenous metabolism, also the hydrolysis of the slowly biodegradable substrates. The equations that describe the kinetics have been reported in Section 3.1, and the parameters that have to be determined are k_h, K_x, $Y_{X/S}$, μ_{max} and b. The parameter K_S can be estimated based on the literature. Similar to what was shown in the previous sections, OUR data can be used to estimate the parameter values, but since the number of parameters to be fitted is quite large, one single experiment might not be enough for a reliable estimation of all the parameters. Better results and more reliable estimations are obtained if multiple experiments are run under different conditions, with the same type of biomass and of wastewater, and are fitted with the same parameters.

Figure 3.18 shows typical OUR profiles that can be expected in experiments with a real wastewater. The tests represent typical results for high and low initial biomass concentration, or, better, for high and low X_0/S_0 ratio, since it is the ratio between the initial concentrations of biomass and substrate that determines the shape of the OUR curve. Note that it is possible to observe a significant increase in the OUR only in the test at low initial X_0/S_0, because if the X_0/S_0 ratio is high, there is not enough substrate to observe an appreciable biomass growth. In Figure 3.18, the presence of a slowly biodegradable COD fraction is shown by the fact that the OUR drops more gradually than in the case of purely readily biodegradable COD. This can be seen by comparing the final part of the OUR

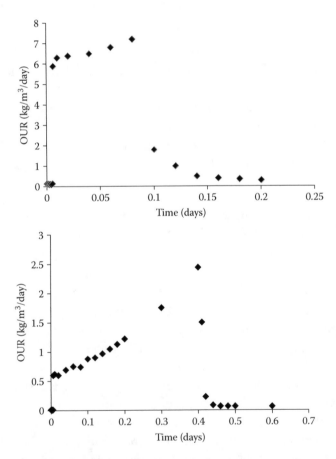

FIGURE 3.18 Typical OUR profiles with real wastewater with two different values of the initial biomass concentration, top: high (0.4 g/l) and bottom: low (0.04 g/l). Initial substrate concentration in both tests: 200 mgCOD/l.

profiles in Figure 3.18 with the final part of the OUR profile in Figure 3.13, where the substrate was entirely readily biodegradable).

In order to find the values of the parameters that best fit the experimental data, Equations 3.7, 3.8 and 3.10 need to be integrated and the resulting OUR values need to be calculated from Equation 3.12. Then, the values of the parameters need to be adjusted to fit the experimental data, using nonlinear regression, as discussed in Appendix C. Note that in this type of tests, an important parameter that needs to be determined on the basis of the experimental data is the fraction of the influent COD that is slowly biodegradable (X_{S0}). We assume that the total COD of the wastewater at the start of the test is known ($X_{S0} + S_0$), so that either X_{S0} or S_0 can be adjusted to fit the experimental data and the other one can be calculated as the difference. By using nonlinear regression, the following values of the parameters have been obtained and the model results are compared with the experimental data in Figure 3.19:

$$\mu_{max} = 4.8 \ \text{day}^{-1}$$

$$b = 0.3 \ \text{day}^{-1}$$

$$k_h = 4.5 \ \text{kgCOD/kg biomass/day}$$

$$K_x = 0.088 \ \text{kg} \, COD / \text{kg biomass}$$

$$Y_{X/S} = 0.32 \ \text{kg biomass/kg glucose}$$

3.3.4 Estimation of Kinetic Parameters on Readily Biodegradable Substrates by Continuous Reactors

All the previous examples use batch tests to determine the model parameters. Instead of batch tests, continuous reactors can also be used. The method described in this section allows the calculation of μ_{max}, K_s, and $Y_{X/S}$ using data from various runs of a continuous reactor, with measurement of residual substrate concentration and biomass concentration at steady state. For example, consider the determination of kinetic parameters for a readily biodegradable substrate. We can feed this substrate to a continuous reactor without recycle where biomass grows and the substrate is removed. At steady state, the mass balances for substrate and biomass are shown below:

$$QS_0 = \frac{\mu_{max} S}{K_S + S} \frac{X}{Y_{X/S}} V + QS$$

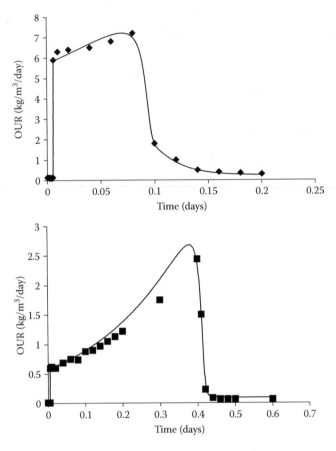

FIGURE 3.19 Fitting of the experimental data in Figure 3.18 with the kinetic model for slowly biodegradable substrates.

$$QX = \frac{\mu_{max} S}{K_S + S} XV - bXV$$

We define the dilution rate as:

$$D(day^{-1}) = \frac{Q}{V} \tag{3.75}$$

The biomass balance can, therefore, be rewritten as:

$$\frac{\mu_{max} S}{K_S + S} = D + b \quad \Rightarrow \quad \frac{1}{S} = \frac{1}{D+b} \frac{\mu_{max}}{K_S} - \frac{K_S}{\mu_{max}} \tag{3.76}$$

and substituting into the substrate balance, the substrate balance can be rearranged as:

$$\frac{(S_0 - S)}{X} = \frac{1}{Y_{X/S}} + \frac{b}{Y_{X/S}} \frac{1}{D} \tag{3.77}$$

Therefore, in order to obtain the model parameters from chemostat experiments, a series of runs at different residence time, i.e. different dilution rate D, needs to be carried out and the steady state has to be achieved in all the runs. The values of the substrate (S) and biomass concentration (X) in the reactor at steady state need to be measured. Then the following plots need to be generated:

$$\frac{(S_0 - S)}{X} \quad \text{vs} \quad \frac{1}{D} \tag{3.78}$$

$$\frac{1}{S} \quad \text{vs} \quad \frac{1}{D + b} \tag{3.79}$$

The first plot will give the values of b and $Y_{X/S}$ from Equation 3.77, while once b is known, the second plot will give the values of μ_{max} and K_s from Equation 3.76.

The procedure is illustrated in Figures 3.20 through 3.22. Figure 3.20 shows typical profiles of biomass and substrate in a chemostat experiment at different dilution rates. Figure 3.21 shows the linearised plots according to Equation 3.77 and Figure 3.22 shows the linearisation of Equation 3.76. From Figure 3.21, the following values of the parameters can be calculated:

$$Y_{X/S} = 0.196 \frac{\text{kg biomass}}{\text{kg COD}}$$

$$b = 0.295 \, \text{day}^{-1}$$

And from Figure 3.22:

$$\mu_{max} = 4.57 \, \text{day}^{-1}$$

$$K_S = 0.0026 \frac{\text{kg COD}}{\text{m}^3}$$

3.3.5 Estimation of Kinetic Parameters under Anoxic Conditions

The determination, or estimation, of the parameters μ_{max}, K_S, b and $Y_{X/S}$ under anoxic conditions, i.e. when nitrate is used, instead of oxygen, as electron acceptor, can be done using experimental procedures that are

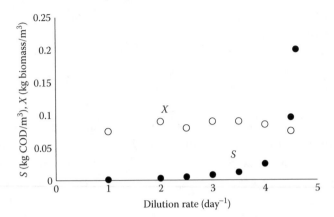

FIGURE 3.20 Biomass and substrate concentration in a chemostat experiment at different dilution rates.

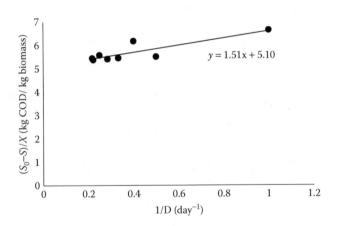

FIGURE 3.21 Linearisation according of Equation 3.77 of the data in Figure 3.20.

absolutely analogous to the ones described in previous sections for aerobic experiments. The only difference is that the experiments have to be designed and carried out so that there is no oxygen available for the microorganisms, which will then use nitrate. Therefore, as for aerobic experiments, the kinetic parameters can be determined by measuring only the OUR, under anoxic conditions, the parameters can be determined just from the nitrate profile.

For example, Figure 3.23 shows the use of nonlinear regression to estimate the value of parameter b from the nitrate profile in an experiment

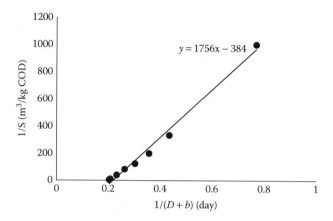

FIGURE 3.22   Linearisation according of Equation 3.76 of the data in Figure 3.20.

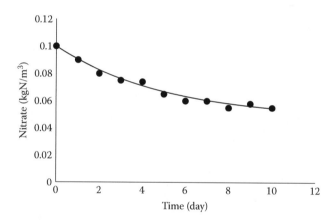

FIGURE 3.23   Experiment without external substrate under anoxic conditions for the determination of the parameter b.

with microorganisms and no external substrate addition. The nitrate profile in Figure 3.23 is described by Equation 3.13, with $S = 0$, and by doing nonlinear regression using this equation, we obtain:

$$b = 0.21 \, \text{day}^{-1}$$

Figure 3.24 shows the nitrate profile in a typical experiment with external substrate and the comparison with the best-fit model, given by Equation 3.13. In this case, the fitted parameters have been μ_{max} and $Y_{X/S}$. The value of b has been taken equal to the value obtained from Figure 3.23 ($b = 0.21 \, \text{day}^{-1}$)

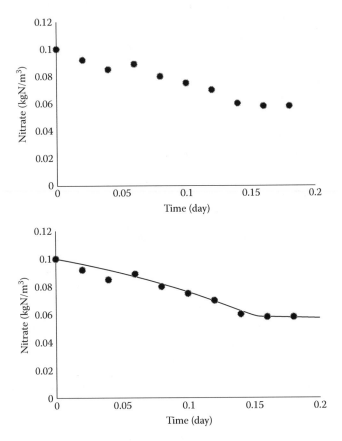

FIGURE 3.24   Typical experimental profile in a test with external COD and nitrate as electron acceptor (top) and comparison with the best-fit model (bottom).

and the value of K_S has been assumed to be 0.004 kgCOD/m³. In this case, the values of the parameters obtained are:

$$\mu_{max} = 4.9 \ \text{day}^{-1}$$

$$Y_{X/S} = 0.44 \frac{\text{kg biomass}}{\text{kg COD}}$$

3.3.6 Estimation of Kinetic Parameters for Anaerobic Microorganisms

The estimation of parameters under anaerobic conditions is usually more complicated than under aerobic conditions because the anaerobic degradation of the organic matter requires the coexistence of various microbial populations, each of them with its own kinetic parameters to be determined. In principle, however, parameter estimation under anaerobic conditions can be done using the same tools as under aerobic conditions, using nonlinear regression to find the values of the parameters that make the model to correspond as close to the experimental data as possible. For example, let us consider the simplest case where the substrate is glucose, and there are three populations of microorganisms: fermentative, which convert glucose into acetic acid and hydrogen, and acetoclastic methanogens and hydrogenotrophic methanogens, which convert acetic acid and hydrogen, respectively, into methane. Assuming we do a batch test, by spiking a certain concentration of glucose at time 0, we should be able to measure the following variables during the test: glucose and acetic acid in the liquid phase, and hydrogen and methane produced. The amount of hydrogen and methane produced can be calculated by measuring the volume of gas produced and the composition of the liquid phase. Figure 3.25 shows the typical profiles that are to be expected in this type of test.

In order to do the parameter fitting, we have to consider that the mass balances for this system are described by Equations 3.14–3.26. From these equations, the profiles of glucose, acetic acid, and produced hydrogen and methane can be calculated. Then, the optimum values of the kinetic parameters can be estimated as the values that make the model predictions to correspond as close to the experimental data as possible. For example,

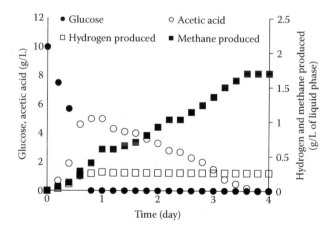

FIGURE 3.25 Typical profiles of glucose, acetic acid, hydrogen and methane produced in a batch test with glucose.

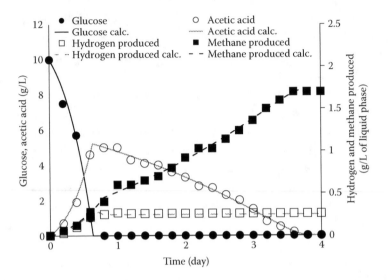

FIGURE 3.26   Best-fit model for the data in Figure 3.25.

with the data in Figure 3.25, the best-fit model is shown in Figure 3.26. The calculated results in Figure 3.26 have been obtained by using fixed values, from the literature, of the following parameters:

$$K_{SGLU} = 0.02 \, \frac{kg}{m^3}$$

$$b_{GLU} = 0.3 \, day^{-1}$$

$$K_{SAc} = 0.14 \, \frac{kg}{m^3}$$

$$b_{Ac} = 0.2 \, day^{-1}$$

$$K_{SH2} = 1.6E-5 \, \frac{kg}{m^3}$$

$$b_{H2} = 0.3 \, day^{-1}$$

The values of the following parameters have been adjusted to fit the experimental data:

$$\mu_{maxGLU} = 2.46 \, day^{-1}$$

$$Y_{X/SGLU} = 0.1 \, \frac{kg \, X_{GLU}}{kg \, GLU}$$

$$X_{0GLU} = 0.26 \frac{kg\ X_{GLU}}{m^3}$$

$$\mu_{maxAC} = 0.29\ day^{-1}$$

$$Y_{X/SAC} = 0.03 \frac{kg\ X_{GLU}}{kg\ Ac}$$

$$X_{0AC} = 0.11 \frac{kg\ X_{GLU}}{m^3}$$

$$\mu_{maxH2} = 2.79\ day^{-1}$$

$$Y_{X/SH2} = 0.45 \frac{kg\ X_{H2}}{kg\ H2}$$

$$X_{0H2} = 0.08 \frac{kg\ X_{H2}}{m^3}$$

Note the large number of parameters that have been necessary to fit. In particular, it has been necessary also to fit the initial concentrations of the microorganisms involved, X_{0GLU}, X_{0Ac} and X_{0H2}. An important limitation of this procedure, common to many other cases of nonlinear regression, is that similar values of the calculated profiles, in agreement with the experimental data, can be calculated with different combinations of parameters. In this case, for example, it is likely that different combinations of μ_{max}, $Y_{X/S}$ and X_0 may give equally good fitting of the experimental data. If this is the case, in order to have a more robust fitting, it is recommended to run multiple tests, under different conditions (e.g. different substrate/biomass ratios) and then simultaneously fit all the data with the same set of parameters.

3.4 KEY POINTS

- Mass balances have the general form:

  Accumulation = input − output + generation − consumption;

- We can write mass balances for each of the relevant species in biological wastewater treatment processes, for example, substrate, ammonia, oxygen and biomass. In writing mass balances, it is important to decide which units to use, for example, whether to express the carbon source as substrate or as COD, and be consistent in their use. Also, in writing mass balances, it is important to specify the type of system we are considering, for example, whether it is a batch reactor or a continuous-flow reactor;

- Enthalpy balances have the general form;

- Enthalpy accumulated = Enthalpy in – Enthalpy out + Enthalpy added to the system – Enthalpy removed from the system;

- In writing the enthalpy balances, the specific enthalpies (as J/mol or J/kg) of all the species entering and leaving the system need to be considered. In enthalpy balances, we need to consider whether the system is batch or continuous and whether it is adiabatic or there is heat transfer with the environment or with an external cooling medium;

- The values of the parameters in the kinetic models are usually to be determined from experimental data or to be taken from the literature (where somebody has determined them for us);

- In the experimental determination of the model parameters, the general procedure is the following: with reference to the experimental setup, write the mass balances for the species of interest using the kinetic models with the unknown parameters; carry out the experiment(s), measuring at least some of the variables that appear in the mass balances; compare the values of the variables simulated by the model with their experimental values, determining the values of the model parameters that minimise the difference between the simulated and the experimental data. In order to minimise the uncertainty in parameter estimation, it is recommended that multiple tests be carried out under the same or different conditions and that all the tests be fitted with the same set of parameters.

Questions and Problems

3.1 A perfectly mixed aerobic continuous reactor without recycle (chemostat) is fed with wastewater with a COD concentration of 300 mgCOD/l, at a flow rate of 100 ml/hr. The COD is entirely soluble and the only processes occurring are microbial growth and endogenous metabolism. It can be assumed that the products of endogenous metabolism are only carbon dioxide and water, with no generation of inert products. The effluent of the reactor has a soluble COD of 20 mgCOD/l and a concentration of microorganisms of 50 mg/l. What is the oxygen consumption rate (g/day) in the reactor?

3.2 Consider the reactor of problem 3.1. It is desired to run the reactor with the same wastewater but using nitrate, instead of oxygen, as electron acceptor. To do so, the reactor is sealed to prevent air from coming in, and sodium nitrate ($NaNO_3$) is added to the feed. Assuming that the growth yield on the COD while using nitrate is the same as that while using oxygen, what is the concentration of sodium nitrate which needs to be added to the feed?

CHAPTER 4

The Activated Sludge Process

4.1 THE ACTIVATED SLUDGE PROCESS FOR CARBON REMOVAL

The activated sludge process for carbon removal can be schematised as shown in Figure 4.1. In this section, we will always assume, unless specified otherwise, that the feed is composed only of readily biodegradable substrates and that the settling tank operates without any biomass losses with the clarified effluent.

We assume that the substrate concentration is expressed as chemical oxygen demand (COD), and we assume the Monod model for the kinetics of biomass growth. According to Chapter 2, we have the following rate equations for microbial growth and endogenous metabolism:

$$r_X \left(\frac{\text{kg biomass}}{\text{m}^3 \text{day}} \right) = \mu X = \frac{\mu_{max} S}{K_S + S} X$$

$$r_{end} \left(\frac{\text{kg biomass}}{\text{m}^3 \text{day}} \right) = -bX$$

Once the rate equations for biomass growth and endogenous metabolism are specified, the rates of production or consumption of all the other

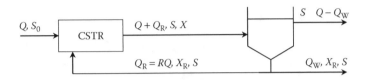

FIGURE 4.1 Scheme of the activated sludge for carbon removal, feed with only readily biodegradable substrates.

species can be derived, as described in Chapter 2. Example: in this case the rate equation of substrate and oxygen are

$$r_S\left(\frac{\text{kg substrate}}{\text{m}^3\text{day}}\right) = -\frac{r_X}{Y_{X/S}} = -\frac{\mu X}{Y_{X/S}} = -\frac{\mu_{\max} S}{K_S + S}\frac{X}{Y_{X/S}}$$

$$r_{O2}\left(\frac{\text{kg O}_2}{\text{m}^3\text{day}}\right) = -\left(\frac{1}{Y_{X/S}} - 1.42\right)r_X + r_{\text{end}}\cdot 1.42 = -\left(\frac{1}{Y_{X/S}} - 1.42\right)\mu X - 1.42bX$$

$$= -\left(\frac{1}{Y_{X/S}} - 1.42\right)\frac{\mu_{\max} S}{K_S + S}\frac{X}{Y_{X/S}} - 1.42bX$$

In these equations, and in all cases in this chapter, $Y_{X/S}$ is considered to have the units of kg biomass/kg COD (for simplicity the subscript COD is omitted in the notation for $Y_{X/S}$). Designing this process means finding the values of all the variables that characterise it. The variables which characterise this process are the following: Q, S_0, V, Q_R (or R), S, X, X_R, Q_W. We assume that Q and S_0 (influent flow rate and composition) are known, so we have six variables that need to be specified. The variables are linked by the following mass balances.

Balance for the biomass in the reactor:

$$\left(r_X + r_{\text{end}}\right)V + Q_R X_R = \left(Q + Q_R\right)X \tag{4.1}$$

Balance for the substrate in the reactor:

$$QS_0 + Q_R S = (Q + Q_R)S + (-r_S)V \Rightarrow Q(S_0 - S) = (-r_S)V \tag{4.2}$$

Balance for the biomass in the whole system (reactor + settling tank):

$$(r_X + r_{\text{end}})V = Q_W X_R \tag{4.3}$$

Note that we do not need to write the mass balance for the substrate in the whole system (reactor + settling tank), because, with the assumption we have made that the substrate is entirely soluble, the substrate concentration does not change in the settling tank.

By introducing the reaction kinetics defined above the mass balances can also be written as:

$$(\mu - b)XV + Q_R X_R = (Q + Q_R)X \tag{4.4}$$

$$Q(S_0 - S) = \frac{\mu XV}{Y_{X/S}} \tag{4.5}$$

$$(\mu - b)XV = Q_W X_R \tag{4.6}$$

The three mass balances can be re-arranged by introducing the recycle ratio R

$$R = \frac{Q_R}{Q} \tag{4.7}$$

and dividing by XV:

$$(\mu - b) + \frac{QRX_R}{XV} = \frac{Q(1 + R)}{V} \tag{4.8}$$

$$\frac{Q(S_0 - S)}{XV} = \frac{\mu}{Y_{X/S}} \tag{4.9}$$

$$(\mu - b) = \frac{Q_W X_R}{XV} \tag{4.10}$$

By introducing the hydraulic retention time (HRT), which is the ratio between the reactor volume and the influent flow rate and therefore represents the 'nominal' residence time of the liquid in the reactor,

$$\text{HRT} = \frac{V}{Q} \tag{4.11}$$

and the solids retention time (SRT), which represents the residence time of the microorganisms in the system,

$$\text{SRT} = \frac{VX}{Q_W X_R} \tag{4.12}$$

we obtain:

$$(\mu - b) + \frac{RX_R}{X \cdot HRT} = \frac{(1+R)}{HRT} \tag{4.13}$$

$$\frac{(S_0 - S)}{X \cdot HRT} = \frac{\mu}{Y_{X/S}} \tag{4.14}$$

$$(\mu - b) = \frac{1}{SRT} \tag{4.15}$$

We have now three Equations 4.13 through 4.15 with the following six variables: HRT, SRT, R, S, X, X_R. Note that from the HRT the volume of the reactor per unit influent flow rate immediately follows, and from the SRT the waste sludge flow rate (again per unit of influent flow rate) can be calculated:

$$\frac{Q_W}{Q} = \frac{HRT}{SRT} \cdot \frac{X}{X_R} \tag{4.16}$$

The system of Equations 4.13 through 4.15 can be solved if the values of three variables are specified by the designer of the process. A good choice is to specify values for SRT, HRT and R and to calculate the values of the remaining variables S, X, X_R by solving the system of equations.

Once all the variables that characterise the process have been calculated, the sludge production and the oxygen consumption can also be calculated. The sludge production in the activated sludge process is simply given by:

$$P_X \left(\frac{kg\ biomass}{day} \right) = Q_W X_R \tag{4.17}$$

and per unit of influent flow rate:

$$\frac{P_X}{Q} \left(\frac{kg\ biomass}{day \cdot \dfrac{m^3}{day}} \right) = \frac{Q_W X_R}{Q} = \frac{V \cdot X}{Q \cdot SRT} = \frac{HRT \cdot X}{SRT} \tag{4.18}$$

The oxygen consumption rate by microorganisms can be calculated using the COD balance on the whole system:

$$\text{Oxygen consumption}\left(\frac{\text{kg O}_2}{\text{day}}\right) = Q_{O2\text{biomass}}$$

$$= Q(S_0 - S) - Q_W X_R \cdot 1.42$$

(4.19)

In Equation 4.19, the factor 1.42 that multiplies the sludge production is needed to convert the biomass concentration in COD units (note that $Q_{O2\text{biomass}}$ can also be calculated as $-r_{O2}V$). The oxygen consumption per unit of influent flow rate is:

$$\text{Oxygen consumption} = \left(\frac{\text{kg O}_2}{\text{day} \cdot \dfrac{\text{m}^3}{\text{day}}}\right)$$

$$= \frac{Q_{O2\text{biomass}}}{Q}$$

(4.20)

$$= (S_0 - S) - \frac{Q_W X_R}{Q} \cdot 1.42$$

$$= (S_0 - S) - \frac{X \cdot \text{HRT}}{\text{SRT}} \cdot 1.42$$

Note that, from the COD balance in the whole process, which has been used to derive Equation 4.19, it follows that, for a given influent flow rate and composition and for a given extent of substrate removal, the sum of the oxygen consumed and biomass produced (converted into COD units) is constant and cannot be altered by changing any design or kinetic parameters:

$$Q(S_0 - S) = Q_{O2\text{biomass}} + P_X \cdot 1.42$$

(4.21)

Equation 4.21 is particularly important considering that usually, for well-designed and well operated processes, $S \ll S_0$ and so $Q(S_0 - S) \cong QS_0$. Therefore, Equation 4.21 shows that, for a well-designed process, the sum of oxygen consumption and biomass production only depends on the flow rate and composition of the influent stream.

In summary, the activated sludge process for carbon removal, in its simplest version of one biological reactor followed by a settling tank, can be designed by specifying the values of the solids residence time (SRT), the hydraulic residence time (HRT) and the recycle ratio from the settling tank to the reactor. Once these three variables have been specified, and assuming certain rate equations for microbial growth, endogenous metabolism and substrate removal, the values of all the variables that characterise the process can be calculated by solving the system of mass balance equations.

4.1.1 Effect of the Choice of the Design Parameters on the Design Results

The design parameters are SRT, HRT and R, and their effect on the design results is discussed below.

The choice of the SRT determines the effluent substrate concentration (i.e. the achieved degree of wastewater treatment). Indeed from the design Equation 4.15:

$$(\mu - b) = \frac{1}{SRT} \tag{4.15}$$

we obtain after re-arrangement:

$$S = \frac{bK_S SRT + K_S}{(\mu_{max} - b)SRT - 1} \tag{4.22}$$

Equation 4.22 shows that the effluent substrate concentration S depends only on the SRT and on the values of the kinetic parameters μ_{max}, b and K_S. S does not depend on the influent substrate concentration S_0. This is typical of biological processes, because, for example, higher values of S_0 give a higher biomass concentration, which removes the substrate at a faster rate, and therefore, the effluent substrate concentration does not change. S decreases by increasing the SRT, because a higher value of the SRT means that the microorganisms have more time to remove the substrate. There is a minimum value of SRT below which there is no substrate removal, that is $S = S_0$. The minimum value of the SRT, SRT_{min}, corresponds to wash out of the microorganisms from the reactor and is given by:

$$SRT_{min} = \frac{1}{\dfrac{\mu_{max} S_0}{K_S + S_0} - b} \tag{4.23}$$

Equation 4.22 also shows that for very high values of the SRT, SRT $\to \infty$, the substrate concentration S approach a minimum value given by:

$$S_{min} = \frac{bK_S}{\mu_{max} - b} \qquad (4.24)$$

Therefore, under the model hypothesis used here, that is biomass growth with endogenous metabolism and completely mixed reactor, the activated sludge process cannot remove the biodegradable substrate completely.

Also, the choice of the SRT determines the total amount of biomass in the biological reactor (XV). Indeed, from the design Equation 4.14

$$\frac{(S_0 - S)}{X \cdot HRT} = \frac{\mu}{Y_{X/S}} \qquad (4.14)$$

we obtain

$$X \cdot HRT = \frac{(S_0 - S)Y_{X/S}}{\mu} = \frac{(S_0 - S)Y_{X/S}SRT}{1 + bSRT} \qquad (4.25)$$

Equation 4.25 shows that, for a given flow rate and composition of the influent wastewater, the total amount of biomass present in the system

$$\frac{XV}{Q} = X \cdot HRT \qquad (4.26)$$

depends only on the SRT, because the substrate concentration S depends only on the SRT. However, note that, differently than the substrate concentration S, the total biomass amount XV also depends on the influent substrate concentration S_0, as expected and as discussed previously. The total amount of biomass in the reactor increases by increasing the SRT, because a higher value of the SRT means that lower amounts of biomass are removed per unit of time.

A very important observation is that, for a given influent flow rate and composition, the SRT is the only design parameter that affects the sludge production and the oxygen consumption. This is evident from Equations 4.18 and 4.20, which can be re-written as:

$$\frac{P_X}{Q} \left(\frac{kg\, biomass}{day \cdot \dfrac{m^3}{day}} \right) = \frac{(S_0 - S)Y_{X/S}}{1 + b \cdot SRT} \qquad (4.27)$$

and

$$\frac{Q_{O2biomass}}{Q} \left(\frac{kg\,O_2}{day \cdot \dfrac{m^3}{day}} \right) = (S_0 - S)\left(1 - \frac{1.42 \cdot Y_{X/S}}{1 + b \cdot SRT} \right) \tag{4.28}$$

Therefore, for given values of the influent and effluent substrate concentration, the relative extent of biomass production and oxygen consumption can be manipulated by changing the SRT of the process. Increasing the SRT increases oxygen consumption and decreases biomass production. However, for given values of the influent and effluent substrate concentration, the sum of oxygen consumption and biomass production does not depend on the SRT, as shown previously by Equation 4.21.

For a given value of the SRT, the value chosen for the HRT determines the biomass concentration in the reactor:

$$X = \frac{(S_0 - S)Y_{X/S}}{\mu \cdot HRT} = \frac{(S_0 - S)Y_{X/S}SRT}{(1 + b \cdot SRT)HRT} \tag{4.29}$$

By increasing the HRT at fixed SRT the biomass concentration in the reactor decreases. This is expected, because, since t0068e the total amount of biomass is fixed by the SRT, having a larger reactor volume (larger HRT) makes the biomass concentration lower. Note that for very large values of SRT, X does not increase indefinitely but approaches an asymptotic value given by (assuming $S \ll S_0$):

$$X_{max} = \frac{S_0 Y_{X/S}}{b \cdot HRT} \tag{4.30}$$

For given values of SRT and HRT, the recycle ratio R determines the concentration of biomass in the recycle stream, X_R. Indeed, Equation 4.13 can be re-arranged as:

$$X_R = \frac{(1 + R)}{R} X - \frac{X \cdot HRT}{R \cdot SRT} \tag{4.31}$$

By increasing R, the biomass concentration in the recycle stream decreases.

In summary, the SRT is the most important design parameter that determines the performance of the activated sludge process. For a given wastewater of a certain flow rate and composition, the SRT determines the effluent substrate concentration, the total biomass amount in the system, the oxygen consumption and the sludge production. For a given value of the SRT, the HRT determines the biomass concentration in the reactor. For given values of the SRT and the HRT, the recycle ratio R determines the concentration of biomass in the recycle stream. Example 4.1 shows numerically the effect of the parameters SRT, HRT and R on the design of an activated sludge process.

Example 4.1: Design of an activated sludge process for carbon removal

A conventional activated sludge process for carbon removal is composed of a reaction tank followed by a settling tank (Figure 4.1). The inlet substrate concentration (readily biodegradable substrate) is:

$$S_0 = 0.5 \text{ kg COD/m}^3$$

Investigate the effect of the design parameters SRT (range 0.8–20 day), HRT (range 0.25–1 day) and R (range 0.5–2) on the values of the variables X, X_R, S, per unit value of the influent flow rate. Discuss the observed trends.

Values of the kinetic parameters:

$$\mu_{max} = 6 \text{ days}^{-1}$$

$$K_S = 0.004 \text{ kg COD/m}^3$$

$$Y_{X/S} = 0.3 \text{ kg biomass/kg COD}$$

$$b = 0.2 \text{ days}^{-1}$$

Also, calculate the values of all the variables that characterise the process for an influent flow rate of 16,000 m³/day and the following values of the design parameters: SRT = 15 day, HRT = 0.3 day, R = 1.5.

Solution

The solution of this problem is obtained by solving the mass balance Equations 4.13, 4.14, 4.15. For fixed values of SRT, HRT and R, the system of equations can be solved to obtain S, X, X_R. First of all, we verify that the given range of SRT (0.8–20 day) is entirely above SRT_{min}. For the given kinetic parameters and influent substrate concentration, from Equation 4.23 we obtain $SRT_{min} = 0.17$ day, therefore the given range of SRT is entirely above SRT_{min}, and we will have substrate removal and presence of microorganisms in the reactor in all cases.

The graphs in Figure 4.2 show the effect of SRT when HRT is set to 0.5 day and R is set to 1. As SRT increases, S concentration decreases, while both X and X_R increase. This is as expected, since

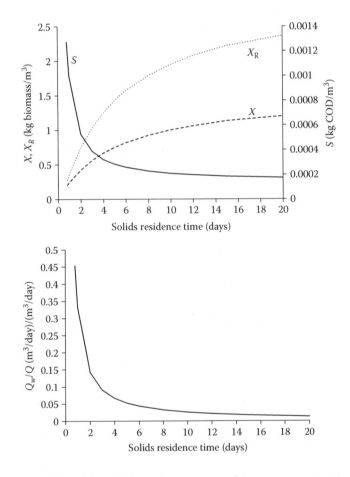

FIGURE 4.2   Effect of the solids residence time on biomass concentration, in the reactor and in the bottom of the settling tank, on substrate concentration in the reactor and on the required sludge waste flow rate. Hydraulic residence time (HRT) = 0.5 day, recycle ratio (R) = 1.

larger SRT means that more solids (biomass) are present in the system and therefore more substrate is removed. Note that, however, in the range of SRT values considered here, the substrate concentration in the reactor, S, is always much lower than the substrate concentration in the influent (S_0). This is typical of all well-designed activated sludge processes. To increase the SRT, the waste flow rate Q_W has to decrease, as shown in the graph.

The effect of HRT on X, X_R, Q_w/Q is shown in the graphs in Figure 4.3 (for $R = 1$). Note that HRT has no effect on S, which is only determined by the SRT. By increasing HRT, X decreases. This is expected, since the total amount of biomass in the system, XV, is only dependent on the SRT. Therefore, by increasing the HRT the reactor becomes larger, and therefore the biomass concentration, X, becomes smaller. Similarly for X_R. By increasing HRT, Q_w increases, due to the fact that X_R decreases and therefore a larger waste flow rate is required to maintain a given value of SRT.

The effect of the recycle ratio R on X_R and Q_w/Q is shown in Figure 4.4, for HRT = 1 day. By increasing R the concentration of biomass in the recycle stream X_R decreases. This is because we are increasing the flow rate from the bottom of the settling tank and therefore obtaining less concentration of the biomass. Q_w/Q increases by increasing R, due to the lower biomass concentration X_R, similarly as what discussed above regarding the effect of HRT on Q_w/Q.

Finally, the effect of SRT on the oxygen consumption and sludge production is shown in Figure 4.5. These two variables only depend on SRT, as shown by Equations 4.27 and 4.28 and are linked by the COD balance, Equation 4.21. The oxygen consumption increases with the SRT, this is because higher SRT means that the biomass stays in the system for a longer time. A longer residence time for the biomass means more substrate removal and more endogenous metabolism (i.e. more self-oxidation by the biomass) and both these phenomena increase the oxygen consumption. Biomass production decreases at higher SRT because endogenous metabolism becomes predominant over substrate removal at high SRT, and this reduces the amount of biomass produced. In other words, at high SRT more biomass decays by self-oxidation than at low SRT and so the biomass production decreases.

For an influent flow rate of 16,000 m³/day and the design parameters of SRT = 15 day, HRT = 0.3 day, $R = 1.5$, by solving the system

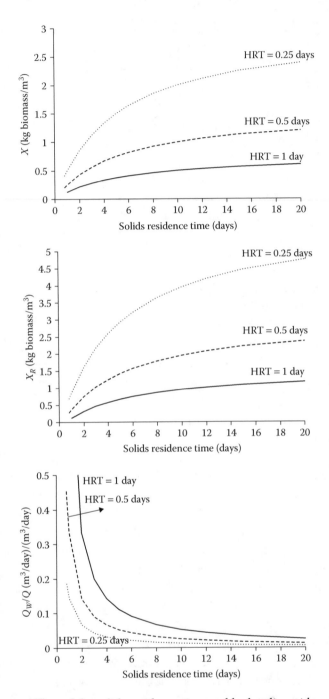

FIGURE 4.3 Effect of the solids residence time and hydraulic residence time on biomass concentration in the reactor and in the bottom of the settling tank and on the required sludge waste flow rate. Recycle ratio $(R) = 1$.

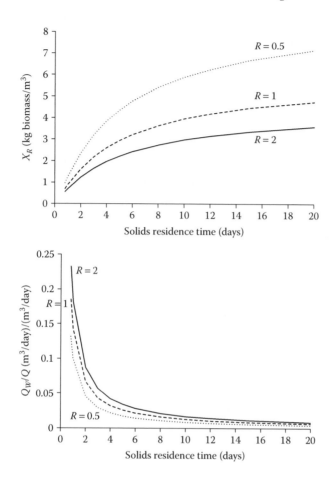

FIGURE 4.4   Effect of the solids residence time and recycle ratio on biomass concentration in the bottom of the settling tank and on the required sludge waste flow rate. Hydraulic residence time (HRT) = 0.25 day.

of Equations 4.13–4.15, and by using Equations 4.7, 4.11, 4.16, 4.17 and 4.19, we obtain:

$$S = 0.00019 \frac{\text{kg COD}}{\text{m}^3}$$

$$X = 1.87 \frac{\text{kg}}{\text{m}^3}$$

$$X_R = 3.10 \frac{\text{kg}}{\text{m}^3}$$

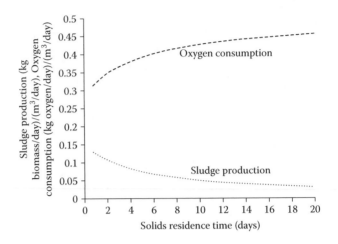

FIGURE 4.5   Effect of the solids residence time on oxygen consumption and sludge production.

$$Q_W = 194 \frac{m^3}{day}$$

$$Q_R = 24,000 \frac{m^3}{day}$$

$$V = 4800 \text{ m}^3$$

$$P_X = 600 \frac{kg}{day}$$

$$Q_{O2biomass} = 7144 \frac{kg \, O_2}{day}$$

It is worth checking that the overall COD balance is satisfied:

$$Q(S_0 - S) = 7997 \frac{kg \, COD}{day}$$

$$= Q_{O2biomass} + P_X \cdot 1.42$$

$$= 7144 + 600 \cdot 1.42 = 7996 \frac{kg \, COD}{day}$$

the very small difference being due to rounding.

4.1.2 Effect of the Values of the Kinetic Parameters on the Design Results

The kinetic parameters which need to be known to design an activated sludge process for carbon removal are, for the case of readily biodegradable substrates considered so far, μ_{max}, K_S, $Y_{X/S}$, b. These parameters describe the rate of microbial growth and substrate removal and of endogenous metabolism. The values of these parameters are highly variable in the literature; therefore, it is important to understand how their values affect the design results, for given values of the design parameters SRT, HRT and R.

The parameter of endogenous metabolism, b, is very important because it affects the substrate effluent concentration, the biomass concentration, the sludge production and the oxygen consumption. As we have seen, the substrate effluent concentration is given by:

$$S = \frac{bK_S \text{SRT} + K_S}{(\mu_{max} - b)\text{SRT} - 1} \tag{4.22}$$

For a given value of SRT, S increases as b increases. This is understandable, since a higher rate of endogenous metabolism means that biomass decays at higher rate, and the rate of substrate removal decreases.

The biomass concentration in the reactor is given by, as shown in previous sections:

$$X = \frac{(S_0 - S)Y_{X/S}\text{SRT}}{(1 + b \cdot \text{SRT})\text{HRT}} \tag{4.29}$$

If b increases, X decreases. This is again due to the fact that higher values of b correspond to a faster decay of the biomass, leading to a lower biomass concentration. Note that if, as it is usually the case, $S \ll S_0$, in the particular case that $b = 0$ (no endogenous metabolism), the biomass concentration increases linearly with the SRT and does not show the asymptotic profile that we observe for $b > 0$.

The biomass production and oxygen consumption are given by Equations 4.27 and 4.28, respectively:

$$\frac{P_X}{Q}\left(\frac{\text{kg biomass}}{\text{day} \cdot \dfrac{\text{m}^3}{\text{day}}}\right) = \frac{(1 + R)Y_{X/S}}{1 + b \cdot \text{SRT}} \tag{4.27}$$

and

$$\frac{Q_{O2biomass}}{Q} \left(\frac{kg\,O_2}{day \cdot \dfrac{m^3}{day}} \right) = (S_0 - S)\left(1 - \frac{1.42 \cdot Y_{X/S}}{1 + b \cdot SRT} \right) \qquad (4.28)$$

These equations show that, assuming $S \ll S_0$, the biomass production decreases and the oxygen consumption increases for higher values of b. This is expected since, by increasing b, the rate of endogenous metabolism increases, causing higher oxygen consumption and lower biomass production.

The value of the growth yield $Y_{X/S}$ affects the biomass concentration, sludge production and oxygen consumption but not the effluent substrate concentration (Equations 4.22, 4.27, and 4.29. Higher values of $Y_{X/S}$ correspond to a higher biomass concentration in the reactor, higher biomass production and lower oxygen consumption. This is as expected if we consider that higher values of $Y_{X/S}$ mean that a larger fraction of the removed substrate is assimilated as biomass, with, consequently, lower oxygen consumption.

The parameters μ_{max} and K_S affect the effluent substrate concentration (Equation 4.22). The effluent substrate concentration increases with increasing K_S and decreases with increasing μ_{max}. If $S \ll S_0$, the parameters μ_{max} and K_S do not affect biomass concentration, sludge production and oxygen consumption.

Example 4.2 shows the effect of the kinetic parameters on the design results.

Example 4.2: Effect of the kinetic parameters on process design

Calculate the effect of the values of the kinetic parameters for an activated sludge with an influent substrate concentration equal to $S_0 = 0.5$ kg COD/m^3. Calculate the values of biomass concentration (X), effluent substrate (S), sludge production and oxygen consumption for a range of parameter values and as a function of the SRT. Assume constant values of HRT = 0.25 day and $R = 1$. Use the following parameter ranges:

$$\mu_{max} = 4 - 8 \text{ day}^{-1}$$

$$K_S = 0.001 - 0.008 \frac{\text{kg COD}}{\text{m}^3}$$

$$b = 0 - 0.4 \text{ day}^{-1}$$

$$Y_{X/S} = 0.15 - 0.45 \frac{\text{kg biomass}}{\text{kgCOD}}$$

For an influent flow rate of 16,000 m³/day, design parameters SRT = 15 day, HRT = 0.3 day and $R = 1.5$ (as in Example 4.1), compare the values of the process variables calculated in Example 4.1 with the values of the variables calculated with the following values of the kinetic parameters:

$$\mu_{max} = 8 \text{ day}^{-1}$$

$$K_S = 0.008 \frac{\text{kg COD}}{\text{m}^3}$$

$$b = 0.1 \text{ day}^{-1}$$

$$Y_{X/S} = 0.40 \frac{\text{kg biomass}}{\text{kg COD}}$$

Solution

First of all, it is important to observe that, for given values of SRT, HRT and R the values of V/Q, Q_R/Q and Q_W/Q do not depend on the kinetic parameters. On the other hand, the performance of the plant, that is, the effluent substrate concentration, the biomass concentration in the reactor and in the settling tank, the sludge production and oxygen consumption will depend on the kinetic parameters, for given values of the design parameters SRT, HRT and R. However, the ratio X/X_R does not depend on the kinetic parameters (this is evident from a simple re-arrangement of Equation 4.31), therefore in this example only the effect of the kinetic parameters on X will be discussed.

In the calculations for this example, one kinetic parameter at the time is changed, and the values of the remaining kinetic parameters are taken as in Example 4.1, that is:

$$\mu_{max} = 6 \ \text{day}^{-1}, \ K_S = 0.004 \frac{\text{kg COD}}{\text{m}^3},$$

$$b = 0.2 \ \text{day}^{-1}, \ Y_{X/S} = 0.30 \frac{\text{kg biomass}}{\text{kg COD}}$$

a. Effect of μ_{max} and K_S

μ_{max} and K_S affect the degree of substrate removal. Assuming that the effluent substrate S is in all cases much lower than the influent substrate S_0, the values of μ_{max} and K_S do not affect the biomass concentration, the sludge production and the oxygen consumption. Figure 4.6 shows the effect of μ_{max} and K_S on the effluent substrate concentration in a range of SRT values.

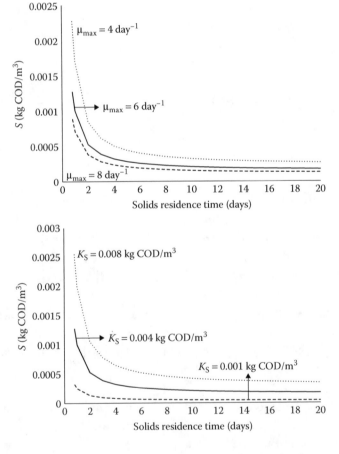

FIGURE 4.6 Effect of the kinetic parameters μ_{max} and K_S on the design results.

In general, lower values of μ_{max} and higher values of K_S give higher effluent substrate concentration. However, for the range of values of μ_{max}, K_S and SRT considered in this example, S is always much lower than S_0. Under these conditions, as discussed above, the values of μ_{max} and K_S have virtually no impact on biomass concentration, sludge production and oxygen consumption.

b. Effect of b

The parameter for endogenous metabolism, b, has a large impact on the results of the simulations. Higher values of b correspond to higher effluent substrate, lower biomass concentration, lower biomass production and higher oxygen consumption. This is shown in Figure 4.7.

Note that for the particular case of $b = 0$ (no endogenous metabolism), the biomass concentration increases linearly with the SRT, instead of reaching an asymptotic value as for $b > 0$, and the sludge production and oxygen consumption do not depend on the SRT. This is because, in the absence of endogenous metabolism, the only contribution to oxygen consumption is substrate removal, but this remains virtually constant as long as $S \ll S_0$, as in all simulations presented in this example.

c. Effect of $Y_{X/S}$

The value of the growth yield $Y_{X/S}$ does not affect the effluent substrate concentration, but it affects biomass concentration, biomass production and oxygen consumption. Higher values of $Y_{X/S}$ give higher biomass concentration, higher biomass production and lower oxygen consumption. This is shown in Figure 4.8.

For an influent flow rate of 16,000 m³/day, design parameters SRT = 15 day, HRT = 0.3 day and R = 1.5 (as in Example 4.1), with the following values of the kinetic parameters:

$$\mu_{max} = 8 \text{ day}^{-1}; \quad K_S = 0.008 \frac{\text{kg COD}}{\text{m}^3};$$

$$b = 0.1 \text{ day}^{-1}; \quad Y_{X/S} = 0.40 \frac{\text{kg biomass}}{\text{kg COD}}$$

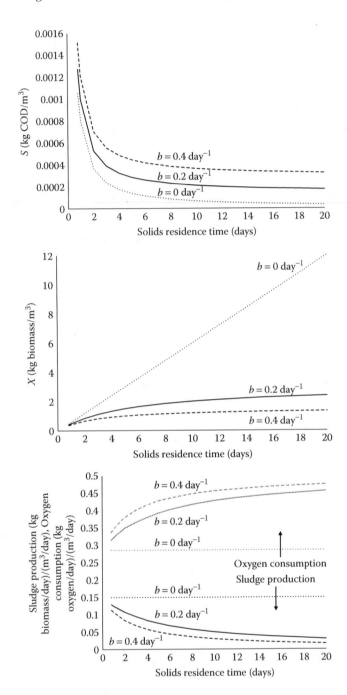

FIGURE 4.7 Effect of the parameter of endogenous metabolism, *b*, on the design results.

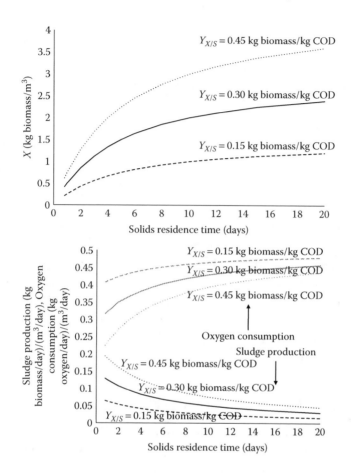

FIGURE 4.8   Effect of the growth yield $Y_{X/S}$ on the design results.

we obtain the following values of the process variables:

$$S = 0.00017 \frac{\text{kg COD}}{\text{m}^3}$$

$$X = 4.00 \frac{\text{kg}}{\text{m}^3}$$

$$X_R = 6.61 \frac{\text{kg}}{\text{m}^3}$$

$$Q_W = 194 \frac{\text{m}^3}{\text{day}}$$

$$Q_R = 24,000 \frac{m^3}{day}$$

$$V = 4800 \, m^3$$

$$P_X = 1280 \frac{kg}{day}$$

$$Q_{O2biomass} = 6179 \frac{kg \, O_2}{day}$$

Note that, as expected, V, Q_W and Q_R do not depend on the kinetic parameters and have the same values as in Example 4.1. The effluent substrate concentration is slightly lower than in Example 4.1, but in both cases it is much lower than in the influent. Biomass concentration X (and consequently X_R) is significantly larger than in Example 4.1, mainly due to the lower value of b and higher value of $Y_{X/S}$. For the same reasons, biomass production is larger and oxygen consumption is lower than in Example 4.1.

Example 4.3: Calculation of ammonia concentration

With reference to the activated sludge process in Example 4.1, assume that the influent stream has a concentration of free ammonia equal to 20 mg N-NH$_3$/L. Calculate the effluent ammonia concentration as a function of the design parameters. Assume that the influent organic matter contains no nitrogen and that the empirical formula for microorganisms is $C_5H_7O_2N$.

Solution

In this process, nitrogen removal is only due to the growth of microorganisms, which remove nitrogen because in their chemical composition nitrogen is 12% by weight ($C_5H_7O_2N$). The biomass production only depends on the SRT; therefore, ammonia removal and the effluent ammonia concentration will only depend on the SRT. The ammonia balance in the whole process is given by:

$$QNH_{30} \left(\frac{kgN - NH_3}{day} \right) = Q_W X_R 0.12 + QNH_3 \qquad (4.32)$$

Where NH_{30} and NH_3 are the ammonia concentration in the influent and in the effluent ($kgN-NH_3/m^3$).

After substitution of the other mass balances and re-arrangements, Equation 4.32 becomes:

$$NH_3 = NH_{30} - \left(S_0 - \frac{bK_S SRT + K_S}{(\mu_{max} - b)SRT - 1} \right) Y_{X/S} \cdot 0.12 \qquad (4.33)$$

Equation 4.33 shows explicitly that the effluent ammonia concentration only depends on the influent ammonia and on the SRT. The effluent ammonia concentration can be therefore calculated (Figure 4.9) as a function of the SRT with the kinetic parameters given in Example 4.1, or, which is the same, from the values of X_R and Q_W calculated for Example 4.1.

Figure 4.9 shows that the effluent ammonia increases as the SRT increases. This is because higher SRT corresponds to a lower production of biomass (e.g. see Figure 4.5). Lower biomass production means that less ammonia is removed, and therefore the effluent ammonia increases.

4.1.3 Aeration Requirements in the Activated Sludge Process

Oxygen needs to be provided to the aeration tank of the activated sludge process, to provide for the requirements of microorganisms growth and of endogenous metabolism. We have seen in previous sections how to

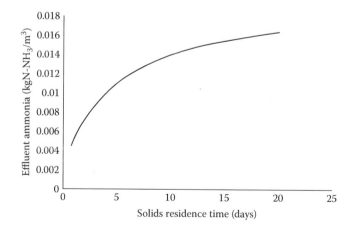

FIGURE 4.9 Effluent ammonia concentration versus solids residence time (SRT) for Example 4.3.

calculate the oxygen consumption by the microorganisms using the COD balance. In this section, we will see how to calculate the oxygen concentration for a given mass transfer coefficient, or, vice versa, how to calculate the required mass transfer coefficient once the required oxygen concentration is known.

Two main types of aerators are used in activated sludge processes: diffusers and mechanical aerators. In aeration by diffusers (Figure 4.10) air is usually introduced at the bottom of the biological reactor through spargers or porous diffusers. The spargers or diffusers produce bubbles which transfer oxygen from the gas phase to the liquid phase in the reactor. Mechanical aerators (Figure 4.11) are essentially agitators, usually mounted above the liquid surface, that agitate the liquid phase and create a good contact with the air in the atmosphere. Therefore, in mechanical aerators air transfers directly from the atmosphere to the liquid phase.

To calculate the oxygen concentration in the liquid phase, or the required $k_L a$ to obtain a certain oxygen concentration, we need to write the oxygen mass balances in the liquid phase. The mass balances refer to the scheme below (Figure 4.12, biological reactor in an activated sludge process).

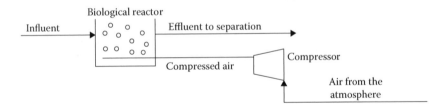

FIGURE 4.10   Aeration by diffusers.

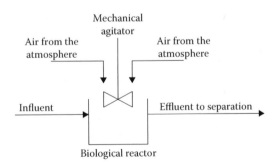

FIGURE 4.11   Mechanical aeration.

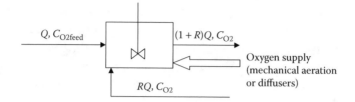

FIGURE 4.12 Scheme of the biological reactor in the activated sludge process, showing the various contributions to the oxygen balance.

Mass balance of oxygen in the liquid phase:

Oxygen in with the inlet liquid stream + oxygen transferred from the gas phase = oxygen consumed by the microorganisms + oxygen out with the outlet liquid stream

$$QC_{O2feed} + RQC_{O2} + k_La\left(k_{eqO2} \cdot p_{O2} - C_{O2}\right)V$$

$$= Q_{O2biomass} + (1+R)QC_{O2} \Rightarrow QC_{O2feed} + k_La\left(k_{eqO2} \cdot p_{O2} - C_{O2}\right)V \quad (4.34)$$

$$= Q_{O2biomass} + QC_{O2}$$

We assume that the process has already been designed, that is, that we know all the values of the flow rates, the substrate and biomass concentration, and, most importantly, the oxygen consumption by the microorganisms $Q_{O2biomass}$ (kg O_2/day).

If aeration is provided by mechanical aerators, the liquid phase is in direct contact with the atmosphere, and therefore p_{O2} can always be assumed to be equal to the partial pressure of oxygen in the atmosphere, that is $p_{O2} = 0.21$ atm. In a similar way if aeration is provided by diffusers using pure oxygen, then p_{O2} is immediately known as it coincides with the total pressure of the bubbles in the aeration tank. If we neglect the hydrostatic head due to the depth of the bubbles inside the liquid phase (this depth is, however, not negligible and may give a not insignificant hydrostatic pressure), we can assume that $p_{O2} = 1$ atm. Therefore, in both cases, mechanical aerators and diffusers with pure oxygen, we can assume p_{O2} is known and Equation 4.34 can easily re-arranged to calculate the concentration of oxygen in the liquid phase, C_{O2}. Alternatively, an important use of Equation 4.34 is to calculate which mass transfer coefficient k_La is required to maintain a certain minimum oxygen concentration in the aeration tank. Once the value of k_La is known, the aeration system can be designed.

The calculation of the oxygen concentration using only Equation 4.34 can also be done if aeration is carried out using diffusers and air, if we assume that the amount of oxygen transferred from the gas to the liquid phase is small compared to the total amount of oxygen fed to the reactor. If this assumption is verified, then the partial pressure of oxygen inside the gas bubbles will be approximately equal to its atmospheric value, that is $p_{O2} = 0.21$ atm. This assumption is often verified in activated sludge process, where the amount of oxygen transferred is usually below 20% of the total oxygen fed to the system.

However, if oxygen is supplied by diffusers using air, in the general case we need to consider that the oxygen partial pressure in the bubbles inside the liquid phase is not necessarily the same as the one in the atmospheric air which is pumped into the system. This is because oxygen transfers from the air bubbles to the liquid phase, and therefore the oxygen partial pressure in the gas phase is usually lower than its value in the atmosphere. Therefore, in the case of aeration with diffusers using air, p_{O2} is in principle not known and needs to be calculated using the mass balance for oxygen in the gas phase. Figure 4.13 shows the scheme of the biological reactor which can be used to calculate the mass balance for oxygen in the gas phase.

In this scheme, all the flow rates are in m³/s (m³ of either gas or liquid), the concentrations in the liquid phase are in kg_{O2}/m^3. We assume that nitrogen gas (N_2) is not transferred to the liquid phase (this is because nitrogen is not consumed by the microorganisms, so we can assume the liquid phase is always in equilibrium with the atmospheric nitrogen); therefore, $Q_{nitrogen}$ does not change between the gas inlet and outlet. We assume that air is composed by 79% nitrogen and 21% oxygen in volume, and we assume that the outlet gas is only composed of nitrogen and oxygen, ignoring in this case the carbon dioxide generation by the

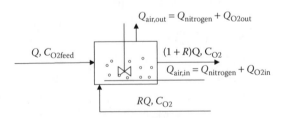

FIGURE 4.13 Scheme of the biological reactor in the activated sludge process, same as Figure 4.12 but including the gas phase.

microorganisms. In the mass balances, we assume that both the liquid phase and the gas phase are perfectly mixed.

Mass balance of oxygen in the gas phase:

Oxygen in with the inlet air flow = oxygen transferred to the liquid phase + oxygen out with the outlet gas stream

$$w_{O2,in} = k_L a \left(k_{eqO2} \cdot p_{O2} - C_{O2} \right) V + w_{O2,out} \tag{4.35}$$

where $W_{O2,in}$ and $W_{O2,out}$ are the mass flow rates of oxygen entering and leaving the system (kgO_2/day). Since the nitrogen does not transfer to the liquid phase, its mass flow rate w_{N2} is the same at the inlet and outlet of the gas phase and the mass flow rates $w_{O2,in}$ and $w_{O2,out}$ can be expressed as:

$$w_{air,in} = w_{N2} + w_{O2,in} \tag{4.36}$$

$$w_{air,out} = w_{N2} + w_{O2,out} \tag{4.37}$$

The mass flow of nitrogen can be expressed as a function of the inlet mass flow rate of air:

$$w_{N2} = \frac{w_{air,in}}{\rho_{air,in}} \cdot \frac{p_{N2,in}}{p_{tot}} \cdot \rho_{N2,in} = w_{air,in} \cdot \frac{y_{N2,in} MW_{N2}}{\left(y_{O2,in} MW_{O2} + y_{N2,in} MW_{N2} \right)} \tag{4.38}$$

Similarly, the inlet mass flow rate of oxygen can be expressed as:

$$w_{O2,in} = w_{air,in} \cdot \frac{y_{O2,in} MW_{O2}}{\left(y_{O2,in} MW_{O2} + y_{N2,in} MW_{N2} \right)} \tag{4.39}$$

The term $y_{O2,in} MW_{O2} / (y_{O2,in} MW_{O2} + y_{N2,in} MW_{N2})$ is the mass fraction of oxygen in atmospheric air and is equal to 0.23 while the term $y_{N2,in} MW_{N2} / (y_{O2,in} MW_{O2} + y_{N2,in} MW_{N2})$ is the mass fraction of oxygen in atmospheric air and is equal to 0.77. So, we can re-write the two Equations 4.38 and 4.39 more simply as:

$$w_{N2} = 0.77 \cdot w_{air,in} \tag{4.40}$$

and

$$w_{O2,in} = 0.23 \cdot w_{air,in} \tag{4.41}$$

The volumetric flow rates of air and oxygen at the outlet of the reactor can be expressed as:

$$\frac{p_{O2}}{p_{tot}} Q_{air,out} = Q_{O2,out} \qquad (4.42)$$

which becomes

$$\frac{p_{O2}}{p_{tot}} \frac{w_{air,out}}{\rho_{air}} = \frac{w_{oxyg,out}}{\rho_{O2}} \qquad (4.43)$$

This can be written as:

$$\frac{p_{O2}}{p_{tot}} \frac{w_{air,out}}{\left(p_{O2} MW_{O2} + p_{N2} MW_{N2} \right)} = \frac{w_{O2,out}}{p_{tot} MW_{O2}}$$

$$\Rightarrow w_{O2,out} = \frac{p_{O2} MW_{O2} w_{air,out}}{\left(p_{O2} MW_{O2} + p_{N2} MW_{N2} \right)} \qquad (4.44)$$

$$\Rightarrow w_{O2,out} = \frac{p_{O2} MW_{O2}}{\left(p_{O2} MW_{O2} + p_{N2} MW_{N2} \right)} \left(w_{N2} + w_{O2,out} \right)$$

and finally

$$w_{O2,out} = w_{N2} \frac{p_{O2} MW_{O2}}{p_{N2} MW_{N2}} = 0.77 w_{air,in} \frac{p_{O2} MW_{O2}}{p_{N2} MW_{N2}} \qquad (4.45)$$

And since $p_{N2} = p_{tot} - p_{O2}$

$$w_{O2,out} = 0.77 w_{air,in} \frac{\dfrac{p_{O2}}{p_{tot}} MW_{O2}}{\left(1 - \dfrac{p_{O2}}{p_{tot}} \right) MW_{N2}} \qquad (4.46)$$

Therefore, the oxygen balance in the gas phase can be re-written as:

$$0.23 w_{air,in} = k_L a \left(k_{eqO2} \cdot p_{O2} - C_{O2} \right) V + 0.77 w_{air,in} \frac{\dfrac{p_{O2}}{p_{tot}} MW_{O2}}{\left(1 - \dfrac{p_{O2}}{p_{tot}} \right) MW_{N2}} \qquad (4.47)$$

In summary, for an activated sludge process aerated with diffusers which use atmospheric air, the oxygen concentration in the aeration tank (or the required $k_L a$) needs to be calculated by solving the mass balances for oxygen in the liquid and in the gas phase, that is the system of Equation 4.48:

$$\begin{cases} QC_{O2feed} + k_L a \left(k_{eqO2} \cdot p_{O2} - C_{O2} \right) V = Q_{O2biomass} + QC_{O2} \\ \\ 0.23 w_{air,in} = k_L a \left(k_{eqO2} \cdot p_{O2} - C_{O2} \right) V + 0.77 w_{air,in} \dfrac{\dfrac{p_{O2}}{P_{tot}} MW_{O2}}{\left(1 - \dfrac{p_{O2}}{P_{tot}} \right) MW_{N2}} \end{cases} \quad (4.48)$$

For a given oxygen consumption rate by the biomass, $Q_{O2biomass}$, and for a given inlet air mass flow rate, $w_{air,in}$, we have a system of two equations and two unknowns, p_{O2} and C_{O2} (or $k_L a$). The solution of the system gives the p_{O2} and C_{O2} values. Obviously, the system of equations can only be solved if the amount of oxygen provided is equal or larger than the amount of oxygen consumed by the biomass, that is it has to be:

$$w_{Oxyg,in} = w_{air,in} \cdot 0.23 \geq Q_{O2biomass} \quad (4.49)$$

Note that the $k_L a$ value can be expressed as a function of the air flow rate by means of correlations, as discussed in Chapter 2.

Example 4.4: Calculation of the oxygen concentration with mechanical aerators

For the activated sludge process designed in Example 4.1 with the design parameters SRT = 15 day, HRT = 0.3 day and R = 1.5, calculate the oxygen concentration in the aeration tank if aeration is provided with a mechanical aerator which provides a $k_L a$ equal to 240 day^{-1}. Assume that the influent to the plant contains no dissolved oxygen. Assume $ke_{qO2} = 0.043$ kg/m^3.atm.

Solution

Since we have a mechanical aerator we only need to use the oxygen balance in the liquid phase, that is Equation 4.34, with $p_{O2} = 0.21$ atm. Equation 4.34, after re-arrangements and assuming the influent contains no oxygen, becomes:

$$C_{O2} = \frac{k_L a \cdot k_{eqO2} \cdot p_{O2} V - Q_{O2biomass}}{Q + k_L aV} \quad (4.50)$$

Since we have a mechanical aerator p_{O2} coincides with the partial pressure of oxygen in the atmosphere, that is 0.21 atm. From the design data in Example 4.1 we have:

$V = 4800 \text{ m}^3$, $Q = 16,000 \text{ m}^3/\text{day}$, $Q_{O2\text{biomass}} = 7144 \text{ kg}_{O2}/\text{day}$.
Substituting these values in Equation 4.50, we obtain:
$C_{O2} = 5.5 \text{ mg/L}$.

Example 4.5: Calculation of the oxygen concentration for aeration with diffusers

For a conventional activated sludge process for carbon removal the following values of the design and operating parameters are known.

$$Q = 1000 \text{ m}^3/\text{day}$$

$$S_0 = 0.250 \text{ kg COD/m}^3$$

$$R = 1$$

$$S = 0.05 \text{ kg COD/m}^3$$

$$Q_w = 50 \text{ m}^3/\text{day}$$

$$V = 1000 \text{ m}^3$$

$$X = 1.1 \text{ kg/m}^3$$

a. Calculate the amount of oxygen consumed by the microorganisms $(Q_{O2\text{biomass}})$
b. Calculate the oxygen concentration in the reactor for a range of air flow rates between 1.1 and 2 times the minimum air flow rate. Assume atmospheric pressure. Assume for $k_L a$ the following correlation holds: $k_L a V(\text{m}^3/\text{day}) = 1.0 \cdot 10^4 \cdot Q_{\text{air,in}}^{0.5}$. Neglect the oxygen concentration in the influent stream. Assume $k_{eqO2} = C_{O2}/p_{O2} = 0.043 \text{ kg/m}^3/\text{atm}$ and a temperature of 20°C

Solution

a. The oxygen consumption by the microorganisms is given by:

$$Q_{O2\text{biomass}}\left(\frac{\text{kg O}_2}{\text{day}}\right) = Q(S_0 - S) - Q_W X_R \cdot 1.42 \qquad (4.19)$$

With the given data, we only miss the value of X_R, which can be easily calculated from a mass balance of the microorganisms on the settling tank, which is (referring to Figure 4.1):

$$(Q+Q_R)X = (Q_W + Q_R)X_R \qquad (4.51)$$

Equation 4.51 can also be obtained from the combination of Equations 4.1 and 4.3.

From Equation 4.51 we obtain:

$$X_R = \frac{Q(1+R)X}{Q_W + RQ} = \frac{1000\,\text{m}^3/\text{day} \cdot 2 \cdot 1.1\,\text{kg/m}^3}{1050\,\text{m}^3/\text{day}} = 2.1\,\text{kg/m}^3$$

Therefore from Equation 4.19 we obtain:

$$Q_{O2biomass}\left(\frac{\text{kg O}_2}{\text{day}}\right) = 1000\ \text{m}^3/\text{day} \cdot (0.25 - 0.05)\text{kg COD/m}^3$$

$$-50\ \text{m}^3/\text{day} \cdot 2.1\,\text{kg/m}^3 \cdot 1.42$$

$$= 50.9\ \text{kg O}_2/\text{day}$$

b. The minimum air flow rate that has to be provided corresponds to the amount of oxygen which is consumed by the microorganisms, that is

$$Q_{airinmin} \cdot 0.21 \cdot \rho_{oxygen} = Q_{O2biomass}$$

ρ_{oxygen} can be calculated assuming ideal gas:

$$PV = nRT,\ \text{that is}\ \frac{n}{V} = \frac{P}{RT} = \frac{1\,\text{atm}}{0.0821\dfrac{1\,\text{atm}}{\text{mol}\,K} \cdot 293\,K} = 0.042\ \text{mol/l}$$

so $\rho_{oxygen} = 0.042\ \text{mol/l} \cdot 32\,\text{g/mol} = 1.33\ \text{g/l}$

therefore,

$$Q_{airinmin} = \frac{Q_{O2biomass}}{0.21 \cdot \rho_{oxygen}} = \frac{50.9\ \text{kgO}_2/\text{day}}{0.21 \cdot 1.33\ \text{kgO}_2/\text{m}^3} = 182.3\ \text{m}^3/\text{day}$$

The calculation of the oxygen concentration in the reactor will be done for the following values of $Q_{air,in}$: 200, 400, 600, 800, 1000 m³/day.

The oxygen concentration in the biological reactor can be calculated for the various values of $Q_{air,in}$ from the mass balances for oxygen in the liquid and in the gas phases, that is from the system of Equations 4.48. The air mass flow rate $w_{air,in}$ (kg/day) in Equations 4.48 can be immediately calculated from $Q_{air,in}$ (which is in m³/day):

$$w_{air,in} = Q_{air,in}\rho_{air}$$

The air density ρ_{air} is obtained as, assuming a molecular weight of 29 for air:

$$\rho_{air} = 0.042 \text{ mol/l} \cdot 29 \text{ g/mol} = 1.22 \text{ g/l}$$

The results of the calculations are tabulated in Table 4.1. Note that a flow rate of 200 m³/day is not enough to maintain the oxygen concentration above 2 mg/L, which is often considered the minimum value for an unrestricted biomass growth rate. For a flow rate of 200 m³/day the oxygen partial pressure in the gas bubbles in the reactor is much lower than its value in the inlet atmospheric air. This indicates that with very low gas flow rates the oxygen fed to the reactor is almost entirely consumed by the microorganisms. When the inlet gas flow rate increases, the dissolved oxygen concentration and the oxygen partial pressure in the gas bubbles increase considerably, indicating that only a minor fraction of the oxygen fed to the system is consumed by the microorganisms.

4.1.3.1 Effect of the Operating Parameters of the Plant on the Aeration Requirements

We have seen in previous sections that the oxygen consumption by the microorganisms, $Q_{O2biomass}$, only depends, for a given influent substrate concentration and flow rate, on the choice of the SRT. We have also seen how to calculate the oxygen concentration or the required mass transfer coefficient k_La using the oxygen balance. But how does the required

TABLE 4.1   Oxygen Concentration in the Biological
Reactor for Different Values of the Inlet Air Flow Rate $Q_{air,in}$

$Q_{air,in}$ (m³/day)	k_La (day⁻¹)	C_{O2} (kg/m³)	p_{O2} (atm)
200	141.4	7.8E–4	0.026
400	200	5.61E–3	0.135
600	244.9	6.83E–3	0.163
800	282.8	7.43E–3	0.176
1000	316.2	7.75E–3	0.183

$k_L a$ depend on the design parameters of the plant, that is SRT, HRT and R? This can be seen by re-arranging the oxygen balance introducing the design parameters. In this section for simplicity we will always assume that the oxygen concentration can be calculated with Equation 4.34 only, and that the mass balance for oxygen in the gas phase, Equation 4.47, is not needed. This implies that, if aeration is carried out with diffusers using air, the amount of oxygen transferred can be considered small compared to the total amount of oxygen fed to the reactor.

Re-arranging Equation 4.34 and introducing the expression for $Q_{O2biomass}$ derived earlier (Equation 4.19), we obtain:

$$k_L a = \frac{\dfrac{Q_{O2biomass}}{V} + \dfrac{Q}{V}(C_{O2} - C_{O2feed})}{k \cdot p_{O2} - C_{O2}}$$

$$= \frac{\dfrac{(S_0 - S)}{HRT} - \dfrac{X}{SRT} \cdot 1.42 + \dfrac{(C_{O2} - C_{O2feed})}{HRT}}{k \cdot p_{O2} - C_{O2}}$$

(4.52)

This equation gives the $k_L a$ that is required to maintain a certain oxygen concentration C_{O2} in the aeration tank and shows that the required $k_L a$ depends on the choice of the SRT and HRT and does not depend on the choice of R. The required $k_L a$ increases by increasing the SRT and by decreasing the HRT. So, in practice increasing the SRT and decreasing the HRT increases the value of the $k_L a$ required. However, it is very important to observe that the required value of the factor $k_L a V$, that is the total mass transfer rate available in the reactor, does not depend on the HRT but only on the SRT. Indeed, re-arranging Equation 4.52, we obtain:

$$k_L a V = \frac{Q_{O2biomass} + Q(C_{O2} - C_{O2feed})}{k \cdot p_{O2} - C_{O2}}$$

(4.53)

Equation 4.53 shows that $k_L a V$ only depends on the SRT, because $Q_{O2biomass}$ depends only on the SRT. This is an important design consideration because, as we have seen in Chapter 2, it is the factor $k_L a V$, and not $k_L a$ per se, which is affected by the value of the inlet air flow rate (for aeration with diffusers) or by the agitator power (for mechanical aerators). This means in practice that the choice of the HRT does not affect the required air flow rate or mechanical power of the aeration system, for a given value of the influent wastewater flow rate, substrate concentration and for a given value of the SRT. In other words, all the said parameters being constant, we can choose to have a smaller or larger reactor without affecting the energy consumption for aeration. This is because the total mass of microorganisms in the reactor is only determined by the SRT, and the HRT only determines the microorganisms' concentration, not their total mass. However, these considerations do not take into account the possible negative effect on $k_L a$ of higher biomass concentrations. If we are under conditions where higher biomass concentration decreases the $k_L a$ value, for example due to the increase in the mixed liquor viscosity, then reducing the HRT, with the consequent increase in biomass concentration, will cause a requirement for larger inlet air flow rate or mechanical power, for the same oxygen consumption rate by the microorganisms.

Example 4.6: Calculation of the aeration requirements for the activated sludge process

For the wastewater treatment plant of Example 4.1, calculate the following, for an influent flow rate of 10,000 m³/day. Assume the concentration of oxygen in the inlet wastewater is 0 mg/L and that the equilibrium equation for oxygen in water is given by the equation $k_{eqO2} = C_{O2}/p_{O2} = 0.043 \, \text{kg/m}^3/\text{atm}$.

a. The $k_L a$ required to maintain an oxygen concentration of 2 mg/L in the reactor, as a function of the HRT and SRT;
b. The $k_L a V$ required to maintain an oxygen concentration of 2 mg/L in the reactor, as a function of the SRT;
c. The required $k_L a$ as a function of the oxygen concentration in the reactor, in the range 2–7 mg/L, for a SRT of 15 days and an HRT of 0.5 days;
d. The required air flow rate, compressor power and corresponding oxygen transfer efficiency for aeration with diffusers as a function of the SRT for an oxygen concentration of 2 mg/L in the reactor. Compare two diffuser models characterised by the following correlations for mass transfer:

$k_L a V = 27.4 Q_{air}^{0.74}$; $k_L a V = 20.1 Q_{air}^{0.89}$ (where $k_L a V$ is in m³/day and Q_{air} is in m³/day)

Assume that the diffusers in the reactor are placed 5 m below the liquid surface and that the fraction of the inlet oxygen that is transferred to the liquid is small;

e. The required agitator speed, power and oxygen transfer efficiency for aeration with mechanical aerators as a function of the SRT for an oxygen concentration of 2 mg/L in the reactor. Compare two mechanical aerators characterised by the following correlations for the power number:

$$P_0 = 0.71\,Fr^{-0.51}; \quad P_0 = 0.78\,Fr^{-0.57}$$

assume that the diameter of the agitator is 2 m in all cases.

Solution

a. The required $k_L a$ is given by Equation 4.52, for an oxygen concentration in the reactor of 2 mg/L:

$$k_L a(\text{day}^{-1}) = \frac{\dfrac{(0.5-S)}{HRT} - \dfrac{X}{SRT} \cdot 1.42 + \dfrac{0.002}{HRT}}{0.043 \cdot 0.21 - 0.002}$$

The values of S and X can be calculated as a function of SRT and HRT as described in Section 4.1.1. Figure 4.14 shows the required $k_L a$ as a function of the SRT and HRT. Note that the required $k_L a$ increases with increasing SRT, because the oxygen consumption by the microorganisms ($Q_{O2biomass}$) increases. Also, the required $k_L a$ increases for decreasing HRT, because the same amount of oxygen needs to be transferred in a smaller volume.

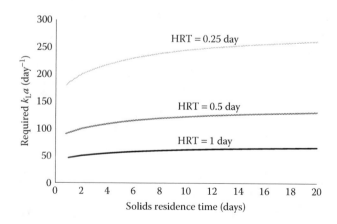

FIGURE 4.14 Effect of solids residence time (SRT) and hydraulic residence time (HRT) on the required $k_L a$ to maintain an oxygen concentration in the reactor equal to 2 mg/L.

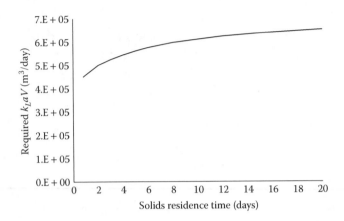

FIGURE 4.15   Effect of solids residence time (SRT) on the required $k_L aV$ to maintain an oxygen concentration in the reactor equal to 2 mg/L.

b. The calculation of the required $k_L aV$ can be done immediately using Equation 4.53, which becomes in this case:

$$k_L aV\left(\frac{m^3}{day}\right) = \frac{Q_{O2biomass} + 10{,}000 \cdot 0.002}{0.043 \cdot 0.21 - 0.002}$$

where $Q_{O2biomass}$ depends only on the SRT and can be calculated using Equation 4.19. The plot of $k_L aV$ versus SRT is shown in Figure (4.15). As discussed above, $k_L aV$ increases with increasing SRT.

c. For SRT = 15 day and HRT = 0.5 day, the values of the steady-state substrate and biomass concentration in the reactor are (calculated by solving the system of Equation (4.13–4.15):

$$S = 0.000185 \frac{kg\ COD}{m^3}$$

$$X = 1.13 \frac{kg}{m^3}$$

The required $k_L a$ as a function of the desired oxygen concentration in the reactor is given by Equation 4.52, which, for HRT = 0.5 day and SRT = 15 day, becomes:

$$k_L a(day^{-1}) = \frac{\dfrac{(0.5 - 0.000185)}{0.5} - \dfrac{1.13}{15} \cdot 1.42 + \dfrac{C_{O2}}{0.5}}{0.043 \cdot 0.21 - C_{O2}}$$

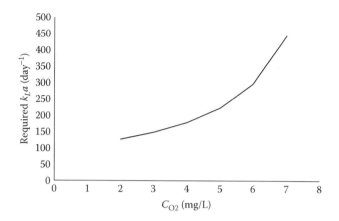

FIGURE 4.16   Dependence of the required $k_L a$ on the desired oxygen concentration in the reactor.

The plot of this equation is shown in Figure 4.16. As expected, the required $k_L a$ increases if a higher oxygen concentration in the reactor is required. This is because the driving force for mass transfer decreases. Therefore, the oxygen concentration should be kept to the minimum value that ensures unrestricted growth (i.e. growth rate not limited by oxygen concentration) of the microorganisms.

d. For aeration with diffusers, the required air flow rate is linked to the required $k_L a V$ by the equation shown in Chapter 2:

$$Q_{air} = \left(\frac{k_L a V}{k_{diff}} \right)^{\frac{1}{b_{diff}}}$$

For the two diffusers considered in this example, this equation becomes:

$$Q_{air} = \left(\frac{k_L a V}{27.4} \right)^{\frac{1}{0.74}} ; \quad Q_{air} = \left(\frac{k_L a V}{20.1} \right)^{\frac{1}{0.89}}$$

where $k_L a V$ and Q_{air} are both in m³/day.

Since $k_L a V$ only depends on the SRT (for a given flow rate and composition of the influent wastewater and for a given concentration of dissolved oxygen in the reactor), the required air flow rate also only depends on the SRT. The plot of the required air flow rate as a function of the SRT is shown in Figure 4.17 for the two models of diffusers.

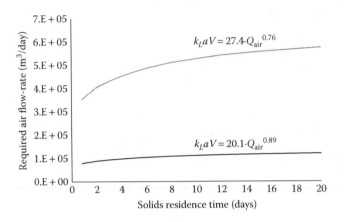

FIGURE 4.17   Effect of the solids residence time (SRT) on the required air flow rate (aeration with diffusers).

For a given air flow rate, the required compressor power is given, as seen in Chapter 2, by:

$$P_{compr}(W) = \frac{\gamma}{\gamma-1} Q_{air}\, p_{inlet} \left[\left(\frac{p_{outlet}}{p_{inlet}} \right)^{\frac{\gamma-1}{\gamma}} - 1 \right]$$

where Q_{air} is in m³/s and p_{inlet} and p_{outlet} are in Pa. p_{inlet} can usually be assumed to be atmospheric (101,325 Pa), while p_{outlet} depends on the depth of the diffusers in the reactor, which in this example is taken equal to 5 m. A depth of 5 m corresponds, assuming the physical properties of water, to a p_{outlet} of 49,050 Pa. Since for air we have: $(\gamma-1)/\gamma = 0.283$ the equation to calculate the required compressor power is, for the two diffusers:

$$P_{compr}(W) = \frac{\dfrac{\left(\dfrac{k_L aV}{27.4} \right)^{\frac{1}{0.74}}}{86,400} 101,325}{0.283} \left[\left(\frac{101,325 + 49,050}{101,325} \right)^{0.283} - 1 \right]$$

$$P_{compr}(W) = \frac{\dfrac{\left(\dfrac{k_L aV}{20.1} \right)^{\frac{1}{0.89}}}{86,400} 101,325}{0.283} \left[\left(\frac{101,325 + 49,050}{101,325} \right)^{0.283} - 1 \right]$$

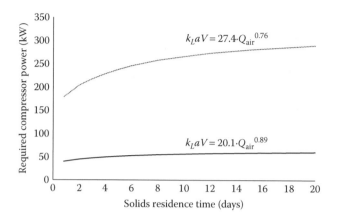

FIGURE 4.18 Effect of the solids residence time (SRT) on the required compressor power (aeration with diffusers).

Since $k_L aV$ only depends on the SRT, for a given wastewater of a certain flow rate and composition, the required compressor power only depends on the SRT. The effect of the SRT on the required compressor power is shown in Figure 4.18.

The efficiency of air transfer is the ratio between the air transferred to the liquid phase in the reactor and the air provided to the system. This ratio is given by (Chapter 2):

$$\eta \left(\frac{kg\, O_2\ \text{transferred}}{kg\, O_2\ \text{provided}} \right) = \frac{k_L aV \left(k_{eqO2}\, p_{O2} - C_{O2} \right)}{Q_{air}\, \rho_{air} \cdot 0.23}$$

$$= \frac{Q_{O2biomass} + Q(C_{O2} - C_{O2feed})}{Q_{air}\, \rho_{air} \cdot 0.23}$$

Where Q_{air} is in m³/day, ρ_{air} is the air density in kg/m³ and 0.23 is the mass fraction of oxygen in air.

Since the term $Q(C_{O2} - C_{O2feed})$ is usually much lower than $Q_{O2biomass}$, for a given diffuser the efficiency of air transfer only depends on the SRT. Indeed, both $Q_{O2biomass}$ and the required Q_{air} only depend on the SRT.

The effect of the SRT on the efficiency of air transfer is shown in Figure 4.19. The efficiency is higher for the diffuser which gives the higher $k_L aV$ for a given air flow rate or, in other words, which gives the same oxygen transfer rate with a lower air flow rate. The efficiency decreases as the SRT increases, because higher

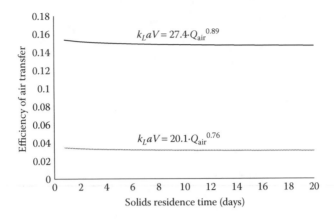

FIGURE 4.19 Effect of the solids residence time (SRT) on the efficiency of air transfer (aeration with diffusers).

SRT values require higher value of $k_L aV$ and the $k_L aV$ increases less than linearly as the air flow rate increases (i.e. the exponent b_{diff} is lower than 1 for the diffuser types used in this example).

e. For mechanical aerators the mass transfer rate $k_L aV$ can be expressed according to what shown in Chapter 2:

$$k_L aV\left(\frac{m^3}{s}\right) = \frac{0.063 k_{mech}^{1.5}}{9.8^{1.5 b_{mech}}} N^{(3+3 b_{mech})} D^{(4.5+1.5 b_{mech})}$$

where N and D are the agitator speed (revs/s) and diameter (m), respectively, and k_{mech} and b_{mech} are the parameters of the power number correlations for the agitator:

$$P_0 = k_{mech}\left(\frac{N^2 D}{g}\right)^{b_{mech}}$$

This equation shows that, for a given agitator of a given diameter D, the required speed can be calculated from the required $k_L aV$. It is also evident that the required speed only depends on the SRT, since the required $k_L aV$ only depends on the SRT (for a given flow rate and composition of the influent wastewater and given oxygen concentration in the aeration tank). Equation 2.101 can be easily re-arranged to give the required agitator speed explicitly:

$$N = \left(k_L aV \frac{9.8^{1.5 b_{mech}}}{0.063 k_{mech}^{1.5} D^{(4.5+1.5 b_{mech})}}\right)^{\frac{1}{3+3 b_{mech}}}$$

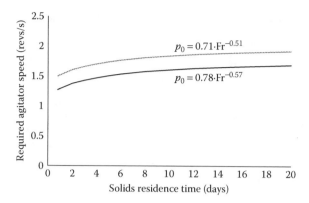

FIGURE 4.20 Effect of the solids residence time (SRT) on the required rotational speed of the mechanical aerator.

where $k_L aV$ is expressed in m^3/s.

It is evident that the required agitation speed increases as the required $k_L aV$ increases (i.e. as the SRT increases) and decreases as the diameter of the agitator decreases. The plot of N versus SRT is shown in Figure (4.20) for the two agitators.

The required power for the mechanical agitators is given, as seen in Chapter 2, by:

$$P(W) = \frac{k_{mech} 1000 \cdot N^{3+2b_{mech}} D^{5+b_{mech}}}{9.8^{b_{mech}}}$$

Therefore, the required agitator power depends on the SRT, as shown in Figure 4.21.

For mechanical aerators the efficiency of air transfer is expressed as the kg of oxygen transferred per unit energy consumed by the agitator (Chapter 2):

$$\eta \left(\frac{kg\, O_2\ transferred}{kWhr\ consumed} \right) = \frac{\dfrac{k_L aV \left(\dfrac{m^3}{day} \right)(k_{eqO2}\, p_{O2} - C_{O2})}{24}}{1000 \dfrac{P(W)}{W}\, 1hr}$$

$$= \frac{\dfrac{Q_{O2biomass} + Q(C_{O2} - C_{O2feed})}{24}}{\dfrac{k_{mech} \cdot N^{3+2b_{mech}} D^{5+b_{mech}}}{9.8^{b_{mech}}}}$$

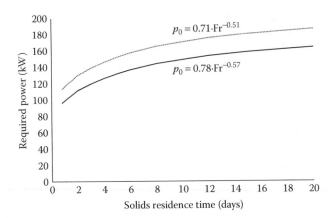

FIGURE 4.21   Effect of the solids residence time (SRT) on the required power of the mechanical aerator.

This equation shows that, for a given agitator, the efficiency depends on the SRT, since $Q_{O2biomass}$ and N both depend on the SRT. The most important effect is that the required N increases at higher SRT; therefore, the efficiency of mechanical aerators decreases at higher SRT. With the data of this example, the plot of the aeration efficiency versus the SRT is shown in Figure 4.22.

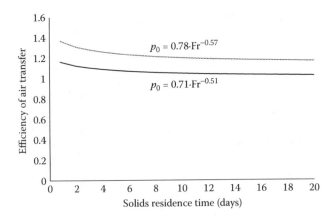

FIGURE 4.22   Effect of the solids residence time (SRT) on the efficiency of air transfer (mechanical aeration).

4.1.4 Calculation of the Required Area of the Settling Tank

Once the activated sludge plant has been designed, and the values for all the concentrations and flow rates are known, we need to make sure that the settling tank is able to settle all the solids that it receives from the biological reactor. Therefore, we need to calculate the required minimum area of the thickening section. This can be calculated as described in Chapter 2. With reference to the thickening zone only, the minimum area of the settling tank, which is able to settle solids entering at concentration X and leaving from the bottom at concentration X_R, with an inlet flow rate to the settling tank of $Q + Q_R$, is given by:

$$A_{min} = (Q + Q_R) X \left[\frac{\frac{1}{X} - \frac{1}{X_R}}{u_C - u_U} \right]_{max} \tag{4.54}$$

where u_c and u_u are the settling velocities of the biomass at the concentration X and X_R, respectively. Generally, the settling velocity decreases as the biomass concentration increases.

Looking at Equation 4.54 it is evident that the minimum required area of the settling tank increases when the inlet biomass concentration X increases. This is because a higher biomass concentration increases the inlet flux $(Q + Q_R)X$ (kg/m².day) and decreases the settling velocity u_c. Also, the required area increases as the inlet flow rate to the settling tank $(Q + Q_R)$ increases.

For a given wastewater of a certain flow rate and composition, it is also possible to evaluate the effect of the design parameters of the process, SRT, HRT and R, on the required area of the settling tank. Increasing the SRT, everything else being the same, causes an increase in A_{min}, because X increases. Increasing the HRT, everything else being constant, decreases X and therefore A_{min} decreases. This shows that, while in terms of reactor volume there is an obvious benefit in working at low HRT, in terms of area of the settling tank the opposite is true and in practice the HRT cannot be chosen at a too low value otherwise the required area of the settling tank would be unfeasibly large. Increasing R increases Q_R and decreases X_R. The increase in Q_R causes an increase in the required area, while the decreases in X_R causes a decrease in the area. Therefore, there will be in general an optimum value of R for which the required area will be at its minimum.

Example 4.7: Minimum area of the settling tank for an activated sludge process

Calculate the minimum area of the settling tank for the activated sludge process of Example 4.1, for an influent flow rate of 10,000 m³/day and the following design parameters: HRT = 0.25 day, SRT = 15 day, R = 1.

Assume that the settling velocity of the sludge is given by the following equation:

$u_C\left(m/h\right) = 5.5e^{-0.64\,X}$ where X is the biomass concentration in kg/m³.

Solution

The minimum area of the settling tank is given by Equation 4.54. From Example 4.1, we have, for the specified design parameters:

$$Q = 10,000\frac{m^3}{day}, \quad Q_R = 10,000\frac{m^3}{day}, \quad X = 2.25\frac{kg}{m^3} \quad X_R = 4.46\frac{kg}{m^3}$$

We have:

$$u_U = 5.5 \cdot 24 e^{-0.64 \cdot 4.46} = 7.6\frac{m}{day}$$

We need to plot the function $(1/X - 1/X_R)/u_C - u_U$ between X and X_R and find its maximum value. This is shown in Figure 4.23.

The maximum value of the function $(1/X - 1/X_R)/u_C - u_U$ between X = 2.25 kg/m³ and X_R = 4.46 kg/m³ is 0.0103 kg/m².day. Using Equation 4.54 this gives a minimum area of 464 m². This corresponds to a diameter of the settling tank of approx. 12.1 m.

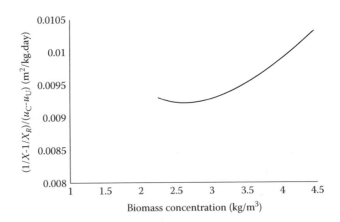

FIGURE 4.23 Calculation of the minimum required area of the settling tank.

Example 4.8: Effect of the design parameters on the minimum area required for settling

For the activated sludge process of Example 4.1, calculate the effect of the design parameters SRT, HRT and R on the minimum area required for settling. Assume an influent flow rate to the plant of 10,000 m³/day.

Solution

The solution is obtained by solving the mass balances 4.13 through 4.15 that define the values of the process variables for various values of the design parameters SRT, HRT and R. Then for each set of design parameters, the corresponding values of the process variables are used in Equation 4.54 to calculate the minimum required area, as shown in Example 4.7.

For fixed values of HRT = 0.25 day and $R = 1$, Figure 4.24 shows the effect of the SRT on the required area of the settling tank. As expected, the required area increases because the biomass concentrations X and X_R increase with the SRT.

For fixed values of SRT = 15 day and $R = 1$, Figure 4.25 shows the effect of the HRT on the required area. The required area decreases when the HRT increases up to a value of HRT approximately equal to 0.5 day, then it slightly increases as the HRT increases. At low values of the HRT the most important effect is the decrease in biomass

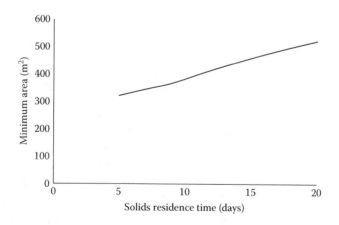

FIGURE 4.24   Effect of the solids residence time (SRT) on the minimum required area of the settling tank. Hydraulic residence time (HRT) = 0.25 day, recycle ratio $(R) = 1$.

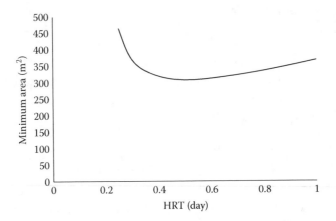

FIGURE 4.25   Effect of the hydraulic residence time (HRT) on the minimum required area of the settling tank. Solids residence time (SRT) = 15 day, recycle ratio (R) = 1.

concentration X as HRT increases, causing a decrease of the area. However, as the HRT increases further the effect of decreasing X_R when HRT increases also becomes important. This counterbalances the effect of the reduction in X and causes a slight increase in the required area.

For fixed values of SRT = 15 day and HRT = 0.25 day, Figure 4.26 shows the effect of the recycle ratio R. At low values of R the required area decreases as R increases, because increasing R causes a decrease in X_R. However, at higher values of R the required area

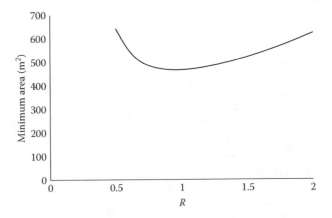

FIGURE 4.26   Effect of the recycle ratio (R) on the minimum required area of the settling tank. Solids residence time (SRT) = 15 day, hydraulic residence time (HRT) = 0.25 day.

increases because the main effect of increasing R becomes the increase in Q_R, which causes an increase in the flux applied to the settling tank.

4.1.5 pH Calculation

The basis for pH calculation is the charge balance. In this section, a simplified procedure to calculate pH in the biological reactor of an activated sludge process for carbon removal is described. The procedure is based on the assumptions that the influent wastewater does not contain weak acid or bases with the exception of carbonic acid and that the only reaction which can alter pH is the carbonic acid equilibrium and carbon dioxide dissolution/stripping. As we have seen in Chapter 2, under these assumptions the charge balance for the biological reactor can be written as follows, where all the concentrations refer to the biological reactor:

$$\left[H_3O^+\right]+\left[\Sigma Cat\right]=\left[HCO_3^-\right]+2\left[CO_3^{2-}\right]+\left[\Sigma An\right]+\left[OH^-\right] \quad (4.55)$$

The charge balance (Equation 4.55), taking into account the equilibrium of carbonic acid as seen in Chapter 2, can be re-written as:

$$\left[H_3O^+\right]+\left[\Sigma Cat-\Sigma An\right]$$

$$= \frac{\left[H_2CO_{3,tot}\right]}{1+K_{CO2}+\dfrac{K_{H2CO3}}{\left[H_3O^+\right]}+\dfrac{K_{HCO3}K_{H2CO3}}{\left[H_3O^+\right]^2}}$$

$$\left(\dfrac{K_{H2CO3}}{\left[H_3O^+\right]}+2\dfrac{K_{HCO3}K_{H2CO3}}{\left[H_3O^+\right]^2}\right)+\dfrac{K_w}{\left[H_3O^+\right]} \quad (4.56)$$

To calculate the pH of the reactor, that is $[H_3O^+]$, we need the values of $\left[\Sigma Cat-\Sigma An\right]$ and of $H_2CO_{3,tot}$. $\left[\Sigma Cat-\Sigma An\right]$ can be assumed equal to its value in the feed, assuming that these ions are inerts, that is they are not consumed nor removed, in the process. $H_2CO_{3,tot}$ is different in the reactor and in the feed, because of the generation and stripping of carbon dioxide due to the growth and endogenous metabolism of the microorganisms, and needs to be calculated by means of mass balances.

The first step of the procedure is to calculate the feed composition in terms of $\left[\Sigma Cat-\Sigma An\right]$ and of $H_2CO_{3,tot}$. This can be done if the values of pH and total alkalinity are known for the feed. As described in Chapter 2, assuming that the only species that contribute to the alkalinity of the feed are the strong acid and bases, bicarbonate and carbonate, if the pH and alkalinity of the feed are known, $[\Sigma Cat - \Sigma An]_{feed}$ and $[H_2CO_{3,totfeed}]$ can be calculated as follows:

$$[\Sigma Cat - \Sigma An]_{feed} = [Alk_{mol}]_{feed} + \frac{K_w}{10^{-4.5}} - 10^{-4.5} \tag{4.57}$$

$$[H_2CO_{3,tot}]_{feed}$$

$$= \frac{10^{-pHfeed} - \dfrac{K_w}{10^{-pHfeed}} + [\Sigma Cat - \Sigma An]_{feed}}{\dfrac{K_{H2CO3}}{10^{-pHfeed}} + 2\dfrac{K_{HCO3}K_{H2CO3}}{10^{-2pHfeed}}} \tag{4.58}$$

$$\left(1 + K_{CO2} + \frac{K_{H2CO3}}{10^{-pHfeed}} + \frac{K_{HCO3}K_{H2CO3}}{10^{-2pHfeed}}\right)$$

Once the values for the feed $[\Sigma Cat - \Sigma An]$ and $H_2CO_{3,totfeed}$ are known, the pH in the reactor can be calculating by simultaneously solving the mass balances for $H_2CO_{3,tot}$ and the charge balance. However, the writing of the mass balance for H_2CO_{3tot} depends on the way the inorganic carbon is transferred from the liquid to the gas phase. We distinguish two cases: aeration with diffusers and mechanical aeration.

a. Aeration with diffusers.

In this case, we assume that carbon dioxide is transferred from the liquid phase to the air bubbles that are present in the reactor. We call p_{CO2} the partial pressure of carbon dioxide in the air bubbles and $[CO_2]$ (mol/L or kmol/m³) the concentration of carbon dioxide in the liquid phase. The rate of carbon dioxide transfer from the liquid to the gas phase is given by:

$$r_{CO2transfer}\left(\frac{mol\,CO_2}{m^3 \cdot day}\right) = k_L a_{CO2}\left([CO_2] - k_{eqCO2}\,p_{CO2}\right) \tag{4.59}$$

where k_{eqCO2} is the equilibrium constant of the gas–liquid partition of carbon dioxide into water:

$$k_{eqCO2}\left(\frac{kmol}{m^3.atm}\right) = \left(\frac{[CO_2]}{p_{CO2}}\right)_{eq} \tag{4.60}$$

In Equation 4.59 we assume, as usual, that the various forms of dissolved carbonic acid are in equilibrium with each other, and therefore, the concentration of dissolved carbon dioxide can be written as (as seen in Chapter 2):

$$[CO_2] = \frac{K_{CO2}\left[H_2CO_{3,tot}\right]}{1 + K_{CO2} + \dfrac{K_{H2CO3}}{10^{-pH}} + \dfrac{K_{HCO3}K_{H2CO3}}{10^{-2pH}}}$$

And so Equation 4.59 becomes:

$$r_{CO2transfer}\left(\frac{\text{kmol } CO_2}{\text{m}^3 \cdot \text{day}}\right) = k_L a_{CO2}$$

$$\left(\frac{K_{CO2}\left[H_2CO_{3,tot}\right]}{1 + K_{CO2} + \dfrac{K_{H2CO3}}{10^{-pH}} + \dfrac{K_{HCO3}K_{H2CO3}}{10^{-2pH}}} - k_{eqCO2}p_{CO2}\right) \tag{4.61}$$

To develop the mass balances for H_2CO_{3tot} in the liquid and in the gas phase we will use the scheme as shown in Figure 4.27. With reference to Figure 4.27, the balance for $H_2CO_{3,tot}$ in the liquid phase can be written as:

$$QH_2CO_{3totfeed} + RQH_2CO_{3,tot} + r_{Co2biomass}V$$

$$= k_L a_{CO2}\left(\frac{K_{CO2}\left[H_2CO_{3,tot}\right]}{1 + K_{CO2} + \dfrac{K_{H2CO3}}{10^{-pH}} + \dfrac{K_{HCO3}K_{H2CO3}}{10^{-2pH}}} - k_{eqCO2}p_{CO2}\right) \tag{4.62}$$

$$V + (1 + R)QH_2CO_{3,tot}$$

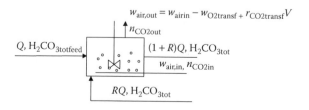

FIGURE 4.27 Scheme of the biological reactor showing the various contribution to the inorganic carbon balance in the gas and in the liquid phase.

where $r_{CO2biomass}V$ is the rate of CO_2 generation (kmol/day) due to the biological processes in the reactor. With the elimination of the term $R \cdot QH_2CO_{3,tot}$, Equation 4.62 immediately becomes:

$$QH_2CO_{3,totfeed} + r_{Co2biomass}V$$

$$= k_L a_{CO2}\left(\frac{K_{CO2}\left[H_2CO_{3,tot}\right]}{1 + K_{CO2} + \dfrac{K_{H2CO3}}{10^{-pH}} + \dfrac{K_{HCO3}K_{H2CO3}}{10^{-2pH}}} - k_{eqCO2}p_{CO2}\right) \quad (4.63)$$

$$V + QH_2CO_{3,tot}$$

Since we are considering a fully designed activated sludge process, we know the values of, or are able to calculate, $r_{co2biomass}V$ (see below), $Q, R, H_2CO_{3,totfeed}, V$. Equation 4.63 needs to be solved simultaneously with the charge balance (4.55). However, these two equations have three unknowns, that is pH, $[H_2CO_{3,tot}]$ and p_{CO2}. We therefore need a third equation, and this is the balance of $H_2CO_{3,tot}$ in the gas phase, which can be written as follows:

$$n_{CO2,in} + k_L a_{CO2}\left(\frac{K_{CO2}\left[H_2CO_{3,\,tot}\right]}{1 + K_{CO2} + \dfrac{K_{H2CO3}}{10^{-pH}} + \dfrac{K_{HCO3}K_{H2CO3}}{10^{-2pH}}} - k_{eqCO2}p_{CO2}\right) \quad (4.64)$$

$$V = n_{CO2,out}$$

The rate with which CO_2 enters and leaves the biological reactor, $n_{CO2,in}$ and $n_{CO2,out}$ (kmol/day) are given by:

$$n_{CO2,in} = \frac{w_{air,in}}{MW_{air}} \cdot \frac{p_{CO2,in}}{p_{tot}} \cdot 1000 \quad (4.65)$$

$$n_{CO2,out} = \frac{w_{air,out}}{MW_{air}} \cdot \frac{p_{CO2}}{p_{tot}} \cdot 1000$$

$$= \left(\frac{\dfrac{w_{air,in}}{MW_{air}} \cdot 1000 - \dfrac{w_{O2,transf}}{MW_{O2}} \cdot 1000 + k_L a_{CO2}}{\left(\dfrac{K_{CO2}\left[H_2CO_{3,tot}\right]}{1 + K_{CO2} + \dfrac{K_{H2CO3}}{10^{-pH}} + \dfrac{K_{HCO3}K_{H2CO3}}{10^{-2pH}}} - k_{eqCO2}p_{CO2}\right)V}\right)\dfrac{p_{CO2,out}}{p_{tot}} \quad (4.66)$$

where, consistently with the notation used in previous sections, $w_{air,in}$ is the inlet air flow rate in kg/day. $w_{O2,transf}$ is the rate of oxygen transfer from the gas to the liquid phase, which is equal to $w_{O2,transf}\left(kg\,O_2/day\right)=k_L a\left(k_{eqO2}\,p_{O2}-C_{O2}\right)V$ and can be immediately calculated if the aeration system has been designed according to what described in Section 4.1.3.

Therefore, the balance for $H_2CO_{3,\,tot}$ in the gas phase can be re-written as:

$$\frac{w_{air,in}}{MW_{air}}\cdot\frac{p_{CO2,in}}{p_{tot}}\cdot 1000+k_L a_{CO2}$$

$$\left(\frac{K_{CO2}\left[H_2CO_{3,tot}\right]}{1+K_{CO2}+\dfrac{K_{H2CO3}}{10^{-pH}}+\dfrac{K_{HCO3}K_{H2CO3}}{10^{-2pH}}}-\beta_{CO2}\,p_{CO2}\right)V$$

$$\text{(4.67)}$$

$$=\frac{\left(\dfrac{w_{air,in}}{MW_{air}}\cdot 1000-\dfrac{w_{O2,transf}}{MW_{O2}}\cdot 1000+k_L a_{CO2}\right)}{\left(\dfrac{K_{CO2}\left[H_2CO_{3,tot}\right]}{1+K_{CO2}+\dfrac{K_{H2CO3}}{10^{-pH}}+\dfrac{K_{HCO3}K_{H2CO3}}{10^{-2pH}}}-k_{eqCO2}\,p_{CO2}\right)V}\cdot\frac{p_{CO2,out}}{p_{tot}}$$

Equations 4.56, the charge balance, (4.63), balance of $H_2CO_{3,tot}$ in the liquid phase, and (4.67), balance of $H_2CO_{3,tot}$ in the gas phase, constitute a system of three equations in the three unknowns $H_2CO_{3,tot}$, pCO_2, pH. The solution of this system of equations gives the pH of the biological reactor when aeration is carried out with diffusers.

The rate of CO_2 generation due to biological processes in the reactor, $r_{CO2,biomass}V$, which is present in Equation 4.63, can be calculated from the rate of substrate removal and biomass production, if the carbon content of the substrate and of the biomass is known. E.g. assuming the influent substrate is glucose ($C_6H_{12}O_6$) and that the biomass composition is, as usual, $C_5H_7O_2N$, $r_{CO2,biomass}V$ can be calculated as follows:

$$r_{CO2,biomass}V\left(\frac{kmol}{day}\right)=\frac{Q\left(S_0-S\right)}{1.067}\times\frac{6}{180}-Q_w X_R\times\frac{5}{113}\qquad\text{(4.68)}$$

The term $Q(S_0 - S)/1.067 \cdot 6/180$ represents the kmol/day of carbon removed by the biological process, where the factor 1.067 is the conversion factor of glucose into COD (S is expressed as kg COD/m³), 180 is the molecular weight of glucose and 6 is the number of carbon atoms in 1 mol of glucose. The term $Q_w X_R \times 5/113$ represents the kmol/day of carbon assimilated into biomass. The difference between these two terms represents the rate (in kmol/day) of carbon dioxide generation by the biological process.

A particular case of the aeration with diffusers is if we assume that the gas and the liquid phase are in equilibrium for the carbon dioxide concentration. This case can be solved by using Equations 4.56, 4.63 and 4.6) with a very large value of the mass transfer coefficient $k_L a_{CO2}$. Another way of calculating the pH in this case is by re-writing the $H_2CO_{3,tot}$ balance in the liquid phase as:

$$QH_2CO_{3,totfeed} + r_{Co2,biomass} V = Q_{gas} \cdot \frac{p_{CO2}}{P_{tot}} \frac{P_{tot}}{0.0821 \cdot T} + QH_2CO_{3,tot} \quad (4.69)$$

where Q_{gas} is the flow rate (m³/day) of gas that leaves the biological reactor, 0.0821 (L.atm/mol. K) is the ideal gas constant and T is the temperature of the system in K. Since carbon dioxide is in equilibrium between the gas and the liquid phase and we always assume the equilibrium of inorganic carbon we have:

$$p_{CO2} = \frac{[CO_2]}{k_{eqCO2}} = \frac{\dfrac{K_{CO2}}{\beta_{CO2}}[H_2CO_{3,tot}]}{1 + K_{CO2} + \dfrac{K_{H2CO3}}{10^{-pH}} + \dfrac{K_{HCO3} K_{H2CO3}}{10^{-2pH}}} \quad (4.70)$$

and Equation 4.69 becomes:

$$QH_2CO_{3totfeed} + r_{Co2biomass} V = Q_{gas} \cdot$$

$$\frac{\dfrac{K_{CO2}}{\beta_{CO2}}[H_2CO_{3,tot}]}{1 + K_{CO2} + \dfrac{K_{H2CO3}}{10^{-pH}} + \dfrac{K_{HCO3} K_{H2CO3}}{10^{-2pH}}} \frac{1}{0.0821 \times T} + QH_2CO_{3,tot} \quad (4.71)$$

Equation 4.71 needs to be solved together with the charge balance (4.56); however, this is not possible yet since we do not have an equation from Q_{gas}. The equation for Q_{gas} can be written as follows:

$$Q_{gas}\left(\frac{m^3}{day}\right) = \frac{w_{air,in} - w_{O2transf}}{\rho_{air}}$$

$$+ \frac{r_{CO2biomass}V + n_{CO2in} + QH_2CO_{3totfeed} - QH_2CO_{3,tot}}{\rho_{CO2}}$$

(4.72)

In Equation 4.72 the term $w_{air,in} - w_{O2transf}/\rho_{air}$ corresponds to the influent air minus the oxygen that has been transferred to the liquid phase, and the term $r_{CO2biomass}V + n_{CO2in} + QH_2CO_{3totfeed} - QH_2CO_{3,tot}/\rho_{CO2}$ corresponds to the inorganic carbon that has been transferred to the gas phase, calculated as the difference between the total inorganic carbon entering the system or generated in it.

$(r_{CO2biomass}V + n_{CO2in} + QH_2CO_{3totfeed})$ and the total inorganic carbon leaving the system (QH_2CO_{3tot}). ρ_{air} and $\rho CO2$ are the density of the air and carbon dioxide at the conditions of the system, in kg/m^3.

Equation 4.72 can be substituted into (4.71) to give:

$$QH_2CO_{3totfeed} + r_{Co2biomass}V$$

$$= \frac{w_{air,in} - w_{O2transf}}{\rho_{air}}$$

$$+ \frac{r_{CO2biomass}V + n_{CO2in} + QH_2CO_{3totfeed} - QH_2CO_{3tot}}{\rho_{CO2}}$$

(4.73)

$$\cdot \frac{\dfrac{K_{CO2}}{k_{eqCO2}}\left[H_2CO_{3,tot}\right]}{1 + K_{CO2} + \dfrac{K_{H2CO3}}{10^{-pH}} + \dfrac{K_{HCO3}K_{H2CO3}}{10^{-2pH}}} \frac{1}{0.0821 \times T} + QH_2CO_{3tot}$$

Equation 4.73 and the charge balance (4.56) constitute a system of two equations with the two unknowns H_2CO_{3tot} and pH.

b. Aeration with mechanical aerators.

When there are no diffusers but instead aeration is carried out by mechanical aerators, carbon dioxide is transferred from the liquid phase to the atmosphere. The concentration of carbon dioxide in the atmosphere can be considered constant and not dependent on the amount of carbon dioxide transferred from the biological reactor. In this case, the rate of carbon dioxide transfer from the biological reactor to the atmosphere can be written as:

$$r_{CO2transfer} \left(\frac{kmol\ CO_2}{m^3 \cdot day} \right) = k_L a_{CO2}$$

$$\left(\frac{K_{CO2} \left[H_2CO_{3,tot} \right]}{1 + K_{CO2} + \dfrac{K_{H2CO3}}{10^{-pH}} + \dfrac{K_{HCO3} K_{H2CO3}}{10^{-2pH}}} - k_{eqCO2}\, p_{CO2atm} \right) \quad (4.74)$$

where p_{CO2atm} is the constant partial pressure of carbon dioxide in the atmosphere, which here will be taken equal to 0.0004 atm. The balance of the inorganic carbon H_2CO_{3tot} in the liquid phase can therefore be written as:

$$QH_2CO_{3totfeed} + r_{Co2biomass}\, V$$

$$= k_L a_{CO2} \left(\frac{K_{CO2} \left[H_2CO_{3,tot} \right]}{1 + K_{CO2} + \dfrac{K_{H2CO3}}{10^{-pH}} + \dfrac{K_{HCO3} K_{H2CO3}}{10^{-2pH}}} - k_{eqCO2}\, p_{CO2atm} \right) \quad (4.63)$$

$$V + QH_2CO_{3tot}$$

Equation 4.63 can be immediately solved together with the charge balance (4.56) without the need of other equations, since the only unknowns are now pH and H_2CO_{3tot}. Note that for a given $k_L a_{CO2}$, the transfer of carbon dioxide to the gas phase is faster with mechanical aerators than with diffusers, because with mechanical aerators the concentration of carbon dioxide in the gas phase will stay at its lowest level, that is the atmospheric value, while with diffusers it will be higher than this value. Therefore, the driving force for mass transfer of carbon dioxide will be larger for mechanical aerators than with diffusers. However, in general the mass transfer coefficient is often lower for mechanical aerators than for diffusers, because of the larger transfer area of the gas bubbles in the latter system, and this counterbalances the effect mentioned above.

Example 4.9: pH calculation in the activated sludge process for carbon removal

Calculate the pH in the biological reactor of an activated sludge process for carbon removal. Assume the substrate is readily bio-degradable (glucose) and assume the values of the kinetic parameters given in Example 4.1. Assume that the influent flow rate is 1000 m^3/day. Consider, as a base case, that the substrate concentration in the feed is 0.5 gCOD/L and that the plant has been designed with an SRT = 10 day, HRT = 0.5 day and $R = 1$. Calculate the pH considering both cases, aeration with diffusers and mechanical aeration, assuming values of $k_L a_{CO2}$ in the range 0–100 day^{-1} and an inlet air flow rate in the range 2500–5000 kg air/day. Calculate the pH for a feed having an alkalinity value in the range 5–1000 mgCaCO$_3$/L and a pH in the range 7–8.

Also, calculate the pH if the plant is designed with an SRT = 2 day (same HRT and R as before). Also, calculate the pH if the substrate concentration in the feed is 0.25 gCOD/L, repeating the design with SRT = 10 day, HRT = 0.5 day and $R = 1$.

Solution

The first step is to calculate the values of $[\Sigma Cat - \Sigma An]$ and $H_2CO_{3totfeed}$ for various values of the feed pH and alkalinity. This is obtained by using Equations 4.57 and 4.58 and the results are tabulated in Table 4.2 (note that in this case we are assuming that the only species that contribute to the alkalinity of the feed are strong acids and bases, bicarbonate and carbonate).

The design of the process can be obtained as shown in Section 4.1, by simultaneously solving the design Equations 4.13, 4.14 and 4.15. For the base case, we obtain:

S = 0.21 mg COD/L; X = 1.0 kg/m^3; X_R = 1.95 kg/m^3; Q_W = 142.7 m^3/day; Biological oxygen consumption = 350 kg/day; V = 500 m^3. From these values, we calculate (Equation 4.68): $r_{CO2biomass} V$ = 13.4 kmol/day.

If SRT = 2 day^{-1}, with the same values of the HRT and R, we obtain:

S = 0.53 mg COD/L; X = 0.43 kg/m^3; X_R = 0.75 kg/m^3; Q_W = 25.6 m^3/day; Biological oxygen consumption = 429 kg/day; V = 500 m^3. From these values, we calculate: $r_{CO2biomass} V$ = 10.9 kmol/day.

First we consider the case of aeration with diffusers.

TABLE 4.2 Calculation of $[\Sigma Cat - \Sigma An]$ and $H_2CO_{3,totfeed}$ for a Feed of Different Values of pH and Alkalinity

Alkalinity (mg CaCO$_3$/L)	$[\Sigma Cat - \Sigma An]$ (mol/L)	H$_2$CO$_{3totfeed}$ (mol/L)
pH 7		
5	6.84E–5	8.44E–5
25	4.68E–4	5.78E–4
50	9.68E–4	1.2E–3
100	1.97E–3	2.43E–3
250	4.97E–3	6.14E–3
500	9.97E–3	1.23E–2
1000	2.00E–2	2.47E–2
pH 8		
5	6.84E–5	6.86E–5
25	4.68E–4	4.68E–4
50	9.68E–4	9.85E–4
100	1.97E–3	2.00E–3
250	4.97E–3	5.06E–3
500	9.97E–3	1.02E–2
1000	2.00E–2	2.03E–2

Note that the values of inlet air flow rate considered in this example, in the range 2500–5000 kg/day are compatible with the oxygen consumption by the microorganisms. Indeed, they correspond to an oxygen mass flow rate of 575 and 1150 kg/day, respectively, which are larger than the oxygen consumption rate by the microorganisms for both values of the SRT. However, it should be verified that the mass transfer coefficient for oxygen, $k_L a$, for these air flow rates is large enough to allow for the required rate of oxygen transfer (as discussed in Section 4.1.3) but this is not considered here. Also, note that in this example the air flow rate and the parameter $k_L a_{CO2}$ are varied independently, while in reality $k_L a_{CO2}$ is dependent on the air flow rate.

Without assuming equilibrium between the carbon dioxide in the gas and the liquid phase, the pH in the biological reactor is obtained by the solution of the system of Equations 4.56, 4.63 and 4.67. Figure 4.28 plots the pH of reactor as a function of the feed alkalinity for different values of the parameter $k_L a_{CO2}$.

Figure 4.28 shows the strong effect that the alkalinity of the feed has on the pH in the biological reactor. Due to the generation of carbon dioxide in the biological reaction, the pH in the reactor tends

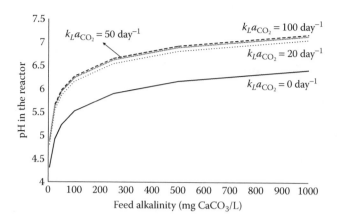

FIGURE 4.28   pH in the reactor at various values of $k_L a_{CO2}$. Aeration with diffusers, $S_0 = 0.5$ kg COD/m³, $w_{air,in} = 5000$ kg/day, $pH_{feed} = 7$, solids residence time (SRT) = 10 day.

to be more acidic than the feed. The presence of cations in the feed, which increases with the alkalinity of the feed (Table 4.2) tends to counterbalance the drop in pH caused by the carbon dioxide generation. Therefore, the higher the alkalinity of the feed, the higher or less acidic is the pH in the reactor. Also, very important is the mass transfer of carbon dioxide to the gas phase. In the absence of any transfer ($k_L a_{CO2} = 0$), all the carbon dioxide generated remains in the liquid phase, and so the pH in the reactor is the most acidic. When carbon dioxide transfers to the gas phase ($k_L a_{CO2} \neq 0$), the pH is higher than in the absence of mass transfer, because an acid (carbon dioxide) is being removed from the system. Note that above a certain value of $k_L a_{CO2}$ (in this example above $k_L a_{CO2} = 50$ day⁻¹), the increase in pH is very low, this is because we tend to the equilibrium condition between the gas and liquid phase, and there is no benefit in increasing the carbon dioxide mass transfer rate further.

Figure 4.29 shows the same calculations shown in Figure 4.28 but showing the effect of the air flow rate fed to the reactor. It is evident that decreasing the air flow rate tends to give a lower pH, because there is less stripping of the carbon dioxide, even though the effect is quite modest, at least with the parameter values used in this example.

Figure 4.30 shows the effect of the pH of the feed on the pH in the reactor. This effect is quite modest, especially for low values of the alkalinity of the feed. The reason is that, when the alkalinity of

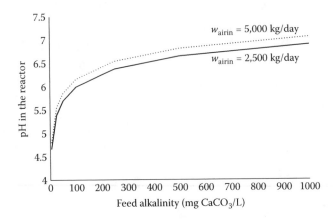

FIGURE 4.29 Effect of the air flow rate on the pH in the reactor. Aeration with diffusers, $S_0 = 0.5$ kg COD/m^3, $k_L a_{CO2} = 20$ day^{-1}, pH$_{feed} = 7$, solids residence time (SRT) = 10 day.

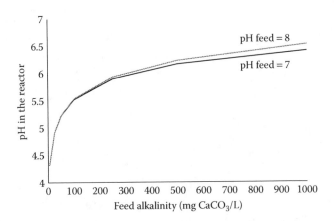

FIGURE 4.30 Effect of the pH of the feed on the pH in the biological reactor. Aeration with diffusers, $S_0 = 0.5$ kg COD/m^3, $k_L a_{CO2} = 0$, $w_{air,in} = 5000$ kg/day, solids residence time (SRT) = 10 day.

the feed is very low, both $[\Sigma Cat - \Sigma An]_{feed}$ and $[H_2 CO_{3tot}]_{feed}$ are very low and so the pH of the reactor is essentially determined by the carbon dioxide generated by the biological reactor, which does not depend on the pH of the feed. For higher values of the alkalinity of the feed, the pH in the reactor is slightly higher when the pH of the feed increases, because this corresponds to less total inorganic carbon, $H_2 CO_{3tot}$, which is a weak acid.

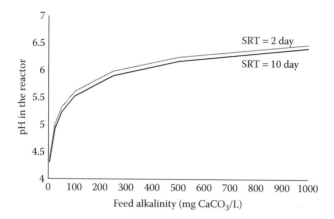

FIGURE 4.31 Effect of the solids residence time (SRT) on the pH in the biological reactor. Aeration with diffusers, $S_0 = 0.5$ kg COD/m^3, pH$_{feed} = 7$, $k_L a_{CO2} = 0$ day^{-1}, $w_{air,in} = 5000$ kg/day.

Figure 4.31 compares the effect of the design SRT. With SRT = 2 day, the pH is slightly higher than with SRT = 10 day, because of the lower generation of carbon dioxide by the microorganisms, which is due to the shorter residence time of the microorganisms, which gives a lower rate of endogenous metabolism and so lower carbon dioxide generation. This effect is however quite modest.

Figure 4.32 shows how the calculated pH changes if we make the hypothesis that the carbon dioxide concentration is in equilibrium

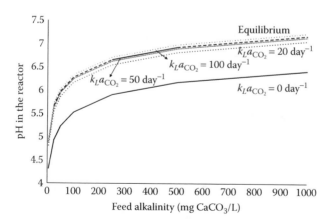

FIGURE 4.32 Effect of the assumption of gas–liquid equilibrium for carbon dioxide on the pH in the reactor. Aeration with diffusers, $S_0 = 0.5$ kg COD/m^3, $w_{air,in} = 5000$ kg/day, pH$_{feed} = 7$, solids residence time (SRT) = 10 day.

between the gas and the liquid phase. In this case, the pH has been calculated as the solution of Equations 4.56 and 4.73. As expected, Figure 4.32 shows that the pH in the equilibrium assumption tends to be equal to the pH obtained without the equilibrium assumption, but with high k_La_{CO2} values. Note however, that, at least with the values used in this example, the equilibrium conditions for carbon dioxide are reached even with relatively modest k_La_{CO2} values, that is equilibrium is approximated quite well even with a k_La_{CO2} equal to 50 day^{-1}. These values of k_La_{CO2} are easily reached in activated sludge processes, and this indicates that equilibrium for carbon dioxide between the gas and the liquid phase is probably reached in many plants.

All the calculations done so far refer to the case of aeration with diffusers. Figure 4.33 shows how the pH changes if we assume aeration is provided by mechanical aerators. It is evident, as expected, that the pH in the reactor is higher with mechanical aerators than with diffusers, for the same value of k_La_{CO2}. The reason is, as explained above, that the driving force for the mass transfer of carbon dioxide is larger with mechanical aerators, because equilibrium is not reached between the liquid and the gas phase. For this reason, the pH in the reactor tends to increase even with large values of the mass transfer coefficient for carbon dioxide, because equilibrium is never reached. On the other hand, with diffusers the air bubbles tend

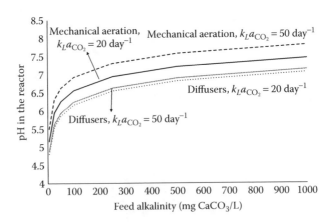

FIGURE 4.33 Comparison of the effect of mechanical aerators and diffusers on the pH in the reactor. $S_0 = 0.5$ kg COD/m^3, pH$_{feed} = 7$, solids residence time (SRT) = 10 day, for aeration with diffusers $w_{air,in} = 5000$ kg/day.

to saturate with carbon dioxide and therefore, above a certain value of $k_L a_{CO2}$, the mass transfer rate cannot increase further.

It is interesting to observe what happens if the design influent concentration for the plant is $S_0 = 0.25$ kg COD/m³ (instead than 0.5 kg COD/m³, as assumed so far). With the design parameters: SRT = 10 day; HRT = 0.5 day; $R = 1$ day, we obtain:

$S = 0.21$ mg COD/L; $X = 0.51$ kg/m³; $X_R = 0.99$ kg/m³; $Q_W = 25.4$ m³/day; biological oxygen consumption = 214 kg/day; $V = 500$ m³. From these values, we calculate: $r_{CO2biomass}V = 6.7$ kmol/day.

With these values, assuming aeration with diffusers, Figure 4.34 compares the pH in the reactor for an influent COD of 0.25 versus 0.5 (previous base case) kg COD/m³. It is evident that the pH in the reactor is higher (less acidic) for the lowest COD concentration in the feed. This is due to the lower generation of carbon dioxide.

4.1.6 Extension to Slowly Biodegradable Substrates

Often, the feed to biological processes is also, or mainly, composed of slowly biodegradable substrate. Here, we will assume that the slowly biodegradable substrate (X_S) is totally soluble and therefore, like the readily biodegradable substrate, its concentration does not change in the settling tank. According to the model seen in Chapter 2, we will assume that the slowly biodegradable substrate is hydrolysed by the biomass to the readily biodegradable substrate, which is then metabolised.

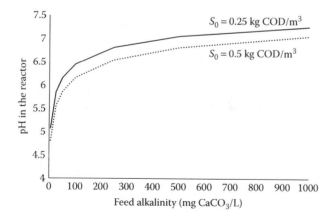

FIGURE 4.34 Effect of the substrate concentration in the feed on the pH in the biological reactor. Aeration with diffusers, $w_{air,in} = 5000$ kg/day, $pH_{feed} = 7$, solids residence time (SRT) = 10 day, $k_L a_{CO2} = 20$ day⁻¹.

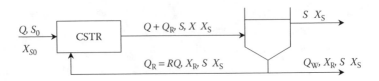

FIGURE 4.35 Scheme of an activated sludge process with both readily and slowly biodegradable substrates in the feed.

With reference to Figure 4.35, the design of this process can be done, similarly as for all the other biological wastewater treatment processes, by writing the mass balances for biomass and substrates. Since in this case we have two substrates, we have to write the mass balances for both the readily and the slowly biodegradable substrates. Therefore, in this case we will have to write four mass balances: biomass in the reactor, biomass in the whole system, readily biodegradable substrate in the reactor, and slowly biodegradable substrate in the whole system. In total, therefore, we have now four mass balances with the seven unknowns V, Q_R (or R), S, X, X_R, Q_W, X_S.

The rate equations for biomass growth, endogenous metabolism and readily biodegradable substrate removal are the same used in Section 4.1 (and seen in Chapter 2). The rate equation for the hydrolysis of slowly biodegradable substrate is the same seen in Chapter 2, that is:

$$r_{hydr}\left(\frac{kg\ COD}{m^3 day}\right) = -k_h \frac{(X_S/X)}{K_X+(X_S/X)} X$$

The mass balances are written below.

Biomass in the reactor:

$$(\mu - b) XV + Q_R X_R = (Q+Q_R) X \qquad (4.4)$$

Slowly biodegradable substrate in the reactor:

$$QX_{S0} + Q_R X_S = (Q+Q_R) X_S + k_h \frac{X_S/X}{K_X+(X_S/X)} XV \qquad (4.75)$$

Readily biodegradable substrate in the reactor:

$$QS_0 + Q_R S + k_h \frac{X_S/X}{K_X+(X_S/X)} XV = (Q+Q_R)S + \frac{\mu XV}{Y_{X/S}} \qquad (4.76)$$

Balance for the biomass in the whole system (reactor + settling tank):

$$(\mu - b) XV = Q_W X_R \tag{4.6}$$

The equations can now be re-arranged introducing the HRT, SRT and R in the same way as it was done for the case with only readily biodegradable substrate.

$$(\mu - b) + \frac{RX_R}{X \cdot HRT} = \frac{(1+R)}{HRT} \tag{4.13}$$

$$\frac{(X_{S0} - X_S)}{X \cdot HRT} = k_h \frac{X_S/X}{K_X + (X_S/X)} \tag{4.77}$$

$$\frac{(S_0 - S)}{X - HRT} + k_h \frac{X_S/X}{K_X + (X_S/X)} = \frac{\mu}{Y_{X/S}} \tag{4.78}$$

$$(\mu - b) = \frac{1}{SRT} \tag{4.15}$$

If the values of the design parameters SRT, HRT and R are chosen, Equations 4.13, 4.15, 4.77 and 4.78 constitute a system of four equations in the four unknowns X, S, X_S and X_R. By solving this system of equations we calculate these values of all the variables that characterise the process.

Once all the variables that define the systems are calculated, for a given influent wastewater flow rate Q we can proceed exactly as done in Section 4.1 (readily biodegradable substrates only in the feed) and use Equations 4.7, 4.11 and 4.16 to calculate the recycle flow rate, reactor volume and sludge waste flow rate. The sludge production is also immediately calculated using Equation 4.17, while for the oxygen consumption rate we need to take into account also the slowly biodegradable substrate and Equation 4.19 becomes:

$$\text{Oxygen consumption} \left(\frac{kg\ O_2}{day} \right) = Q_{O2biomass}$$

$$= Q(S_0 + X_{S0} - S - X_S) - Q_W X_R \cdot 1.42 \tag{4.79}$$

The effect of the design parameters on the performance of the process is analogous to what discussed in Section 4.1.1 for a feed composed only of readily biodegradable substrate. The effluent concentration of readily biodegradable substrate, S, is still dependent only on the SRT, and it is still given by Equation 4.22. However, an important difference with the case of only readily biodegradable substrates in the feed is that the effluent concentration of slowly biodegradable substrate, X_S, depends not only on the SRT but also on the HRT. This is evident by combining Equations 4.77 and 4.78 and re-arranging them to express X_S as:

$$X_S = X_{S0} + (S_0 - S) - \frac{\left(\frac{1}{SRT} + b\right) X \cdot HRT}{Y_{X/S}} \tag{4.80}$$

Equation (4.80) shows that the effluent concentration of slowly biodegradable substrate depends not only on the SRT but also on the HRT. X_S decreases by increasing the SRT and the HRT. The reason for the decrease of X_S with increasing HRT is that the rate of hydrolysis (which is the rate of X_S removal) per unit of biomass, r_{hydr}/X, increases with decreasing biomass concentration. Therefore, at higher values of HRT, where the biomass concentration is lower, the specific rate of hydrolysis will be higher and overall the rate of hydrolysis will be higher than at lower values of HRT. However, the extent of this effect is dependent on the parameters k_h and K_X of the rate equation of hydrolysis. Note also that the minimum required value of SRT to avoid biomass washout, SRT_{min}, is different, and higher, than in the case of readily biodegradable substrates only in the feed. The value of SRT_{min} can be calculated by solving Equation 4.80 for $X_S = X_{S0}$ and will be different than the value given by Equation 4.23.

In summary, the design of an activated sludge process for carbon removal, treating a wastewater containing also slowly biodegradable substrates in the feed, can be done with the same approach described for a feed composed only of readily biodegradable substrates. By choosing the values of the design parameters SRT, HRT and R the process is completely defined, and all the variables can be calculated by solving the appropriate mass balances. The effect of the design parameters on the performance of the process will be similar, but not identical, to the case of readily biodegradable substrates only in the feed.

Example 4.10: Design of an activated sludge process with slowly biodegradable substrates in the feed

A biological wastewater treatment plant (activated sludge process) for carbon removal has a feed entirely composed of slowly biodegradable substrates:

$$X_{S0} = 0.5 \text{ kg COD/m}^3$$

Investigate the effect of the design parameters SRT (range 0.8–20 days), HRT (range 0.25–1 days) and R (range 0.5–2) on the values of the variables X, X_R, S, per unit value of the influent flow rate. Discuss the observed trends.

Values of the kinetic parameters:

$\mu_{max} = 6$ days^{-1}

$K_S = 0.004$ kg COD/m^3

$Y_{X/S} = 0.3$ kg biomass/kg COD

$b = 0.2$ days^{-1}

$k_h = 4$ kg COD/kg biomass/day

$k_x = 0.07$ kg COD/kg biomass

Solution

The solution of this problem is obtained by solving the mass balance Equations 4.13, 4.15, 4.77 and 4.78. For fixed values of SRT, HRT and R, the system of equations can be solved to obtain S, X, X_S and X_R.

Figure 4.36 shows the effect of the SRT on the biomass and substrate concentrations. The trends are similar to what observed in Example 4.1 for readily biodegradable substrates in the feed. However, the residual substrate concentration in the effluent ($X_S + S$, almost coincident with X_S, which is $>> S$) is larger than in Example 4.1, because the rate limiting process is the hydrolysis of the slowly biodegradable substrate. Also, comparing Figure 4.36 with

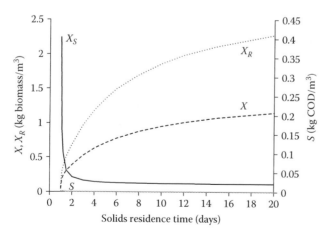

FIGURE 4.36 Effect of the solids residence time on substrate and biomass concentration in the reactor and at the bottom of the settling tank (hydraulic residence time, HRT = 0.5 day, recycle ratio, R = 1).

Figure 4.2 it can be noted that the minimum value of SRT is larger when slowly biodegradable substrates are present in the feed, as discussed above.

Figure 4.37 shows the calculated values of the waste sludge flow rate, oxygen consumption and sludge production. The trends are similar as in Example 4.1, however, as also shown in Figure 4.36, the minimum value of SRT, which corresponds to biomass washout and therefore no sludge production or oxygen consumption, is higher than for a feed only composed of readily biodegradable substrates.

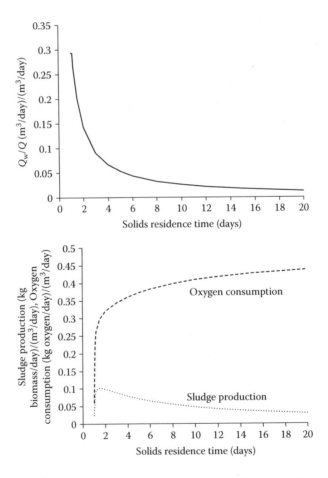

FIGURE 4.37 Effect of the solids residence time on the required waste sludge flow rate and on the sludge production and oxygen consumption rate (hydraulic residence time, HRT = 0.5 day, recycle ratio R = 1).

Figure 4.38 shows the effect of the HRT on biomass concentration and effluent substrate (X_S) concentration. The effect of HRT on the biomass concentration is similar to what discussed in Example 4.1 for readily biodegradable substrates. The effect of HRT on X_S shows, as discussed above, that X_S in the effluent decreases with increasing HRT at fixed SRT, as discussed above. Figure 4.39 shows the effect of the HRT on the biomass concentration in the recycle stream and on the required sludge waste flow rate. Figure 4.40 shows the effect of the recycle ratio. These effects are analogous to what was observed in Example 4.1.

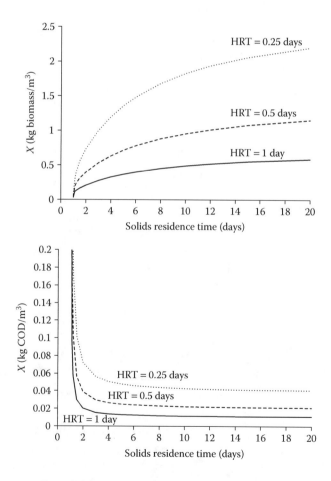

FIGURE 4.38 Effect of the solids residence time and hydraulic residence time on the biomass and substrate concentration in the biological reactor (recycle ratio, $R = 1$).

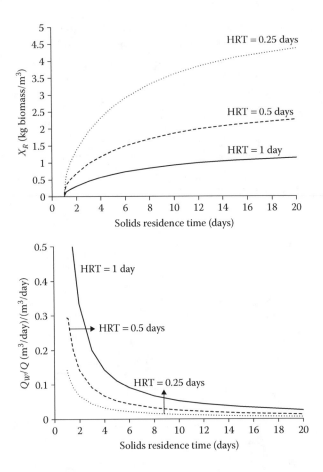

FIGURE 4.39   Effect of the solids residence time and hydraulic residence time on the biomass concentration in the recycle stream and on the required sludge waste flow rate (recycle ratio, $R = 1$).

4.1.7 The Activated Sludge Process as a Series of CSTRs

So far we have modelled the activated sludge process as one single continuous stirred tank reactor (CSTR). Sometimes, however, still focussing our attention on aerobic processes for carbon removal only, the biological process may be composed of two reactors or more reactors in series. Even though the process has only one biological reactor, the fluid dynamic profiles inside the reactor might give a condition different than perfect mixing and the reactor might be modelled better by two (or more) CSTRs in series than by a single CSTR. Figure 4.41 shows the scheme of an activated sludge process with two CSTRs in series.

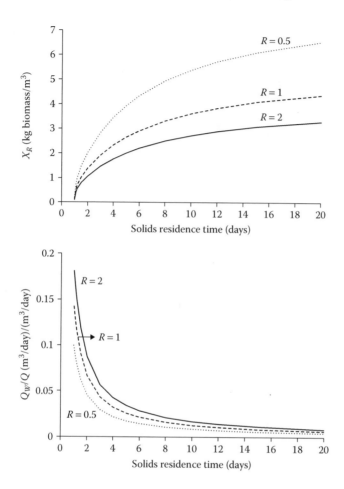

FIGURE 4.40 Effect of the solids residence time and of the recycle ratio on the biomass concentration in the recycle stream and on the required sludge waste flow rate (hydraulic residence time, HRT = 0.25 day).

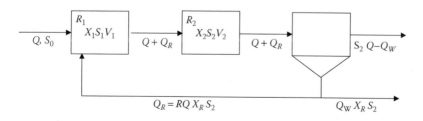

FIGURE 4.41 Scheme of an activated sludge process with two continuous stirred tank reactors in series.

If the process is composed of more than one CSTR in series, the design procedure is essentially the same seen for one single tank with the appropriate differences in the definition of HRT and SRT and in the mass balances. For a system of two tanks in series, will have two values of the HRT:

$$\text{HRT}_1 = \frac{V_1}{Q}, \quad \text{HRT}_2 = \frac{V_2}{Q} \tag{4.81}$$

And the definition of SRT is:

$$\text{SRT} = \frac{V_1 X_1 + V_2 X_2}{Q_W X_R} \tag{4.82}$$

In this case obviously the rate equations for biomass growth and endogenous metabolism and substrate removal are different in the two reactors. Indicating with subscript 1 and 2 the values for the first (R_1) and second (R_2) reactor, respectively, we have:

$$r_{X1}\left(\frac{\text{kg biomass}}{\text{m}^3\text{day}}\right) = \mu_1 X_1 = \frac{\mu_{\max} S_1}{K_S + S_1} X_1 \tag{4.83}$$

$$r_{X2}\left(\frac{\text{kg biomass}}{\text{m}^3\text{day}}\right) = \mu_2 X_2 = \frac{\mu_{\max} S_2}{K_S + S_2} X_2 \tag{4.84}$$

$$r_{S1}\left(\frac{\text{kg substrate}}{\text{m}^3\text{day}}\right) = -\frac{\mu_1 X_1}{Y_{X/S}} = -\frac{\mu_{\max} S_1}{K_S + S_1} \frac{X_1}{Y_{X/S}} \tag{4.85}$$

$$r_{S2}\left(\frac{\text{kg substrate}}{\text{m}^3\text{day}}\right) = -\frac{\mu_2 X_2}{Y_{X/S}} = -\frac{\mu_{\max} S_2}{K_S + S_2} \frac{X_2}{Y_{X/S}} \tag{4.86}$$

$$r_{\text{end1}}\left(\frac{\text{kg biomass}}{\text{m}^3\text{day}}\right) = -bX_1 \tag{4.87}$$

$$r_{\text{end2}}\left(\frac{\text{kg biomass}}{\text{m}^3\text{day}}\right) = -bX_2 \tag{4.88}$$

We need to write the mass balances for biomass and substrate in the two reactors and for biomass in the whole system. The mass balances are reported below.

Biomass in R_1:

$$\left(r_X + r_{end}\right)_1 V_1 + Q_R X_R = \left(Q + Q_R\right) X_1 \qquad (4.89)$$

Biomass in R_2:

$$\left(r_X + r_{end}\right)_2 V_2 + \left(Q + Q_R\right) X_1 = \left(Q + Q_R\right) X_2 \qquad (4.90)$$

Biomass in the whole system:

$$\left(r_X + r_{end}\right)_1 V_1 + \left(r_X + r_{end}\right)_2 V_2 = Q_W X_R \qquad (4.91)$$

Substrate in R1:

$$QS_0 + Q_R S_2 = \left(Q + Q_R\right) S_1 + \frac{r_{X1}}{Y_{X/S}} V_1 \qquad (4.92)$$

Substrate in R2:

$$\left(Q + Q_R\right) S_1 = \left(Q + Q_R\right) S_2 + \frac{r_{X2}}{Y_{X/S}} V_2 \qquad (4.93)$$

By introducing the definitions of R, SRT, HRT_1 and HRT_2, the mass balances can be rewritten as follows.
Biomass in R1:

$$\left(\mu - b\right)_1 X_1 HRT_1 + R X_R = \left(1 + R\right) X_1 \qquad (4.94)$$

Biomass in R_2:

$$\left(\mu - b\right)_2 X_2 HRT_2 + \left(1 + R\right) X_1 = \left(1 + R\right) X_2 \qquad (4.95)$$

Biomass in the whole system:

$$\frac{\left(\mu - b\right)_1}{1 + \dfrac{X_2 HRT_2}{X_1 HRT_1}} + \frac{\left(\mu - b\right)_2}{1 + \dfrac{X_1 HRT_1}{X_2 HRT_2}} = \frac{1}{SRT} \qquad (4.96)$$

Substrate in R1:

$$S_0 + R S_2 = \left(1 + R\right) S_1 + \frac{\mu_1 X_1}{Y_{X/S}} HRT_1 \qquad (4.97)$$

Substrate in R2:

$$\left(1 + R\right) S_1 = \left(1 + R\right) S_2 + \frac{\mu_2 X_2}{Y_{X/S}} HRT_2 \qquad (4.98)$$

In summary, to design an activated sludge process for carbon removal we need to set the values of SRT, HRT_1, HRT_2 and R and then solve the five Equations 4.94–4.98 obtaining the values of the five unknowns X_1, X_2, S_1, S_2 and X_R. Once these values are known, the design can be completed by calculating the values of V_1, V_2, Q_R and Q_W using Equations 4.7, 4.81 and 4.82. Example 4.11 shows the design of an activated sludge process with two CSTRs in series, comparing the results with the case of only one CSTR.

Example 4.11: Design of an activated sludge process with two reactors in series

Design an activated sludge process with two reactors in series, assuming readily biodegradable substrates in the feed and assuming the kinetic parameters of Example 4.1. Evaluate the effect of the design parameters HRT_1, HRT_2 and SRT and compare with the case of having only one reactor of a volume equal to the sum of the volumes of the two CSTRs.

Solution

The design can be done solving the system of Equations 4.94–4.98. Assuming a total HRT ($HRT_1 + HRT_2$) equal to 0.5 day, Figure 4.42 shows the effect of the relative size of the two reactors, ranging from $V_1 = 10\%$ (i.e. $HRT_1 = 0.05$ day, $HRT_2 = 0.45$ day) to $V_1 = 80\%$ (i.e. $HRT_1 = 0.40$ day, $HRT_2 = 0.10$ day) of the total reaction volume. For low value of SRT and a low volume fraction of R_1, there is a considerable residual substrate concentration in R_1, which is however, almost entirely degraded in R_2 (note that S_2 is always much lower than S_1). It is important to observe that the effluent substrate concentration from two reactors in series is always lower than the effluent concentration from one reactor, with the same SRT and total HRT. The reason is that with two (or more) reactors in series there is a spatial substrate gradient which is obviously not present with only one CSTR. This means that the substrate concentration in the first reactor is always higher than in the effluent, and so on average substrate removal occurs at a higher concentration, and therefore at higher rate, than in the case with only one reactor. So, in general, we can conclude that having two or more reactors in series (or having substrate gradients in the process, e.g. plug flow reactor) has a beneficial effect on reducing the effluent substrate concentration.

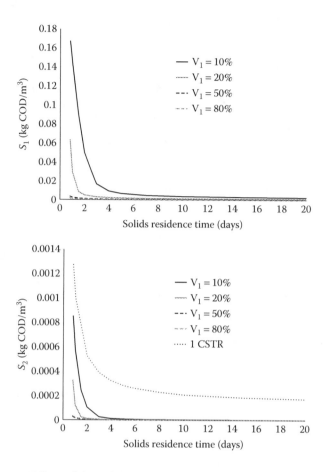

FIGURE 4.42 Effect of the solids residence time (SRT) and of the fraction of the total volume taken by reactor 1 on the effluent substrates from R_1 and R_2. Hydraulic residence time $(HRT)_{total} = 0.5$ day, recycle ratio $(R) = 1$.

Figure 4.43 shows the effect of the volume fraction of R_1 and of the SRT on the biomass concentration in R_2 and on the ratio X_1/X_2. Note that X_2 is virtually independent on the volume fraction of the two reactors and that for most values of the SRT X_1 is slightly larger than X_2. The reason is that in most cases the vast majority of the substrate is removed in R_1, and therefore, the main phenomenon happening in R_2 is biomass metabolism, which causes a slight reduction in biomass concentration. The exception to this trend are the cases of low SRT and low volume fraction of R_1 (10% or 20%), when $X_1 < X_2$. The reason is that, as mentioned above, under these conditions there is a high concentration of

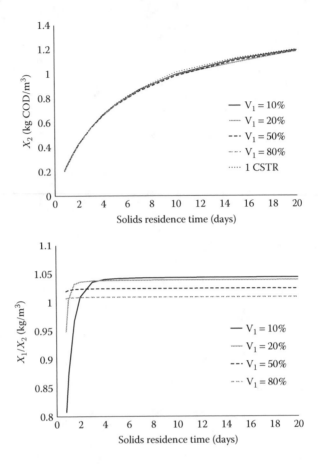

FIGURE 4.43 Effect of the solids residence time (SRT) and of the fraction of the total volume taken by reactor 1 on the effluent substrates from R_1 and R_2. Hydraulic residence time $(HRT)_{total} = 0.5$ day, recycle ratio $(R) = 1$.

residual substrate in R_1 which is then metabolised in R_2, allowing a significant growth in the second reactor.

4.2 THE ACTIVATED SLUDGE PROCESS FOR CARBON AND NITROGEN REMOVAL

So far, we have only considered the activated sludge process for carbon removal. If nitrogen also needs to be removed from the wastewater, a typical process is the activated sludge process with denitrification followed by nitrification (described in Chapter 1). This scheme is shown in Figure 4.44, and the first reactor operates in the absence of oxygen (anoxic reactor) while in the second reactor oxygen is provided (aerobic reactor).

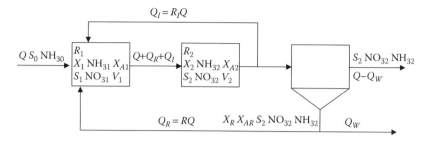

FIGURE 4.44 Scheme of an activated sludge process for nitrogen removal with anoxic tank followed by an aerobic tank and by the final settling tank.

Here, we will assume that the only source of nitrogen in the feed waste-water is soluble ammonia.

The rate equations for biomass growth can be written as shown in Chapter 2 and are summarised below. Note that the rate equations have to be written for each type of microorganisms (heterotrophic and auto-trophic) and for each of the two reactors. The subscripts 1 and 2 refer to the anoxic and aerobic reactors, respectively. The units of the growth rate terms need to be mass/volume·time, for example $kg/m^3 \cdot day$.

$$r_{X1} = \mu_{max} \frac{S_1}{K_S + S_1} \frac{NO_{31}}{K_{SNO3} + NO_{31}} X_1 \qquad (4.99)$$

$$r_{end1} = -b \frac{NO_{31}}{K_{SNO3} + NO_{31}} X_1 \qquad (4.100)$$

$$r_{X2} = \mu_{max} \frac{S_2}{K_S + S_2} X_2 \qquad (4.101)$$

$$r_{end2} = -b X_2 \qquad (4.102)$$

$$r_{XA2} = \mu_{maxA} \frac{NH_{32}}{K_{SNH3} + NH_{32}} X_2 \qquad (4.103)$$

$$r_{endA2} = -b_A X_{A2} \qquad (4.104)$$

Once the rate equations for biomass growth and endogenous metabolism have been written, the rate equations for the production or removal of all the other species, which are reagents or products of the biological reactions, can be written easily according to what shown in Chapter 2. These rate equations are written below.

Rate equation for organic substrate removal (kg COD/m³·day):

$$r_{S1} = -\frac{r_{X1}}{Y_{X/S}} \tag{4.105}$$

$$r_{S2} = -\frac{r_{X2}}{Y_{X/S}} \tag{4.106}$$

Rate equation for ammonia removal (kgN/m³·day):

$$r_{NH31} = -\left(r_{X1} + r_{end1}\right) \cdot 0.12 \tag{4.107}$$

$$r_{NH32} = -r_{XA2}\left(0.12 + \frac{1}{Y_{XA/NO3}}\right) - r_{endA2} \cdot 0.12 - \left(r_{X2} + r_{end2}\right) \cdot 0.12 \tag{4.108}$$

Rate equation for nitrate production and removal (kgN/m³·day):

$$r_{NO31} = \frac{-r_{X1}\left(\dfrac{1}{Y_{X/S}} - 1.42\right) + 1.42 \cdot r_{end1}}{2.86} \tag{4.109}$$

$$r_{NO32} = \frac{r_{XA}}{Y_A} \tag{4.110}$$

Rate equation for oxygen consumption (kg O₂/m³·day) (only in reactor 2, because there is no oxygen in reactor 1):

$$r_{O22} = -r_{XA}\left(\frac{4.57}{Y_{XA/NO3}} - 1.42\right) - r_{X1}\left(\frac{1}{Y_{X/S}} - 1.42\right) - \left(r_{end2} + r_{endA2}\right) \cdot 1.42 \tag{4.111}$$

Designing the process means determining the values of all the variables that characterise it as a function of the design parameters. The process variables are linked by the mass balances reported below. These mass balances are based on the scheme in Figure 4.44 and on the definition of the internal recycle ratio $R_I = Q_I/Q$ and of the recycle ratio from the settling tank, defined as usual as $R = Q_R/Q$.

Heterotrophic biomass in reactor 1:

$$\left(r_{X1} + r_{end1}\right) \cdot V_1 + RQX_R + R_I QX_2 = Q\left(1 + R + R_I\right)X_1 \tag{4.112}$$

Autotrophic biomass in reactor 1:

$$RQX_{AR} + R_I QX_{A2} = Q\left(1 + R + R_I\right)X_{A1} \tag{4.113}$$

Heterotrophic biomass in reactor 2:

$$\left(r_{X2}+r_{end2}\right)V_2+Q\left(1+R+R_I\right)X_1=Q\left(1+R+R_I\right)X_2 \qquad (4.114)$$

Autotrophic biomass in reactor 2:

$$\left(r_{XA2}+r_{end2}\right)V_2+Q\left(1+R+R_I\right)X_{A1}=Q\left(1+R+R_I\right)X_{A2} \qquad (4.115)$$

Heterotrophic biomass in the whole system:

$$\left(r_{X1}+r_{end1}\right)V_1+\left(r_{X2}+r_{end2}\right)V_2=Q_W X_R \qquad (4.116)$$

Autotrophic biomass in the whole system:

$$\left(r_{XA2}+r_{endA2}\right)V_2=Q_W X_{AR} \qquad (4.117)$$

Carbon substrate in reactor 1:

$$QS_0+RQS_2+R_IQS_2+r_{S1}V_1=Q\left(1+R+R_I\right)S_1 \qquad (4.118)$$

Carbon substrate in reactor 2:

$$Q\left(1+R+R_I\right)S_1+r_{S2}V_2=Q\left(1+R+R_I\right)S_1 \qquad (4.119)$$

Ammonia in reactor 1:

$$QNH_{30}+RQNH_{32}+R_IQNH_{32}+r_{NH31}V_1=Q\left(1+R+R_I\right)NH_{31} \qquad (4.120)$$

Ammonia in reactor 2:

$$Q(1+R+R_I)NH_{31}+r_{NH32}V_2=Q(1+R+R_I)NH_{32} \qquad (4.121)$$

Nitrate in reactor 1:

$$RQNO_{32}+R_IQNO_{32}+r_{NO31}V_1=Q\left(1+R+R_I\right)NO_{31} \qquad (4.122)$$

Nitrate in reactor 2:

$$r_{XA}\frac{V_2}{Y_A}+Q\left(1+R+R_I\right)NO_{31}=Q\left(1+R+R_I\right)NO_{32} \qquad (4.123)$$

The mass balances can be re-written using a procedure which is analogous to what done for the activated sludge process for carbon removal only. Each mass balance equation can be re-written using the rate Equations 4.105–4.111 and divided by XV (i.e. either X_1V_1, or X_2V_2, or $X_{A2}V_2$, depending on the specific equation). Then the new variables HRT_1, HRT_2 and SRT, defined below, can be introduced.

$$HRT_1 = \frac{V_1}{Q} \qquad (4.124)$$

$$HRT_2 = \frac{V_2}{Q} \qquad (4.125)$$

$$SRT = \frac{V_1(X_1 + X_{A1}) + V_2(X_2 + X_{H2})}{Q_w(X_R + X_{AR})} \qquad (4.126)$$

Note that from the definition of SRT, HRT_1, HRT_2, it follows that:

$$\frac{Q_w}{Q} = \frac{HRT_1(X_1 + X_{A1}) + HRT_2(X_{A2} + X_2)}{SRT(X_R + X_{AR})} \qquad (4.127)$$

The mass balances can therefore be re-written as follows.

Heterotrophic biomass in reactor 1:

$$(r_{X1} + r_{end1}) \cdot \frac{NO_{31}}{K_{SNO3} + NO_{31}} + \frac{RX_R}{X_{H1}HRT_1} + \frac{R_I X_2}{X_1 HRT_1} = \frac{(1 + R + R_I)}{HRT_1} \qquad (4.128)$$

Autotrophic biomass in reactor 1:

$$\frac{RX_{AR}}{X_{A1}HRT_1} + \frac{R_I X_{A2}}{X_{A1}HRT_1} = \frac{(1 + R + R_I)}{HRT_1} \qquad (4.129)$$

Heterotrophic biomass in reactor 2:

$$(r_{X2} + r_{end2}) + \frac{(1 + R + R_I)X_{H1}}{X_{H2}HRT_2} = \frac{(1 + R + R_I)}{HRT_2} \qquad (4.130)$$

Autotrophic biomass in reactor 2:

$$(r_{XA2} + r_{endA2}) + \frac{(1 + R + R_I)X_{A1}}{X_{A2}HRT_2} = \frac{(1 + R + R_I)}{HRT_2} \qquad (4.131)$$

Heterotrophic biomass in the whole system:

$$(r_{X1} + r_{end1}) \cdot \frac{NO_{31}}{K_{SNO3} + NO_{31}} + (r_{X2} + r_{end2}) \frac{X_2 HRT_2}{X_1 HRT_1} =$$

$$= \frac{X_R}{X_1 HRT_1} \left(\frac{HRT_1(X_1 + X_{A1}) + HRT_2(X_{A2} + X_2)}{SRT(X_R + X_{AR})} \right) \qquad (4.132)$$

Autotrophic biomass in the whole system:

$$\left(r_{XA2}+r_{endA2}\right)=\frac{X_{AR}}{X_{A2}HRT_2}\left(\frac{HRT_1\left(X_1+X_{A1}\right)+HRT_2\left(X_{A2}+X_2\right)}{SRT\left(X_R+X_{AR}\right)}\right) \quad (4.133)$$

Carbon substrate in reactor 1:

$$\frac{S_0+RS_2+R_1S_2}{X_1HRT_1}=r_{X1}\frac{1}{Y_{X/S}}\cdot\frac{NO_{31}}{K_{SNO3}+NO_{31}}+\frac{\left(1+R+R_1\right)S_1}{X_1HRT_1} \quad (4.134)$$

Carbon substrate in reactor 2:

$$\frac{\left(1+R+R_1\right)S_1}{X_2HRT_2}=r_{X2}\frac{1}{Y_{X/S}}+\frac{\left(1+R+R_1\right)S_2}{X_2HRT_2} \quad (4.135)$$

Ammonia in reactor 1:

$$\left(r_{X1}+r_{end1}\right)\cdot\frac{NO_{31}}{K_{SNO3}+NO_{31}}0.12+\frac{\left(1+R+R_1\right)NH_{31}}{X_1HRT_1}$$

$$=\frac{NH_{30}+RNH_{32}+R_1NH_{32}}{X_1HRT_1} \quad (4.136)$$

Ammonia in reactor 2:

$$\left(r_{X2}+r_{end2}\right)\frac{X_2}{X_{A2}}0.12+\left(r_{XA2}+r_{endA2}\right)0.12+r_{XA2}\frac{1}{Y_A}$$

$$+\frac{\left(1+R+R_1\right)NH_{31}}{X_{A2}HRT_2}=\frac{\left(1+R+R_1\right)NH_{32}}{X_{A2}HRT_2} \quad (4.137)$$

Nitrate in reactor 1:

$$\frac{RNO_{32}+R_1NO_{32}}{X_1HRT_1}=\left[r_{X1}\left(\frac{1}{Y_{X/S}}-1.42\right)+r_{end1}\right]\cdot\frac{NO_{31}}{K_{SNO3}+NO_{31}}$$

$$+\frac{\left(1+R+R_1\right)NO_{31}}{X_1HRT_1} \quad (4.138)$$

Nitrate in reactor 2:

$$r_{XA2}\frac{1}{Y_A}+\frac{\left(1+R+R_1\right)NO_{31}}{X_{A2}HRT_2}=\frac{\left(1+R+R_1\right)NO_{32}}{X_{A2}HRT_2} \quad (4.139)$$

In conclusion, for the activated sludge process for nitrogen removal we have a system of 12 Equations, 4.128–4.139, with the 17 unknowns, X_1, X_2,

X_{A1}, X_{A2}, S_1, S_2, NH_{31}, NH_{32}, NO_{31}, NO_{32}, X_R, X_{AR}, HRT_1, HRT_2, R, R_1, SRT. The system can be solved if the values of 5 variables are fixed, with the remaining 12 variables being calculated as the solution of the 12 mass balance equations. It is convenient to use HRT_1, HRT_2, R, R_1, SRT as design variables, that is to fix their values and to obtain the values of the other 12 variables from the solution of the system of equations.

Once the system has been designed, that is the values of all the variables in the system have been determined, the sludge production and the oxygen consumption can be calculated. The sludge production is given by:

$$P_X\left(\frac{kg\,biomass}{day}\right) = Q_W\left(X_R + X_{AR}\right) \tag{4.140}$$

The oxygen consumption is given by the sum of the oxygen consumed by the heterotrophic and autotrophic microorganisms.

$$Q_{O2biomass}\left(\frac{kgO_2}{day}\right) = Q_{O2biomass,het} + Q_{O2biomass,aut} \tag{4.141}$$

The oxygen consumed by the heterotrophic microorganisms can be calculated as usual from the COD balance over the whole process, but in this case we need to subtract the COD that has been oxidised using nitrate as electron acceptor in the anoxic reactor, because this COD obviously has been removed without any oxygen consumption. The COD oxidised by nitrate in the anoxic reactor can be calculated by multiplying the nitrate removed (as N) by the factor 2.86 used to convert nitrate into COD. Therefore, the COD removed by nitrate in the anoxic reactor is $\left[QNO_{32}\left(R_1 + R\right) - QNO_{31}\left(1 + R + R_1\right)\right] \cdot 2.86$, and the oxygen consumption by the heterotrophic microorganisms is:

$$Q_{O2biomass,het} = Q\left(S_0 - S\right) - Q_W X_R \times 1.42$$
$$-\left[QNO_{32}\left(R_1 + R\right) - \left(1 + R + R_1\right)QNO_{31}\right] \cdot 2.86 \tag{4.142}$$

The oxygen consumption by the heterotrophic biomass can be calculated from the nitrate balance in the aerobic reactor, adding the contribution of the endogenous metabolism of the autotrophic biomass:

$$Q_{O2biomass,aut} = \left(4.57 - 1.42Y_{XA/NO3}\right)\left(NO_{32} - NO_{31}\right)Q(1 + R + R_1)$$
$$+ b_A X_{A2} V_2 1.42 \tag{4.143}$$

The total oxygen consumption $Q_{O2biomass}$ can be calculated according to Equation (4.141) by adding up the contributions of Equations 4.142 and 4.143.

The total nitrogen removal is given by:

$$\text{Total nitrogen removal}\,(\%)=\left(\frac{QNH_{30}-Q\left(NO_{32}+NH_{32}\right)}{QNH_{30}}\right)\cdot100 \quad (4.144)$$

Example 4.12: Design of an activated sludge process for carbon and nitrogen removal

It is desired to design an activated sludge process for carbon and nitrogen removal. The reaction tanks are composed of an anoxic reactor followed by an aerobic one with internal recycle.

Inlet conditions:

$S_0 = 0.5$ kg COD/m³

$NH_{30} = 0.05$ kgN-NH₃/m³

Investigate the effect of the design parameters SRT (range 0.8–20 days), HRT_1, HRT_2 (range 0.25–1 days), R (range 0.5–2) and R_I (range 1–7) on the values of the calculated variables. Discuss the observed trends.

Values of the kinetic parameters:

$\mu_{max} = 6$ day⁻¹

$K_S = 0.004$ kg COD/m³

$Y_{X/S} = 0.3$ kg biomass/kg COD

$b = 0.2$ days⁻¹

$\mu_{maxA} = 1$ day⁻¹

$K_{SA} = 0.001$ kg N-NH₃/m³

$Y_A = 0.17$ kg biomass/kg N-NO₃

$b_A = 0.1$ day⁻¹

$K_{SNO3} = 0.001$ kg N-NO₃/m³

Solution

The solution of this example is obtained by solving the mass balances that describe the system, Equations (4.128 through 4.139). The system has five degrees of freedom, that is we need to set the values of five parameters to be able to solve the mass balance and to design the process. Initially, we will see the effect of the SRT, at fixed values of HRT_1 (= 0.33 day), HRT_2 (= 0.66 day), R (= 1) and R_I (= 3). The results are reported below in Figures 4.45 and 4.46. The main effect

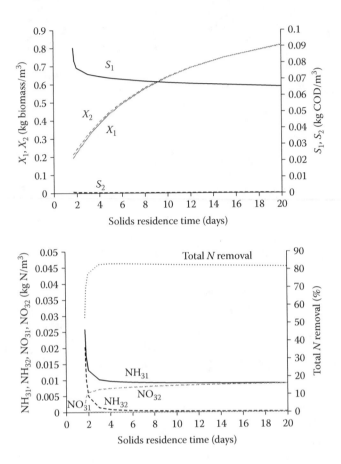

FIGURE 4.45 Solution of Example 4.12. Effect of the solids residence time (SRT) on the design of an activated sludge process for carbon and nitrogen removal. Hydraulic residence time $(HRT)_1 = 0.33$ day, $HRT_2 = 0.66$ day, recycle ratio $(R) = 1$, $R_I = 3$.

of the SRT in a process for carbon and nitrogen removal is on nitrogen removal. Indeed, washout of nitrifying microorganisms happens for SRT slightly lower than two days and this obviously corresponds to a sharp increase in ammonia concentration and a sharp decrease in nitrogen removal. When the autotrophic microorganisms are washed out due to a too short SRT, nitrogen removal occurs only due to the growth of heterotrophic microorganisms. It is interesting to observe that when autotrophic microorganisms are washed out from the process, heterotrophic microorganisms are still present, as also shown from the fact that the carbon substrate is always virtually

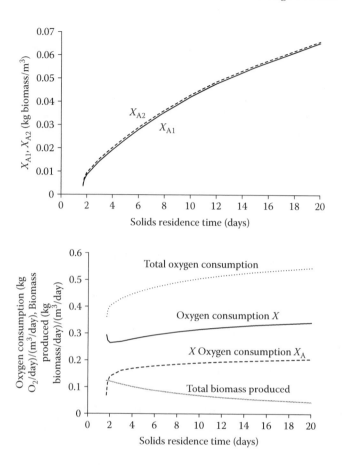

FIGURE 4.46 Solution of Example 4.12. Effect of the solids residence time (SRT) on the design of an activated sludge process for carbon and nitrogen removal. Hydraulic residence time $(HRT)_1 = 0.33$ day, $HRT_2 = 0.66$ day, recycle ratio $(R) = 1$, $R_I = 3$.

completely removed from the system in the aerobic tank (S_2 concentration is virtually 0 for any value of the SRT in Figure 4.45). When autotrophic microorganisms are washed out, nitrate production stops and this cause the increase in the concentration S_1 in the anoxic reactor, because there is not enough nitrate for the degradation of the organic matter in the anoxic reactor. The total oxygen consumption (Figure 4.46) is due to the sum of the oxygen consumption of the heterotrophs and of the nitrifiers and increases at increasing SRT, while the total biomass production decreases, and this is a similar trend as already observed for the process with carbon removal only.

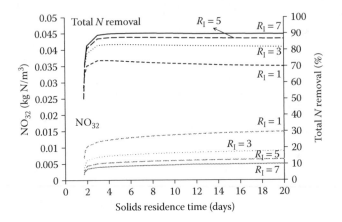

FIGURE 4.47 Solution of Example 4.12. Effect of the solids residence time (SRT) and of the internal recycle R_I on the design of an activated sludge process for carbon and nitrogen removal. Hydraulic residence time $(HRT)_1 = 0.33$ day, $HRT_2 = 0.66$ day, recycle ratio $(R) = 1$.

The effect of the internal recycle flow rate on the calculated variables is shown in Figure 4.47. Higher values of R_I give lower concentration of effluent nitrate and therefore higher total nitrogen removal, because the internal recycle dilutes the influent wastewater and therefore decreases the concentration of ammonia at the inlet of the aerobic reactor. Changing the recycle flow rate from the bottom of the settling tank has a similar effect as changing the internal recycle (Figure 4.48). Therefore, in principle, a high nitrogen removal could also be obtained without the internal recycle, by increasing the recycle from the bottom of the clarifier. This is however often not possible because it would cause an increase in the area required for settling (Section 4.1.4).

The effect of changing the HRT on the total N removal is shown in Figure 4.49. In these simulations, the total HRT in the plant has been kept constant ($HRT_{tot} = 1$ day) and this means that increasing HRT_1 corresponds to a decrease in HRT_2 and vice versa. In practice, for a given influent flow rate, changing the HRTs corresponds to changing the volume of the two reactors. It can be observed that by increasing HRT_1, which corresponds to a decrease in HRT_2, nitrogen removal decreases. This is because the autotrophic biomass is only active in reactor 2 (the aerobic one) and therefore a reduction in the aerobic volume corresponds to a decrease in the aerobic SRT and therefore to a decrease in the nitrification efficiency.

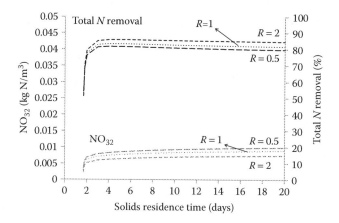

FIGURE 4.48 Solution of Example 4.12. Effect of the solids residence time (SRT) and of the recycle R on the design of an activated sludge process for carbon and nitrogen removal. Hydraulic residence time $(HRT)_1 = 0.33$ day, $HRT_2 = 0.66$ day, $R_I = 3$.

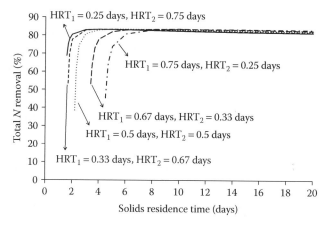

FIGURE 4.49 Solution of Example 4.12. Effect of the solids residence time (SRT) and of the hydraulic residence time (HRT) in the two reactors on the design of an activated sludge process for carbon and nitrogen removal. $R_I = 3$, $R = 1$.

4.2.1 pH Calculation in the Activated Sludge Process for Carbon and Nitrogen Removal

In the activated sludge process for nitrogen removal the pH can be calculated with the same approach used in Section 4.1.5 for the activated sludge process for carbon removal. The additional complication is that in this case we have two reactors and more species which affect the pH, that is we have ammonia and nitrate in addition to carbonic acid.

For the anoxic reactor (reactor 1), the charge balance is:

$$\left[H_3O^+\right]_1 + \left[\Sigma Cat - \Sigma An\right]_1 + \frac{K_{NH3}\left[H_3O^+\right]_1\left[NH_{3tot}\right]_1}{K_w + K_{NH3}\left[H_3O^+\right]_1}$$

$$= \frac{\left[H_2CO_{3,tot}\right]_1}{1 + K_{CO2} + \dfrac{K_{H2CO3}}{\left[H_3O^+\right]_1} + \dfrac{K_{HCO3}K_{H2CO3}}{\left[H_3O^+\right]_1^2}} \qquad (4.145)$$

$$\left(\frac{K_{H2CO3}}{\left[H_3O^+\right]_1} + 2\frac{K_{HCO3}K_{H2CO3}}{\left[H_3O^+\right]_1^2}\right) + \left[NO_3\right]_1 + \frac{K_w}{\left[H_3O^+\right]_1}$$

For the aerobic reactor (reactor 2) the charge balance is, with the same approach:

$$\left[H_3O^+\right]_2 + \left[\Sigma Cat - \Sigma An\right]_2 + \frac{K_{NH3}\left[H_3O^+\right]_1\left[NH_{3tot}\right]_2}{K_w + K_{NH3}\left[H_3O^+\right]_2}$$

$$= \frac{\left[H_2CO_{3,tot}\right]_2}{1 + K_{CO2} + \dfrac{K_{H2CO3}}{\left[H_3O^+\right]_2} + \dfrac{K_{HCO3}K_{H2CO3}}{\left[H_3O^+\right]_2^2}} \qquad (4.146)$$

$$\left(\frac{K_{H2CO3}}{\left[H_3O^+\right]_2} + 2\frac{K_{HCO3}K_{H2CO3}}{\left[H_3O^+\right]_2^2}\right) + \left[NO_3\right]_2 + \frac{K_w}{\left[H_3O^+\right]_2}$$

In these equations, we know in both reactors $\left[\Sigma Cat - \Sigma An\right]$, $[NO_3]$, $[NH_{3,tot}]$. Indeed, we are calculating the pH for a fully designed a specified process, so we assume that the main design concentrations are known. $\left[\Sigma Cat - \Sigma An\right]$ is known from the characterisation of the feed (pH and alkalinity) as shown in previous sections. We assume, as we usually do, that the concentration $\left[\Sigma Cat - \Sigma An\right]$ is the same in the feed and in both reactors, since we ignore any possible adsorption or precipitation reactions. To solve Equations 4.145 and 4.146, we need additional equations to calculate $[H_2CO_{3,tot}]$ in both reactors. Similarly, as for the process for carbon removal only (4.1.5) the total concentration of inorganic carbon depends on its concentration in the feed (which is linked to the alkalinity of the feed), on its generation due to the biological reactions and to the stripping of carbon dioxide.

For the anoxic reactor, there are no air diffusers or mechanical aeration; therefore, we will assume that no stripping of carbon dioxide occurs (although some carbon dioxide might leave with the nitrogen gas generated by denitrification). Under this hypothesis, the balance of the inorganic carbon for reactor 1 is:

$$QH_2CO_{3totfeed} + (R+R_I)QH_2CO_{3tot2} + r_{CO2biomass1}V_1$$
$$= (1+R+R_I)QH_2CO_{3tot1}$$
(4.147)

Where $r_{CO2biomass1}V_1$ is the generation of inorganic carbon due to microorganisms' activity and can be calculated from a carbon balance on the anoxic reactor. For example assuming the carbon substrate is glucose:

$$r_{CO2biomass1}V_1\left(\frac{kmol}{day}\right) = \frac{(QS_0 + Q(R+R_I)S_2)}{1.067}\frac{6}{180}$$
$$+ Q(R+R_I)X_{H2}\frac{5}{113} - \frac{Q(1+R+R_I)S_1}{1.067}\frac{6}{180}$$
(4.148)
$$- Q(1+R+R_I)X_{H1}\frac{5}{113}$$

For the aerobic reactor, the balance of inorganic carbon can be written exactly as we did in Section 4.1.5 for the case of carbon removal only

$$Q(1+R+R_I)H_2CO_{3tot1} + r_{CO2biomass2}V_2$$

$$= k_La_{CO2}\left(\frac{K_{CO2}[H_2CO_{3,tot}]_2}{1+K_{CO2}+\frac{K_{H2CO3}}{10^{-pH}}+\frac{K_{HCO3}K_{H2CO3}}{10^{-2pH}}} - k_{eqCO2}p_{CO2}\right)V_2 \quad (4.149)$$

$$+ Q(1+R+R_I)H_2CO_{3tot2}$$

Which needs to be coupled with the balance of carbon dioxide in the gas phase:

$$\frac{w_{air,in}}{MW_{air}} \times \frac{p_{CO2,in}}{p_{tot}} \times 1000$$

$$+ k_L a_{CO2} \left(\frac{K_{CO2}\left[H_2CO_{3,tot}\right]}{1+K_{CO2}+\dfrac{K_{H2CO3}}{10^{-pH}}+\dfrac{K_{HCO3}K_{H2CO3}}{10^{-2pH}}} - k_{eqCO2}\, p_{CO2} \right) V$$

$$(4.150)$$

$$= \left(\begin{array}{c} \dfrac{w_{air,in}}{MW_{air}} \times 1000 - \dfrac{w_{O2,transf}}{MW_{O2}} \times 1000 \\[2em] + k_L a_{CO2} \left(\dfrac{K_{CO2}\left[H_2CO_{3,tot}\right]}{1+K_{CO2}+\dfrac{K_{H2CO3}}{10^{-pH}}+\dfrac{K_{HCO3}K_{H2CO3}}{10^{-2pH}}} - k_{eqCO2}\, p_{CO2} \right) V \end{array} \right) \dfrac{p_{CO2,out}}{p_{tot}}$$

The generation of inorganic carbon due to biological activity in reactor 2 can be calculated from a carbon balance around reactor 2:

$$r_{CO2biomass2}V_2 = \frac{Q(1+R+R_I)S_1}{1.067}\frac{6}{180} + Q(1+R+R_I)(X_{H1}+X_{A1})\frac{5}{113}$$

$$- \frac{Q(1+R+R_I)S_2}{1.067}\frac{6}{180} - Q(1+R+R_I)(X_{H2}+X_{A2})\frac{5}{113}$$

$$(4.151)$$

Equations 4.145, 4.146, 4.147, 4.149 and 4.150 constitute a system of five equations in the five unknowns pH_1, pH_2, H_2CO_{3tot1}, H_2CO_{3tot2}, pCO_2 and therefore allow the calculation of pH in the two reactors.

Example 4.13: pH calculation in the activated sludge process for carbon and nitrogen removal

Calculate the pH in the anoxic and aerobic reactors of an activated sludge process for carbon and nitrogen removal. Assume the kinetic parameters given in Example 4.12. Assume that the feed has an influent COD of 0.5 kg COD/m³, entirely composed of glucose, and an ammonia concentration of 50 mgN-NH₃/L. Calculate the pH of the reactors for the following values of the design parameters:

$$HRT_1 = 0.33 \text{ day}; HRT_2 = 0.67 \text{ day}; R = 1; R_I = 3; SRT = 10$$

Calculate the pH in a range of feed alkalinity 5–1000 mgCaCO$_3$/L and with a pH of 7. Assume aeration in the aerobic reactor is achieved by diffusers and that the inlet air flow rate is 5000 kg/day.

Solution

The first step is the calculation of $[\Sigma\text{Cat} - \Sigma\text{An}]$ and the inorganic carbon of the feed. In this case, since we have ammonia in the feed, we assume that ammonia also contributes to the alkalinity of the feed, and therefore $[\Sigma\text{Cat} - \Sigma\text{An}]$ and $[\text{H}_2\text{CO}_{3\text{tot}}]$ are calculated according to Equations 2.137 and 2.138. The results are shown in Table 4.3.

For this case, the design of the plant gives the following values:

$S_1 = 68$ gCOD/m^3; $S_2 = 0.17$ gCOD/m^3; NH$_{31} = 9.1$ gN/m^3;
NH$_{32} = 0.33$ gN/m^3; NO$_{31} = 0.04$ gN/m^3; NO$_{32} = 8.4$ gN/m^3,
$XH_1 = 0.639$ kg/m^3, $XH_2 = 0.642$ kg/m^3, $XH_R = 1.22$ kg/m^3, total biological oxygen consumption = 505 kg oxygen/m^3, $Q_W = 52$ m^3/day.

With these values the pH in the reactors can be calculated using Equations 4.145, 4.146, 4.147, 4.149 and 4.15). Figure 4.50 shows the pH in the two reactors, assuming no transfer of carbon dioxide to the gas phase in the aerobic reactor ($k_L a_{\text{CO}2} = 0$). Figure 4.51 shows the effect of carbon dioxide stripping in the aerobic reactor. As expected, the pH is higher if carbon dioxide is removed from the liquid phase by stripping.

TABLE 4.3   Calculation of $[\Sigma\text{Cat} - \Sigma\text{An}]$ and the Inorganic Carbon of the Feed for as a Function of the Feed Alkalinity

Alkalinity (mg CaCO$_3$/L)	$[\Sigma\text{Cat} - \Sigma\text{An}]$ (mol/L)	H$_2$CO$_{3\text{totfeed}}$ (mol/L)
5	−3.50E−3	6.02E−5
25	−3.1E−3	5.54E−4
50	−2.6E−3	1.17E−3
100	−1.60E−3	2.41E−3
250	1.40E−3	6.11E−3
500	6.40E−3	1.23E−2
1000	1.64E−2	2.46E−2

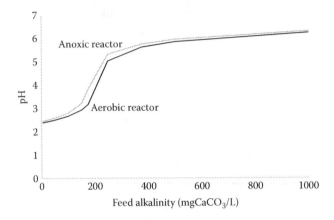

FIGURE 4.50 pH calculation in the anoxic and aerobic reactors.

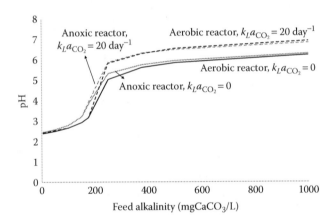

FIGURE 4.51 Effect of carbon dioxide stripping on the aerobic reactor on pH values.

4.3 THE ACTIVATED SLUDGE PROCESS FOR FILAMENTOUS BULKING CONTROL

So far the design of the activated sludge process for carbon removal (not including nitrogen removal) has been based on the assumption of one single population of microorganisms. However, we have seen in previous chapters that a potentially very serious problem in activated sludge processes is filamentous bulking. Filamentous bulking can be modelled as the competition between two populations of microorganisms, floc-forming (desired population) and filamentous (undesired) microorganisms, for a single carbon source. As discussed in Chapter 2, we will assume that floc-forming microorganisms have higher values of both μ_{max} and K_S than filamentous microorganisms, while the kinetic parameter for endogenous metabolism is the same.

In this section, we will see how the design of the activated sludge plant can affect the competition between the two populations.

With these assumptions, the rate equations for growth and endogenous metabolism of the two populations are (Chapter 2):

$$r_{XFLOC} = \frac{\mu_{maxFLOC} S}{K_{SFLOC} + S} X_{FLOC}, \quad r_{endFLOC} = -b X_{FLOC}$$

$$r_{XFIL} = \frac{\mu_{maxFIL} S}{K_{SFIL} + S} X_{FIL}, \quad r_{endFIL} = -b X_{FIL}$$

and for the rate of substrate removal:

$$r_S = -\frac{r_{XFLOC} V}{Y_{X/S}} - \frac{r_{XFIL} V}{Y_{X/S}}$$

In the presence of the two populations, the design of the activated sludge process for carbon removal can be done in the usual way by writing the mass balances for the substrate and the two populations. The mass balances are shown below, with reference to the scheme in Figure 4.52.

To understand the effect of the design parameters on the competition of the two populations we have to write the mass balances.

Balance of floc-forming microorganisms in the reactor:

$$\left(r_{XFLOC} + r_{endFLOC} \right) V + Q_R X_{FLOCR} = \left(Q + Q_R \right) X_{FLOC} \tag{4.152}$$

Balance of filamentous microorganisms in the reactor:

$$\left(r_{XFIL} + r_{ENDFIL} \right) V + Q_R X_{FILR} = \left(Q + Q_R \right) X_{FIL} \tag{4.153}$$

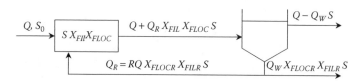

FIGURE 4.52 Scheme of a conventional activated sludge process for carbon removal with competition of filamentous and floc-forming microorganisms.

Balance of substrate in the reactor:

$$QS_0 + Q_R S = (Q + Q_R S) + \frac{r_{\text{XFLOC}} V}{Y_{X/S}} + \frac{r_{\text{XFIL}} V}{Y_{X/S}} \tag{4.154}$$

Balance of floc-forming microorganisms in the whole system:

$$\left(r_{\text{XFLOC}} + r_{\text{ENDFLOC}} \right) V = Q_W X_{\text{FLOCR}} \tag{4.155}$$

Balance of filamentous microorganisms in the whole system:

$$\left(r_{\text{XFIL}} + r_{\text{ENDFIL}} \right) V = Q_W X_{\text{FILR}} \tag{4.156}$$

We can define the HRT and SRT in the usual way:

$$\text{HRT} = \frac{V}{Q} \tag{4.11}$$

$$\text{SRT} = \frac{V \left(X_{\text{FLOC}} + X_{\text{FIL}} \right)}{Q_W \left(X_{\text{FLOCR}} + X_{\text{FILR}} \right)} \tag{4.157}$$

Since we assume that the two populations settle together their relative concentration in the reactor is the same as in the settled sludge and this means that:

$$\frac{X_{\text{FLOC}}}{X_{\text{FLOCR}}} = \frac{X_{\text{FIL}}}{X_{\text{FILR}}} \Rightarrow \text{SRT} = \frac{V X_{\text{FLOC}}}{Q_W X_{\text{FLOCR}}} = \frac{V X_{\text{FIL}}}{Q_W X_{\text{FILR}}} \tag{4.158}$$

The mass balances written above can be rearranged introducing the definitions of HRT and SRT.

Floc-forming microorganisms in the reactor:

$$\left(\mu_{\text{FLOC}} - b \right) X_{\text{FLOC}} \text{HRT} + R X_{\text{FLOCR}} = (1 + R) X_{\text{FLOC}} \tag{4.159}$$

Filamentous microorganisms in the reactor:

$$\left(\mu_{\text{FIL}} - b \right) X_{\text{FIL}} \text{HRT} + R X_{\text{FILR}} = (1 + R) X_{\text{FIL}} \tag{4.160}$$

Balance of substrate in the reactor:

$$S_0 + RS = (1 + RS) + \frac{\mu_{\text{FLOC}} X_{\text{FLOC}} \text{HRT}}{Y_{X/S}} + \frac{\mu_{\text{FIL}} X_{\text{FIL}} \text{HRT}}{Y_{X/S}} \tag{4.161}$$

Balance of floc-forming microorganisms in the whole system:

$$\left(\mu_{\text{FLOC}}-b\right)=\frac{1}{\text{SRT}} \tag{4.162}$$

Balance of filamentous microorganisms in the whole system:

$$\left(\mu_{\text{FIL}}-b\right)=\frac{1}{\text{SRT}} \tag{4.163}$$

From the latter two Equations 4.162 and 4.163 we observe that coexistence of the two populations, with the modelling assumptions made here, is not possible. Indeed, since the kinetic parameters of the two populations are different, it is not possible to satisfy the two Equations 4.162 and 4.163 simultaneously. In other words, for a generic value of the SRT the two equations

$$\mu_{\text{FLOC}}=\frac{1}{\text{SRT}}+b; \quad \mu_{\text{FIL}}=\frac{1}{\text{SRT}}+b \tag{4.164}$$

cannot be simultaneously verified, except for the single value of SRT given by:

$$\frac{1}{\text{SRT}}+b=\mu_{\text{FIL}}=\mu_{\text{FLOC}} \tag{4.165}$$

Except for this single value of SRT, which corresponds to the intersection point of the two growth curves, coexistence of the two populations is not possible. Therefore, the model used here says that, in order to verify all the mass balances for a generic value of the SRT, the concentration of one of the two microbial populations will have to be equal to zero. If the concentration of one population is zero, it will be no longer possible to re-arrange either Equation 4.155 or Equation 4.156 to Equation 4.162 or 4.163 and the mass balances (4.152–4.156) will be satisfied for any values of the SRT. For example assuming that the population that prevails are the floc-formers, the mass balances that characterise the process, and which are required for process design, are the following:

Balance of floc-forming microorganisms in the reactor:

$$\left(r_{\text{XFLOC}}+r_{\text{ENDFLOC}}\right)V+Q_R X_{\text{FLOCR}}=\left(Q+Q_R\right)X_{\text{FLOC}} \tag{4.152}$$

Balance of substrate in the reactor:

$$QS_0 + Q_R S = (Q + Q_R S) + \frac{r_{XFLOC} V}{Y_{X/S}} \qquad (4.166)$$

Balance of floc-forming microorganisms in the whole system:

$$\left(r_{XFLOC} + r_{ENDFLOC}\right) V = Q_W X_{FLOCR} \qquad (4.155)$$

Which can be rearranged as:

$$\left(\mu_{FLOC} - b\right) X_{FLOC} HRT + R X_{FLOCR} = (1 + R) X_{FLOC} \qquad (4.159)$$

$$S_0 + RS = (1 + RS) + \frac{\mu_{FLOC} X_{FLOC} HRT}{Y_{X/S}} \qquad (4.167)$$

$$\left(\mu_{FLOC} - b\right) = \frac{1}{SRT} \qquad (4.162)$$

Equations 4.159, 4.167 and 4.162 are the equations that define the system if the population that prevails is the floc-forming one. If, on the other hand, the population that prevails is the filamentous one, the equations that define the system are the same one, just changing the subscript from 'FLOC' to 'FIL'.

However, which population will grow in the process and which population will be washed out? The answer to this question is that the population that will grow is the population that has the highest growth rate for the substrate concentration that corresponds to $\mu = (1/SRT) + b$. This is evident looking at Figure 4.53. For any value of SRT, either equation: $(1/SRT) + b = \mu_{FLOC}$ or $(1/SRT) + b = \mu_{FIL}$ will have to be satisfied. The population that prevails is the one for which the specific growth rate is higher at the substrate concentration that satisfies these two equations.

It is evident that, with the model for the competition of the two populations considered here, filaments are favoured by higher values of SRT, which give lower substrate concentration in the effluent.

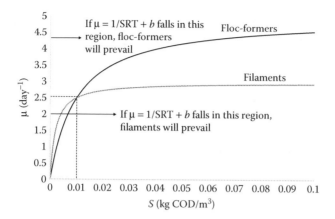

FIGURE 4.53 Comparison of growth rate of filamentous and floc-forming microorganisms for the determination of the species that prevails.

Example 4.14: Calculation of an activated sludge process with two populations

Determine which species (floc-formers or filaments) will prevail as a function of the SRT for an activated sludge process with two microbial populations characterised by the following values of the kinetic parameters:

$$\mu_{max,floc} = 5 \ day^{-1}$$

$$K_{S,floc} = 0.01 \frac{kg \ COD}{m^3}$$

$$\mu_{max,fil} = 3 \ day^{-1}$$

$$K_{S,floc} = 0.002 \frac{kg \ COD}{m^3}$$

$$b = 0.08 \ day^{-1}$$

$$Y_{X/S} = 0.45 \frac{kg \ biomass}{kg \ VSS}$$

Solution

The growth curves of the two populations intersect for $S = 0.01$ kg COD/m³, for which $\mu = 2.5$ day⁻¹. So, if $(1/SRT) + b > 2.5 day^{-1}$ floc-formers will prevail and filaments will be washed out, while if

$(1/SRT) + b < 2.5$ day^{-1} filaments will prevail and floc-formers will be washed out. Therefore, the steady state of the system will be described by Equations 4.159, 4.162 and 4.167, where the kinetic parameters will be the one of floc-formers when $(1/SRT) + b < 2.5$ day^{-1} (i.e. SRT < 0.41 day) and the ones of filaments when $(1/SRT) + b < 2.5$ day^{-1} (i.e. SRT > 0.41 day). The equations can be solved as a function of the SRT, for given values of the HRT and R. This example shows that, using the present model, in a conventional activated sludge process filaments are favoured by longer values of the SRT and, as a consequence, by lower values of the substrate concentration.

4.3.1 Activated Sludge with Selector for Bulking Control

We have seen that, according to the kinetic theory of filamentous bulking, a low substrate concentration in the reactor favours the growth and proliferation of filamentous microorganisms. Obviously, this is a big problem since we want to achieve a process with both high removal of the effluent substrate and good settling properties, that is prevalence of the floc-forming microorganisms. In order to achieve both aims of having a low substrate concentration in the effluent and suppressing the growth of filamentous microorganisms, one strategy is to use two tanks in series, instead than one single tank. The first tank is usually smaller than the second one and is called 'selector'. The rationale behind using two tanks is that in the first tank most of the influent substrate can be removed but still the substrate concentration in the selector is high enough to favour the growth of floc-forming microorganisms. Then in the second tank the substrate concentration is further reduced to the desired low value. In the second tank, the growth of filamentous microorganisms is favoured, but the substrate available for their growth is low, so they will not be able to proliferate and the majority of the microorganisms will still be floc-formers.

In the design of an activated sludge process with selector it is important to size the selector correctly. Indeed, if the selector is too small, its substrate concentration will be too large and, while floc-formers will be favoured in it, there will be enough substrate available for the filamentous microorganisms to proliferate in the second tank. On the other hand, if the selector is too large, its substrate concentration may be too low and favour the growth of filamentous microorganisms. In order to determine the optimum size of the selector, the design of an activated sludge process with selector can be done as usual using the mass balances for the various species in the two tanks and in the whole system. This will be done with reference to Figure 4.54.

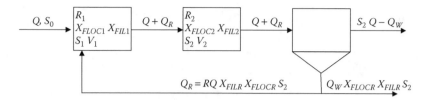

FIGURE 4.54 Scheme of an activated sludge process with selector (R_1) for bulking control.

Balance of floc-forming microorganisms in the selector (R1):

$$\left(r_{\text{XFLOC}} + r_{\text{ENDFLOC}} \right)_1 V_1 + Q_R X_{\text{FLOCR}} = \left(Q + Q_R \right) X_{\text{FLOC1}} \qquad (4.168)$$

Balance of filamentous microorganisms in the selector:

$$\left(r_{\text{XFIL}} + r_{\text{ENDFIL}} \right)_1 V_1 + Q_R X_{\text{FILR}} = \left(Q + Q_R \right) X_{\text{FIL1}} \qquad (4.169)$$

Balance of substrate in the selector:

$$Q S_0 + Q_R S_2 = \left(Q + Q_R \right) S_1 + \frac{r_{\text{XFLOC1}} V_1}{Y_{X/S}} + \frac{r_{\text{XFIL1}} V_1}{Y_{X/S}} \qquad (4.170)$$

Balance of floc-forming microorganisms in R2:

$$\left(r_{\text{XFLOC}} + r_{\text{ENDFLOC}} \right)_2 V_2 + \left(Q + Q_R \right) X_{\text{FLOC1}} = \left(Q + Q_R \right) X_{\text{FLOC2}} \qquad (4.171)$$

Balance of filamentous microorganisms in R2:

$$\left(r_{\text{XFIL}} + r_{\text{ENDFIL}} \right)_2 V_2 + \left(Q + Q_R \right) X_{\text{FIL1}} = \left(Q + Q_R \right) X_{\text{FIL2}} \qquad (4.172)$$

Balance of substrate in R2:

$$\left(Q + Q_R \right) S_1 = \left(Q + Q_R \right) S_1 + \frac{r_{\text{XFLOC2}} V_2}{Y_{X/S}} + \frac{r_{\text{XFIL2}} V_2}{Y_{X/S}} \qquad (4.173)$$

Balance of floc-forming microorganisms in the whole system:

$$\left(r_{\text{XFLOC}} + r_{\text{ENDFLOC}} \right)_1 V_1 + \left(r_{\text{XFLOC}} + r_{\text{ENDFLOC}} \right)_2 V_2 = Q_W X_{\text{FLOCR}} \qquad (4.174)$$

Balance of filamentous microorganisms in the whole system:

$$\left(r_{\text{XFIL}} + r_{\text{ENDFIL}}\right)_1 V_1 + \left(r_{\text{XFIL}} + r_{\text{ENDFIL}}\right)_2 V_2 = Q_W X_{\text{FILR}} \qquad (4.175)$$

We can introduce the definitions of HRT in each reactor and of SRT:

$$\text{HRT}_1 = \frac{V_1}{Q}, \quad \text{HRT}_2 = \frac{V_2}{Q} \qquad (4.81)$$

$$\text{SRT} = \frac{V_1 \left(X_{\text{FLOC1}} + X_{\text{FIL1}}\right) + V_2 \left(X_{\text{FLOC2}} + X_{\text{FIL2}}\right)}{Q_W \left(X_{\text{FLOCR}} + X_{\text{FILR}}\right)} \qquad (4.176)$$

With these definitions and after rearrangements the mass balances can be written as follows.

Balance of floc-forming microorganisms in the selector (R1):

$$\left(\mu_{\text{FLOC}} - b\right)_1 X_{\text{FLOC1}} \text{HRT}_1 + R X_{\text{FLOCR}} = \left(1 + R\right) X_{\text{FLOC1}} \qquad (4.177)$$

Balance of filamentous microorganisms in the selector:

$$\left(\mu_{\text{FIL}} - b\right)_1 X_{\text{FIL1}} \text{HRT}_1 + R X_{\text{FILR}} = \left(1 + R\right) X_{\text{FIL1}} \qquad (4.178)$$

Balance of substrate in the selector:

$$S_0 + R S_2 = \left(1 + R\right) S_1 + \frac{\mu_{\text{FLOC1}} X_{\text{FLOC1}} \text{HRT}_1}{Y_{X/S}} + \frac{\mu_{\text{FIL1}} X_{\text{FIL1}} \text{HRT}_1}{Y_{X/S}} \qquad (4.179)$$

Balance of floc-forming microorganisms in R2:

$$\left(\mu_{\text{FLOC}} - b\right)_2 X_{\text{FLOC2}} \text{HRT}_2 + \left(1 + R\right) X_{\text{FLOC1}} = \left(1 + R\right) X_{\text{FLOC2}} \qquad (4.180)$$

Balance of filamentous microorganisms in R2:

$$\left(\mu_{\text{FIL}} - b\right)_2 X_{\text{FIL2}} \text{HRT}_2 + \left(1 + R\right) X_{\text{FIL1}} = \left(1 + R\right) X_{\text{FIL2}} \qquad (4.181)$$

Balance of substrate in R2:

$$\left(1 + R\right) S_1 = \left(1 + R\right) S_2 + \frac{\mu_{\text{FLOC2}} X_{\text{FLOC2}} \text{HRT}_2}{Y_{X/S}} + \frac{\mu_{\text{FIL2}} X_{\text{FIL2}} \text{HRT}_2}{Y_{X/S}} \qquad (4.182)$$

Balance of floc-forming microorganisms in the whole system:

$$\frac{\left(\mu_{FLOC}-b\right)_1 X_{FLOC1}HRT_1 + \left(\mu_{FLOC}-b\right)_2 X_{FLOC2}HRT_2}{HRT_1\left(X_{FLOC1}+X_{FIL1}\right)+HRT_2\left(X_{FLOC2}+X_{FIL2}\right)}$$

$$=\frac{X_{FLOCR}}{SRT\left(X_{FLOCR}+X_{FILR}\right)}$$

(4.183)

Balance of filamentous microorganisms in the whole system:

$$\frac{\left(\mu_{FIL}-b\right)_1 X_{FIL1}HRT_1 + \left(\mu_{FIL}-b\right)_2 X_{FIL2}HRT_2}{HRT_1\left(X_{FLOC1}+X_{FIL1}\right)+HRT_2\left(X_{FLOC2}+X_{FIL2}\right)}$$

$$=\frac{X_{FILR}}{SRT\left(X_{FLOCR}+X_{FILR}\right)}$$

(4.184)

In summary, we have a system of eight equations (Equations 4.177 through 4.184) which can be solved to calculate the eight unknowns X_{FLOC1}, X_{FLOC2}, X_{FIL1}, X_{FIL2}, S_1, S_2, X_{FLOCR}, X_{FILR}, once the values of SRT, R, HRT$_1$ and HRT$_2$ have been fixed. Note that from the solution of the equations, all the variables that characterise the system can be calculated.

In designing an activated sludge process with selector for bulking control, the choice of the operating parameters SRT, R, HRT$_1$, HRT$_2$ has to be made so that the concentration of filamentous microorganisms in the second reactor, X_{FIL2}, is minimised, in order to minimise filamentous bulking. The most important parameters are the SRT, the relative size of the two reactors, that is the ratio between HRT$_1$ and HRT$_2$. The SRT determines the extent of substrate removal (even though, since we have two reactors, this is also affected by the relative size of the reactors as shown in Section 4.1.7). The relative size of the two reactors mainly affects the substrate concentration in the first reactor, which is the main factor that determines which population will prevail.

Example 4.15: Design of an activated sludge process with selector for bulking control

Assume a wastewater with COD $S_0 = 0.5$ kg COD/m³ has to be treated. We plan to design an activated sludge process with selector for bulking control. Calculate the optimum size of the selector to maximise the fraction of the biomass composed by floc-forming microorganisms. Assume that the kinetic parameters of the two populations are:

$$\mu_{max,floc} = 5 \ day^{-1}$$

$$K_{S,floc} = 0.01 \frac{kg \ COD}{m^3}$$

$$\mu_{max,fil} = 3 \ day^{-1}$$

$$K_{S,floc} = 0.002 \frac{kg \ COD}{m^3}$$

$$b = 0.08 \ day^{-1}$$

$$Y_{X/S} = 0.45 \frac{kg \ biomass}{kg \ VSS}$$

Solution

This problem can be solved by solving the system of Equations 4.177–4.184. In all the calculations presented here, the total HRT, $HRT_{total} = HRT_1 + HRT_2$, has been maintained equal to one day and the recycle ratio has been fixed to $R = 1$. Figure 4.55 shows the substrate concentration in the two reactors as a function of the SRT for various values of the relative volume of the reactors. In all cases, the substrate concentration decreases by increasing the SRT, as expected. However, while in all cases the effluent substrate concentration S_2 is very low, and much lower than S_0, the substrate concentration in the selector (R_1), is very dependent on the relative size of R1. If the selector is small (V_1 10% or less, i.e. $HRT_1 = 0.1$ day or less) the substrate in R1 is quite high, especially at relatively low SRT values. The substrate concentration profiles in Figure 4.55 have obviously effect on the competition of the two populations, and Figure 4.56 shows the relative fraction of the floc-forming population on the total biomass in the second reactor. The value of the ratio $X_{FLOC2}/(X_{FLOC2} + X_{FIL2})$ determines the settling properties of the biomass. If the selector is very small (V_1 5% or less), prevalence of floc-formers is only possible if the SRT is large enough. This is because for low SRT the substrate concentration in the selector is too high, and filaments will have a lot of substrate to grow on in the second reactor, where they are

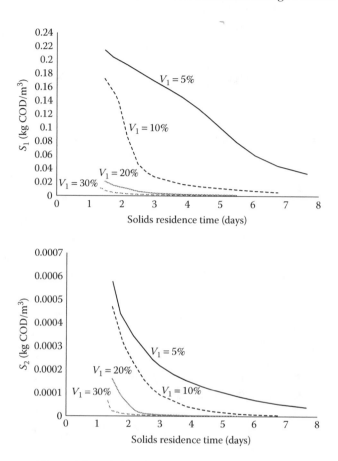

FIGURE 4.55   Effect of the solids residence time (SRT) on the substrate concentration in the two reactors for various values of their relative size.

at a competitive advantage due to the low substrate concentration. A very small selector is only effective if the SRT is large enough, so that the substrate concentration in the first reactor is low enough not to provide too much substrate for filaments in the second reactor, but large enough to give a competitive advantage for floc-formers in the selector. As an opposite behaviour, large selectors (V_1 20% or more) are only effective in selecting for the floc-formers if the SRT is small enough, so that the substrate concentration in the selector does not decrease to a value for which the filaments would prevail. Intermediate volumes of the selector (e.g. $V_1 = 10\%$) give an intermediate behaviour with floc-formers prevailing at intermediate values of the SRT.

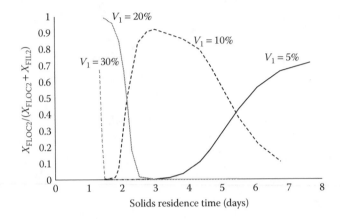

FIGURE 4.56   Effect of the solids residence time (SRT) and of the volume fraction of the selector on the competition between floc-formers and filaments.

4.4 REMOVAL OF XENOBIOTICS IN THE ACTIVATED SLUDGE PROCESS

Let us consider an activated sludge process for carbon removal, where in the feed a xenobiotic substance is also present, in addition to readily biodegradable substrates. We assume, as it is very often the case especially for municipal wastewaters, that in the feed the concentration of the xenobiotic is much lower than the concentration of the readily biodegradable COD. In general (Chapter 2), xenobiotics can be removed by biodegradation, adsorption and stripping. According to the model presented in Chapter 2, we will call X_{XOC} the concentration of biomass that is able to grow on the xenobiotic as only carbon source, and we will assume that the concentration of X_{XOC} is negligible compared to X, the heterotrophic biomass which remove the organic readily biodegradable substrate. The assumption that $X_{XOC} << X$ is justified if, as we are assuming, the concentration of xenobiotic in the feed is much lower than the concentration of readily biodegradable substrates.

Here, we want to determine the effect of the operating parameters of the process on the removal of the xenobiotics. The rate equations for the biodegradation, adsorption and stripping of xenobiotics have been presented in Chapter 2 and are summarised below:

$$r_{X_{XOC}} \left(\frac{\text{kg biomass}}{\text{m}^3 \text{day}} \right) = \frac{\mu_{XOC} S_{XOC}}{K_{SXOC} + S_{XOC}} X_{XOC}$$

$$r_{\text{end}\,\text{XOC}}\left(\frac{\text{kg biomass}}{\text{m}^3\text{day}}\right)=-b_{\text{XOC}}X_{\text{XOC}}$$

$$r_{\text{bio}}\left(\frac{\text{kg}}{\text{m}^3\text{day}}\right)=-\frac{r_{X\text{XOC}}}{Y_{X/S\text{XOC}}}$$

$$r_{\text{strip}}\left(\frac{\text{kg}}{\text{m}^3\text{day}}\right)=k_L a_{\text{XOC}} S_{\text{XOC}}$$

$$r_{\text{ads}}\left(\frac{\text{kg}}{\text{m}^3\text{day}}\right)=k_{\text{ads}}\left(K_P S_{\text{XOC}}-S_{\text{XOC,biom}}\right)X$$

In order to calculate the effect of the design parameters on the extent of removal of the xenobiotics, we need to write and solve the mass balances that characterise the system. We will do this with reference to Figure 4.57.

The mass balances for the readily biodegradable substrate S and for the heterotrophic biomass are the same presented in Section 4.1.

The mass balance for the xenobiotic-degrading microorganisms X_{XOC} in the whole process is:

$$\left(r_{X\text{XOC}}+r_{\text{end}\,\text{XOC}}\right)V=Q_W X_{\text{XOCR}}\ \Rightarrow\left(\mu_{\text{XOC}}-b_{\text{XOC}}\right)=\frac{Q_W X_{\text{XOCR}}}{VX_{\text{XOC}}}=\frac{1}{\text{SRT}}\quad(4.185)$$

where we have used the definition of SRT which in this case is:

$$\text{SRT}=\frac{V\left(X+X_{\text{XOC}}\right)}{Q_W\left(X_R+X_{\text{XOCR}}\right)}=\frac{VX}{Q_W X_R}=\frac{VX_{\text{XOC}}}{Q_W X_{\text{XOCR}}}\quad(4.186)$$

We also need to write the mass balances for the xenobiotic in the liquid phase and in the solid phase.

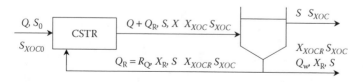

FIGURE 4.57 Scheme of an activated sludge process with xenobiotics and readily biodegradable substrates in the feed.

The mass balance for the xenobiotic in the liquid phase is:

$$QS_{XOC0} + Q_R S_{XOC} = (Q + Q_R) S_{XOC} + (-r_{strip}) V + (-r_{ads}) V + \frac{r_{X_{XOC}}}{Y_{X/SXOC}} V$$

$$= (Q + Q_R) S_{XOC} + k_L a_{XOC} S_{XOC} V \qquad (4.187)$$

$$+ k_{ads} (K_P S_{XOC} - S_{XOC,biom}) XV + \frac{\mu_{XOC}}{Y_{X/SXOC}} X_{XOC} V$$

Which can be rewritten as:

$$S_{XOC0} = S_{XOC} + k_L a_{XOC} S_{XOC} HRT + k_{ads} (K_P S_{XOC} - S_{XOC,biom}) XHRT$$

$$\qquad (4.188)$$

$$+ \frac{\mu_{XOC}}{Y_{X/SXOC}} X_{XOC} HRT$$

The mass balance for the xenobiotic in the solid phase is:

$$(-r_{ads}) V = k_{ads} (K_P S_{XOC} - S_{XOC,biom}) XV = Q_W X_R S_{XOC,biom} \qquad (4.189)$$

which becomes:

$$k_{ads} (K_P S_{XOC} - S_{XOC,biom}) SRT = S_{XOC,biom} \qquad (4.190)$$

In summary, we have a system of three Equations 4.185, 4.188, 4.190 in the three unknowns S_{XOC}, X_{XOC} and $S_{XOC,biom}$. The solution of this system of equations allows to calculate the removal of the xenobiotic as a function of the kinetic parameters and of the operating parameters of the plant. Since the adsorption of the xenobiotic is dependent on the concentration of the heterotrophic biomass X, the solution of the mass balance equations for the xenobiotic requires the previous, or simultaneous, solution of the mass balances for the readily biodegradable substrate S and for the heterotrophic biomass X (described in Section 4.1.1).

4.4.1 Effect of the Process Operating Parameters on the Xenobiotics Removal

Looking at Equations 4.185, 4.188 and 4.190, we can discuss the effect of the operating parameters of the process, mainly the HRT and SRT, on the removal of the xenobiotic. It is important to distinguish which is the main removal mechanism for the considered xenobiotic.

If the only removal mechanism is biodegradation, we only need Equations 4.185 and 4.188 to calculate S_{XOC} and X_{XOC} (since there is no adsorption Equation 4.190 is not needed). In Equation 4.188, $k_L a_{XOC}$ and k_{ads} are both equal to 0. The two equations show that the removal of a biodegradable xenobiotic can be treated exactly as the removal of any other biodegradable substance. In particular, as discussed in Section 4.1.1 for readily biodegradable substrates, the effluent concentration of the xenobiotic, S_{XOC}, is only determined by the SRT, while the HRT and SRT together determine the concentration of the xenobiotic-degrading biomass X_{XOC}. So, if the xenobiotic is only removed by biodegradation, its effluent concentration will decrease as the SRT increases.

If the only removal mechanism is stripping, the only equation that matters is Equation 4.188, which simplifies to:

$$S_{XOC0} = S_{XOC} + k_L a_{XOC} S_{XOC} \text{HRT} \qquad (4.191)$$

Equation 4.191 shows that, for a given value of $k_L a_{XOC}$, the effluent xenobiotic concentration S_{XOC} is inversely proportional to the HRT. In other words, for a given value of $k_L a_{XOC}$, the removal of the xenobiotic increases with as the reactor volume increases. However, as seen in Chapter 2 and in the previous sections on the design of the aeration system, the product $k_L a_{XOC} V$ depends on the air (or oxygen) flow rate, therefore keeping $k_L a_{XOC}$ constant while V increases (i.e. while the HRT increases) can only be achieved by increasing the air flow rate. In reality what determines the effluent concentration of a xenobiotic removed only by stripping is the air flow rate, as shown by combining Equation 1.191 with the expression of the $k_L a$ as a function of the air flow rate:

$$\frac{S_{XOC}}{S_{XOC0}} = \frac{1}{1 + \dfrac{k_{diff} Q_{air}^{b_{diff}}}{Q}} \qquad (4.192)$$

Similarly, if aeration is provided by mechanical aerators we obtain:

$$\frac{S_{XOC}}{S_{XOC0}} = \frac{1}{1 + \dfrac{\dfrac{0.63 k_{mech}^{1.5}}{9.8^{1.5 b_{mech}}} N^{(3+3b_{mech})} D^{(4.5+1.5 b_{mech})}}{Q}} \qquad (4.193)$$

Equations 4.192 and 4.193 show that the removal of a xenobiotic for which the only removal mechanism is stripping can be increased by increasing the air flow rate (for aeration with diffusers) or increasing the agitator speed and/or diameter (for mechanical aerators).

If the only removal mechanism is adsorption on the activated sludge flocs, S_{XOC} and $S_{XOC,biom}$ need to be calculated by solving Equations 4.188 (with k_La_{XOC} and μ_{XOC} both equal to 0) and (4.190). In this case, the only process parameter that determines the effluent concentration of the xenobiotic is the SRT. Increasing the SRT decreases the removal of the xenobiotic, that is increases S_{XOC}, because the lower production of biomass means that the xenobiotic concentration on the biomass, $S_{XOC,biom}$ will be higher. Therefore, in order to enhance the removal of a xenobiotic which is only removed by adsorption on the biomass, the process needs to operate at low SRT. Note that in the case of infinite SRT, that is no biomass removal from the system, no removal of xenobiotics due to adsorption can occur.

Example 4.16: Removal of a xenobiotic in the activated sludge process

Assume that the feed of the wastewater treatment plant of Example 4.1 also contains, in addition to the readily biodegradable substrate S, a xenobiotic at a concentration $S_{XOC0} = 0.01$ kg/m³. Calculate the effluent xenobiotic concentration, S_{XOC}, for the following cases:

a. The only removal mechanism for the xenobiotic is biodegradation, with the following parameters:

$$\mu_{maxXOC} = 1 \text{ day}^{-1}$$

$$K_{SXOC} = 0.001 \frac{kg}{m^3}$$

$$b_{XOC} = 0.1 \text{ day}^{-1}$$

$$Y_{X/SXOC} = 0.3 \frac{kg\,biomass}{kg\,xenobiotic}$$

b. The only removal mechanism is stripping, and aeration is provided with air and with diffusers for which the following correlations has been determined:

$$k_La_{XOC}V\left(\frac{m^3}{day}\right) = 10 \cdot Q_{air}^{0.8} \text{ where } Q_{air} \text{ is in m}^3/\text{day};$$

c. The only removal mechanism is adsorption on the activated
sludge flocs, with the following parameters:

$$k_{ads} = 200 \text{ day}^{-1}$$

$$K_P = 20 \frac{m^3}{kg}$$

Solution

a. If the only removal mechanism for the xenobiotic is biodegradation, the effluent xenobiotic concentration is only determined
by the SRT, while the concentration of the xenobiotic-degrading
microorganisms depends on both the SRT and the HRT.
Figure 4.58 reports the calculated concentrations as a function
of the SRT. As expected, the effluent xenobiotic concentration
S_{XOC} decreases as the SRT increases, exactly as described for
readily biodegradable substrates. The figure also confirms that
the concentration of xenobiotic-degrading microorganisms,
X_{XOC}, is much lower than the heterotrophic biomass concentration (calculated in Example 4.1);

b. If the only removal mechanism is stripping, the effluent xenobiotic concentration only depends on the air flow rate, and
the results obtained in this case are plotted in Figure 4.59.

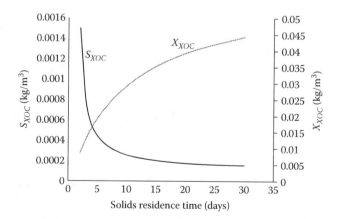

FIGURE 4.58 Xenobiotic removed by biodegradation only. Effluent xenobiotic
concentration and concentration of the xenobiotic degrading biomass as a function of the solids residence time (SRT). Hydraulic residence time (HRT) = 0.5 day.

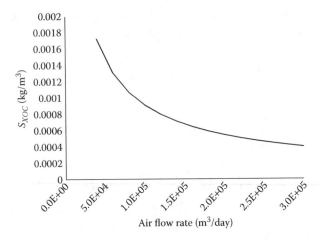

FIGURE 4.59 Xenobiotic removed by stripping only. Effect of the air flow rate on the effluent xenobiotic concentration.

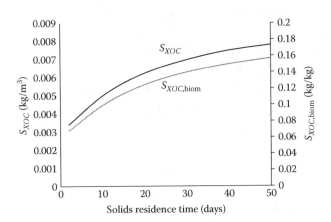

FIGURE 4.60 Xenobiotic removed by adsorption on biomass only. Effluent xenobiotic concentration and concentration of adsorbed xenobiotic as a function of the solids residence time (SRT).

The figure shows that the effluent substrate concentration decreases as the air flow rate increases, as expected.

c. If the only removal mechanism is adsorption, Figure 4.60 reports the solution of Equations 4.188 and 4.190, showing, as expected, that the effluent xenobiotic concentration increases at higher SRT.

4.5 FURTHER EXAMPLES ON THE ACTIVATED SLUDGE PROCESS

In this section, additional examples are proposed to show the applications of mass balances and reaction kinetics to the operation of the activated sludge process.

Example 4.17: Design of an activated sludge process for carbon removal

Design an activated sludge for carbon removal to treat a wastewater having an influent concentration of 300 mgCOD/L and a flow rate of 25,000 m³/day. The required effluent concentration is 1 mgCOD/L. Use an HRT value equal to 5% of the SRT and a recycle ratio equal to 1. Calculate the reactor volume, the recycle and sludge waste flow rate, the sludge production rate, the aeration requirements and the minimum area of the settling tank.

Kinetic parameters:

$$\mu_{max}=3 \text{ day}^{-1}; \quad K_S=0.01 \text{ kg COD/m}^3;$$

$$b=0.15 \text{ day}^{-1}; \quad Y_{X/S}=0.4\frac{\text{kg biomass}}{\text{kg COD}}$$

Aeration with a mechanical aerator with the power number given by:

$$P_0=0.6\,\text{Fr}^{-0.45}$$

Assume the oxygen concentration in the reactor is 2 mg/L and that the oxygen concentration in the influent stream is 0.

Settling velocity expressed as:

$$u_C\left(\frac{m}{h}\right)=3.9e^{-0.79X} \text{ where } X \text{ is the biomass concentration in kg/m}^3.$$

Solution

The SRT required for an effluent concentration of 1 mgCOD/L = 0.001 kg COD/m³, is given by:

$$\text{SRT}=\frac{S+K_S}{S(\mu_{max}-b)bK_S}=8.1 \text{ day}$$

So the HRT will be 0.4 day.

The required reactor volume will be:

$$V = \text{HRT} \cdot Q = 10,125\,\text{m}^3$$

The biomass concentration in the reactor will be:

$$X = \frac{Q(S_0 - S)Y_{X/S}}{\mu V} = 1.08\frac{\text{kg biomass}}{\text{m}^3}$$

With $R = 1$, the recycle flow rate is $Q_R = 25,000\,\text{m}^3/\text{day}$. We still need to calculate the biomass concentration in the recycle stream and the required sludge waste flow rate. These can be calculated by combining the biomass balance on the setting tank and the definition of SRT:

$$(Q + Q_R)X = (Q_R + Q_W)X_R$$

$$\text{SRT} = \frac{VX}{Q_W X_R}$$

After re-arrangements we obtain:

$$X_R = \left(\frac{1+R}{R} - \frac{\text{HRT}}{R \cdot \text{SRT}}\right)X = 2.1\frac{\text{kg}}{\text{m}^3}$$

$$\text{and } Q_W = \frac{VX}{\text{SRT} \cdot X_R} = 643\frac{\text{m}^3}{\text{day}}$$

the sludge production rate is:

$$P_X = Q_W X_R = 1350\frac{\text{kg}}{\text{day}}$$

And the oxygen consumption by the microorganisms can be calculated from the COD balance:

$$Q_{\text{O2biomass}} = Q(S_0 - S) - P_X \times 1.42 = 5558\frac{\text{kg}_{\text{O2}}}{\text{day}}$$

For the design of the aeration system, the first step is to calculate the required $k_L a$, which can be done by the mass balance for oxygen in the aeration tank:

$$RQC_{O2}+k_L a\left(k \times p_{O2}-C_{O2}\right)V=Q_{O2biomass}+(1+R)QC_{O2}$$

$$\Rightarrow k_L a=\frac{Q_{O2biomass}+QC_{O2}}{\left(k\cdot p_{O2}-C_{O2}\right)V}=78 \text{ day}^{-1}$$

The required speed of the mechanical aerator is given by:

$$N=\left(k_L a V \frac{9.8^{1.5 b_{mech}}}{0.063 k_{mech}^{1.5} D^{(4.5+1.5 b_{mech})}}\right)^{\frac{1}{3+3 b_{mech}}}=1.54 s^{-1}$$

Where $k_{mech} = 0.6$, $b_{mech} = -0.45$ and $D = 2.5$ m. Note that in this formula $k_L a$ has to be in the units of s^{-1} and V in m^3.

The required power draw of the mechanical aerator is given by:

$$P(W)=\frac{k_{mech} 1000 \cdot N^{3+2 b_{mech}} D^{5+b_{mech}}}{9.8^{b_{mech}}}=268 kW$$

And the aeration efficiency is:

$$\eta\left(\frac{kg\, O_2 \text{ transferred}}{kWhr \text{ consumed}}\right)=\frac{\dfrac{Q_{O2biomass}+QC_{O2}}{24}}{\dfrac{k_{mech}\cdot N^{3+2 b_{mech}} D^{5+b_{mech}}}{9.8^{b_{mech}}}}=0.87 \frac{kg\, O_2}{kWhr}$$

The minimum required area for the settling tank is

$$A_{min}=\left(Q+Q_R\right)X\left[\frac{\dfrac{1}{X}-\dfrac{1}{X_R}}{u_C-u_U}\right]_{max}$$

$$=50,000\frac{m^3}{day}\times 1.08\frac{kg}{m^3}\left[\frac{\dfrac{1}{X}-\dfrac{1}{2.1}}{3.9\times 24\times e^{-0.79X}-3.9\times 24\times e^{-0.79\times 2.1}}\right]_{max}$$

The maximum value of the function

$$\left[\frac{\dfrac{1}{X} - \dfrac{1}{2.1}}{3.9 \cdot 24 \cdot e^{-0.79X} - 3.9 \cdot 24 \cdot e^{-0.79 \times 2.1}} \right]$$

between $X = 1.08$ and $X = 2.1$ kg/m^3 is 0.0204 m^2.day/kg, which gives:

$$A_{min} = 1100 \text{ m}^2$$

Which corresponds to a diameter equal to about 37 m.

Example 4.18: Comparison of process design

Compare and discuss the choice of the design parameters for two activated sludge processes having the same organic load (kg COD/day), but different influent flow rate and concentration. Assume that plant A is to be designed to treat a wastewater of flow rate Q_A and substrate concentration S_{0A} and plant B needs to treat a wastewater of flow rate $Q_B = 2Q_A$ and substrate concentration $S_{0B} = 0.5 \cdot S_{0A}$. It is required to have in both cases the same effluent concentration S, much lower than the influent concentration.

Solution

Since the effluent concentration S must be the same for the two plants, the SRT has to be the same, that is

$$SRT = SRT_A = SRT_B$$

According to the mass balances developed in Section 4.1.1, the oxygen consumption of an activated sludge process can be written as:

$$Q_{O2biomass} \left(\frac{\text{kg O}_2}{\text{day}} \right) = Q(S_0 - S)\left(1 - \frac{Y_{X/S}}{\mu \cdot SRT} \right)$$

which, assuming $S \ll S_0$ in both cases, becomes:

$$Q_{O2biomass} \left(\frac{\text{kg O}_2}{\text{day}} \right) = QS_0 \left(1 - \frac{Y_{X/S}}{\mu \cdot SRT} \right)$$

The product QS_0 is the same for plant A and plant B, the specific growth rate μ is also the same, because the substrate concentration in the reactor, S, is the same; therefore, the oxygen consumption in the two plants will be the same. The sludge production will also be the same in the two plants:

$$P_X\left(\frac{\text{kg biomass}}{\text{day}}\right) = Q_W X_R = \frac{Q(S_0 - S) - Q_{O2biomass}}{1.42}$$

Which, using again the assumption that $S \ll S_0$, becomes:

$$P_X\left(\frac{\text{kg biomass}}{\text{day}}\right) = Q_W X_R = \frac{QS_0 - Q_{O2biomass}}{1.42}$$

Which shows that the sludge production will be the same for the two plants.

We need to compare the values of the reactor volume, recycle flow rate and sludge waste flow rate for the two plants.

We note that the total mass of biomass in the reactor will also be the same for the two plants:

$$XV \cong \frac{QS_0 Y_{X/S}}{\mu}$$

Therefore: $X_A V_A = X_B V_B$. The values of the biomass concentrations X_A and X_B will depend on the choice of the reactor volumes V_A and V_B. If we choose to have the same biomass concentration for the two plants, that is $X_A = X_B$, then the two reactor volumes will have to be the same, that is $V_A = V_B$. This means that, since $Q_B = 2Q_A$, it will be:

$$\text{HRT}_A = 2\text{HRT}_B$$

The required sludge flow rate depends on the values of SRT, X, X_R and V:

$$Q_W = \frac{VX}{\text{SRT} \cdot X_R}$$

The ratio X/X_R can be calculated from the mass balance in the settling tank (assuming no solid losses with the effluent):

$$(Q + Q_R)X = (Q_R + Q_W)X_R$$

which, after re-arrangements becomes:

$$\frac{X}{X_R} = \frac{1}{1 + \dfrac{1}{R} - \dfrac{HRT}{R \cdot SRT}}$$

Therefore, we have:

$$\frac{\left(\dfrac{X}{X_R}\right)_A}{\left(\dfrac{X}{X_R}\right)_B} = \frac{1 + \dfrac{1}{R_B} - \dfrac{HRT_B}{R_B \cdot SRT}}{1 + \dfrac{1}{R_A} - \dfrac{HRT_A}{R_A \cdot SRT}}$$

One reasonable design criteria is to choose the same ratio X/X_R for the two plants. With this choice of X/X_R, we can calculate the required value of the recycle ratio for plant B as a function of the recycle ratio for plant A. We obtain:

$$R_B = R_A \frac{SRT - HRT_B}{SRT - 2HRT_B}$$

With typical values of the design parameters for activated sludge processes, that is with SRT considerably larger than the SRT, this shows that the required recycle ratio will be very similar for the two plants and slightly higher for plant B.

With the design decisions made so far, that is of having $X_A = X_B$ (which requires $V_A = V_B$), and $X_{RB} = X_{RA}$, it follows that $Q_{WA} = Q_{WB}$.

In summary, the two activated sludge plants can be designed with the same value of reactor volume, biomass concentration in the reactor and in the recycle stream and the same waste flow rate. The required recycle ratio will not be the same, but it will be very similar. The oxygen consumption and sludge production will be the same.

However, it is important to observe that, with the chosen design criteria, the biomass load to the settling tank will not be the same. Indeed, the product $(Q + Q_R)X$ will be higher for plant B, that is for the plant with higher influent flow rate. Indeed, we will have $Q_B = 2Q_A$ and since, the recycle ratio R will be very similar in the two plants, also Q_{RB} will be larger than Q_{RA}. Since X is the same in the two plants, the biomass load to the settling tank for plant B will be almost twice value for plant A. Therefore, plant B will require a larger settling tank than plant A. If this is a problem, the design criteria can be changed, e.g. we can decide to have a lower biomass concentration in plant B than in plant A, which will reduce the solid loading to the settling tank for plant B. However, this will require a larger reactor volume for plant B than for plant A.

Example 4.19: Activated sludge process design

A conventional activated sludge process for COD removal is composed of a reaction tank, to be assumed perfectly mixed, and of a settling tank. The plant treats a municipal wastewater and the influent COD can all be assumed to be readily biodegradable. The plant operates satisfactorily with the following parameters:

$Q = 10,000$ m³/day (flow rate of the influent wastewater)

$Q_R = 15,000$ m³/day (recycle flow rate from the settling tank to the reaction tank)

$S_0 = 400$ Mg COD/L (concentration of the influent wastewater)

$S = 30$ Mg COD/L (concentration of the treated effluent stream)

$V = 8000$ m³ (volume of the aeration tank)

$Q_w = 250$ m³/day (flow rate of the waste sludge stream)

$X = 2$ g/L (biomass concentration in the recycle and waste sludge streams)

It is foreseen that the plant will soon have to cope with an increase of 30% of the influent flow rate, with no change to the influent COD concentration. It is desired to maintain the same effluent quality (i.e. the same concentration of effluent COD). Which changes in the operating parameters of the plant will need to be made in order to maintain the desired effluent quality?

Solution

Since the volume of the reactor cannot be changed, the only options are to change the flow rate of either the recycle (Q_R) or the waste sludge (Q_W) streams.

First of all, under the initial conditions we need to calculate the biomass concentration in the recycle and waste sludge lines, which is not given. This can be immediately calculated with a biomass balance on the settling tank:

$$(Q+Q_R)X = (Q_R+Q_W)X_R \implies$$

$$X_R = \frac{(Q+Q_R)}{(Q_R+Q_W)}X = \frac{25{,}000\dfrac{m^3}{day}}{15{,}250\dfrac{m^3}{day}}2\frac{g}{l} = 3.28\frac{g}{l} = 3.28\frac{kg}{m^3}$$

We can now analyse the two options of changing Q_R or Q_W.

a. Change Q_R (at fixed Q_w).

According to the biomass balance in the whole system:

$$(\mu - b) = \frac{Q_W X_R}{XV}$$

Under the new conditions the substrate concentration S in the reaction tank has to stay constant, therefore, μ cannot change and:

$$\left(\frac{Q_W X_R}{XV}\right)_{\text{old flow rate}} = \left(\frac{Q_W X_R}{XV}\right)_{\text{new flow rate}}$$

This also means that the SRT has to stay the same under the new conditions. Since Q_W and V are constant, under the new conditions the ratio X_R/X has to be the same as under the initial conditions. Therefore, $X_R/X = 1.64$ also under the new conditions.

Now, from the biomass balance in the reactor under the new conditions (Q_{Rnew} and Q_{new}) we have:

$$Q_{Rnew}\frac{X_R}{X} + (\mu - b)V = (Q_{new} + Q_{Rnew})$$

Which gives:

$$Q_{Rnew} = \frac{Q_{new} - \dfrac{Q_W X_R}{X}}{\dfrac{X_R}{X} - 1} = \frac{13{,}000\dfrac{m^3}{day} - 250\dfrac{m^3}{day}1.64}{1.64 - 1} = 19{,}672\frac{m^3}{day}$$

This indicates that if we want to keep the same COD removal with a higher influent flow rate, that is with a lower HRT, we need to increase the recycle flow rate in order to maintain a concentration of microorganisms in the reaction tank that is high enough.

The new biomass concentration X can be calculated from the substrate balance in the reactor under the new conditions:

$$Q_{new} S_0 + Q_{Rnew} S = \frac{\mu}{Y} X_{new} V + (Q_{new} + Q_{Rnew}) S$$

The factor μ/Y in this equation has the same value as in the initial conditions, because Y is a constant parameter and μ is also constant, because S does not change. So the factor μ/Y can be calculated from the same substrate balance under the initial conditions:

$$\frac{\mu}{Y} = \frac{Q(S_0 - S)}{XV} = \frac{10,000 \dfrac{m^3}{day}(0.400 - 0.030)\dfrac{kg\ COD}{m^3}}{2\dfrac{kg}{m^3} \cdot 8000\ m^3} = 0.23 \frac{kg\ COD}{kg \cdot day}$$

And substituting in the substrate balance under the new conditions we obtain X_{new}:

$$X_{new} = \frac{Q_{new}(S_0 - S)}{\dfrac{\mu}{Y} V} = \frac{13,000 \dfrac{m^3}{day}(0.400 - 0.030)\dfrac{kg\ COD}{m^3}}{0.23 \dfrac{kg\ COD}{kg \times day} \cdot 8000\ m^3} = 2.61 \frac{kg}{m^3}$$

And:

$$X_{Rnew} = 1.64 X_{new} = 4.28 \frac{kg}{m^3}$$

b. Change Q_W (at fixed Q_R)

Similarly to what shown in case a), the SRT has to stay the same since S will not change, and therefore, as in case a):

$$\left(\frac{Q_W X_R}{XV} \right)_{old\ flow\ rate} = \left(\frac{Q_W X_R}{XV} \right)_{new\ flow\ rate} = \frac{1}{SRT}$$

Therefore:

$$Q_{Wnew} = \frac{X_{new} V}{SRT \cdot X_{Rnew}}$$

In this equation, SRT is known from the initial conditions:

$$\text{SRT} = \left(\frac{XV}{Q_W X_R} \right)_{\text{old flow rate}} = \frac{2\frac{g}{1} 8000\,\text{m}^3}{250\frac{\text{m}^3}{\text{day}} 3.28\frac{g}{1}} = 19.5\,\text{day}$$

X_{new} can be calculated with the same procedure used for case a), from the substrate balance in the reactor and we obtain the same value, since the value of X_{new} does not depend on Q_R or Q_w:

$$X_{new} = \frac{Q_{new}(S_0 - S)}{\frac{\mu}{Y}V} = \frac{13{,}000\frac{\text{m}^3}{\text{day}}(0.400 - 0.030)\frac{\text{kgCOD}}{\text{m}^3}}{0.23\frac{\text{kgCOD}}{\text{kg}\cdot\text{day}}\cdot 8000\,\text{m}^3} = 2.61\frac{\text{kg}}{\text{m}^3}$$

To calculate Q_{wnew} now we only need to find X_{Rnew}, and this can be done with the biomass balance on the settling tank:

$$(Q_{new} + Q_R)X_{new} = (Q_R + Q_{Wnew})X_{Rnew}$$

$$\Rightarrow X_{Rnew} = \frac{(Q_{new} + Q_R)}{(Q_R + Q_{Wnew})}X_{new} = \frac{(13{,}000 + 15{,}000)\frac{\text{m}^3}{\text{day}}2.61\frac{\text{kg}}{\text{m}^3}}{(15{,}000 + Q_{Wnew})\frac{\text{m}^3}{\text{day}}}$$

$$= \frac{73{,}080}{15{,}000 + Q_{Wnew}}\frac{\text{kg}}{\text{m}^3}$$

This expression can be entered into the equation that gives Q_{wnew}:

$$Q_{Wnew} = \frac{X_{new}V}{\text{SRT}\cdot X_{Rnew}} = \frac{2.61\cdot 8000\cdot(15{,}000 + Q_{wnew})}{19.5\cdot 73{,}080}$$

This equation can be easily solved to find:

$$Q_{Wnew} = 223\frac{\text{m}^3}{\text{day}}$$

And also:

$$X_{Rnew} = 4.8\frac{\text{kg}}{\text{m}^3}$$

We can now compare the two options of changing the recycle flow rate or the sludge waste flow rate to maintain the same process performance with a higher influent flow rate. In both cases, the biomass concentration in the reaction tank will have to increase, to the same value. This is because we need a higher substrate removal rate per unit volume $Q(S_0 - S)/V$ which, if S is constant, can only be obtained with a higher biomass concentration. If we change the recycle flow rate at fixed Q_w, the recycle flow rate will have to increase compared to the initial value, in order to maintain the higher biomass concentration. If we change Q_W at fixed Q_R, Q_W will have to decrease compared to the initial value, because we will need to remove less biomass to obtain the higher biomass concentration in the reactor. In both cases, the biomass concentration at the bottom of the settling tank will increase compared to the initial value.

Example 4.20: pH calculation in the activated sludge process

Calculate the pH in an activated sludge process treating 5000 m^3 of wastewater per day. The volume of the reactor is 800 m^3. Assume that the process removes 95% of the influent substrate and that the sludge production is 300 kg/day. Assume that the pH of the influent is 7.0 and that the feed has no alkalinity, that is it has no carbonic acid in any forms. Assume that the partial pressure of carbon dioxide in contact with the liquid phase in the reactor is equal to the atmospheric partial pressure of carbon dioxide. Assume that $k_L a_{CO2} = 500$ day^{-1}. Consider two cases:

a. The substrate is acetic acid;
b. The substrate is ethanol.

Also, calculate the amount of acid or base to be added in order to maintain the pH in the reactor to the value of 7.50.
  Physical properties:

$$K_{CH3COOH} = \frac{[CH_3COO^-][H_3O^+]}{[CH_3COOH]} = 1.8 \times 10^{-5}$$

$$K_{CO2} = \frac{[CO_2]}{[H_2CO_3]} = 588$$

$$K_{H2CO3} = \frac{\left[HCO_3^-\right]\left[H_3O^+\right]}{\left[H_2CO_3\right]} = 2.5 \ 10^{-4}$$

$$K_{HCO3} = \frac{\left[CO_3^{2-}\right]\left[H_3O^+\right]}{\left[HCO_3\right]} = 4.7 \times 10^{-11}$$

Partial pressure of carbon dioxide in the atmosphere: 0.0004 atm

Solution

a. Substrate is acetic acid

The pH in the reactor is obtained from the charge balance:

$$\left[\Sigma Cat\right] + 10^{-pH} = \left[CH_3COO^-\right] + \left[HCO_3^-\right] + 2\left[CO_3^{2-}\right] + \frac{K_w}{10^{-pH}} + \left[\Sigma An\right]$$

Where the concentrations of all the species refer to the biological reactor. The charge balance can be re-written as:

$$\left[\Sigma Cat - \Sigma An\right] + 10^{-pH} = \frac{\left[CH_3COOH_{tot}\right]}{1 + \dfrac{10^{-pH}}{K_{CH3COOH}}}$$

$$+ \frac{\left[H_2CO_{3tot}\right]}{1 + K_{CO2} + \dfrac{K_{H2CO3}}{10^{-pH}} + \dfrac{K_{HCO3}K_{H2CO3}}{10^{-2pH}}}\left(\frac{K_{H2CO3}}{10^{-pH}} + 2\frac{K_{HCO3}K_{H2CO3}}{10^{-2pH}}\right) + \frac{K_w}{10^{-pH}}$$

The concentration of total acetic acid in the reactor, CH_3COOH_{tot}, is given by:

$$\left[CH_3COOH_{tot}\right] = \frac{0.2\dfrac{kg}{m^3}\dfrac{1}{2}0.05}{60\dfrac{kg}{kmol}} = 1.67 \cdot 10^{-4}\frac{kmol}{m^3}$$

The value of $\left[\Sigma Cat - \Sigma An\right]$ is equal to its value in the influent wastewater, which can be calculated from a charge balance since we know that the pH of the influent wastewater is 7.0, and it has no alkalinity. The charge balance for the influent wastewater is:

$$\left[\Sigma Cat - \Sigma An\right] + 10^{-pH} = \frac{\left[CH_3COOH_{tot}\right]_{feed}}{1 + \dfrac{10^{-pH}}{K_{CH3COOH}}} + \frac{K_w}{10^{-pH}}$$

where $[CH_3COOH_{tot}] = \dfrac{0.2\dfrac{kg}{m^3}}{60\dfrac{kg}{kmol}} = 3.33 \times 10^{-3} \dfrac{kmol}{m^3}$

We obtain:

$$[\Sigma Cat - \Sigma An] = 3.31 \times 10^{-3} \frac{kmol}{m^3}$$

We need to determine the concentration of total carbonic acid in the reactor, H_2CO_{3tot}. We need to write the mass balance for total carbonic acid in the process, which is:

$$r_{CO2}V = k_L a_{CO2}V \left(\frac{K_{CO2}[H_2CO_{3tot}]}{1 + K_{CO2} + \dfrac{K_{H2CO3}}{10^{-pH}} + \dfrac{K_{HCO3}K_{H2CO3}}{10^{-2pH}}} - k_{eqCO2}p_{CO2} \right)$$

$$+ QH_2CO_{3tot}$$

Where $r_{CO2}V$ is the net rate of carbon dioxide generation due to the biological reaction which can be calculated as:

$$r_{CO2}V\left(\frac{kmol}{day}\right) = Q\left(CH_3COOH_{totfeed} - CH_3COOH\right)\cdot 2 \cdot - Q_w X_R \cdot \frac{5}{113}$$

$$= 25.03\frac{kmol}{day}$$

In summary, we have now a system of two equations with the two unknowns pH and $[H_2CO_{3tot}]$. Solving the system gives:

$$pH = 8.3$$

$$[H_2CO_{3tot}] = 3.15 \times 10^{-3}M$$

Clearly, the pH is not ideal for an activated sludge process. In order to maintain the pH to the optimum value of 7.50, we need to add an acid. The concentration of the acid to be added can be obtained again by writing the charge balance as:

$$\left[\Sigma Cat - \Sigma An\right] + 10^{-pH} = \frac{\left[CH_3COOH_{tot}\right]}{1 + \dfrac{10^{-pH}}{K_{CH3COOH}}}$$

$$+ \frac{\left[H_2CO_{3tot}\right]}{1 + K_{CO2} + \dfrac{K_{H2CO3}}{10^{-pH}} + \dfrac{K_{HCO3}K_{H2CO3}}{10^{-2pH}}} \left(\frac{K_{H2CO3}}{10^{-pH}} + 2\frac{K_{HCO3}K_{H2CO3}}{10^{-2pH}}\right)$$

$$+ \frac{K_W}{10^{-pH}} + \left[Acid\right]$$

Where [Acid] is the unknown concentration of acid in the reactor. We can solve the system of two equations in two unknowns [Acid] and $[H_2CO_{3tot}]$, for a fixed pH = 7.5 and we obtain:

$$\left[Acid\right] = 2.28 \times 10^{-3} M$$

$$\left[H_2CO_{3tot}\right] = 9.28 \times 10^{-4} M$$

The daily amount of acid to be added is given by

$$Q \cdot \left[Acid\right] = 11.4 \frac{kmol}{day}$$

b. Substrate is ethanol

The procedure is analogous to the one described above for acetic acid. Ethanol is not an acid, so the fact that the pH of the feed is equal to 7.0 means that $\left[\Sigma Cat - \Sigma An\right] = 0$. Also, in the charge balance, there is no contribution of the substrate. The charge balance for ethanol as substrate can be written therefore as:

$$10^{-pH} = \frac{\left[H_2CO_{3tot}\right]}{1 + K_{CO2} + \dfrac{K_{H2CO3}}{10^{-pH}} + \dfrac{K_{HCO3}K_{H2CO3}}{10^{-2pH}}} \left(\frac{K_{H2CO3}}{10^{-pH}} + 2\frac{K_{HCO3}K_{H2CO3}}{10^{-2pH}}\right)$$

$$+ \frac{K_W}{10^{-pH}}$$

This means that the pH in the reactor is only determined by the equilibrium of carbonic acid. The balance of total carbonic acid in the system can be written in the same way as for acetic acid:

$$r_{CO2}V = k_L a_{CO2} V \left(\frac{K_{CO2}[H_2CO_{3tot}]}{1+K_{CO2}+\dfrac{K_{H2CO3}}{10^{-pH}}+\dfrac{K_{HCO3}K_{H2CO3}}{10^{-2pH}}} - k_{eqCO2}p_{CO2} \right)$$

$$+ QH_2CO_{3tot}$$

Where $r_{CO2}V$ in this case is equal to:

$$r_{CO2}V\left(\frac{kmol}{day}\right) = Q(CH_3CH_2OH_{feed} - CH_3CH_2OH) \cdot 2 - Q_w X_R \cdot \frac{5}{113}$$

$$= 34.66 \frac{kmol}{day}$$

By solving the system of two equations as described above we obtain:

$$pH = 5.19$$

$$[H_2CO_{3tot}] = 1.05 \times 10^{-4} M$$

Clearly, the pH is too low for a good operation of the process. In order to maintain the pH at 7.50, we need to add a base. The calculation of the concentration of base to be added is given, as described above, by the charge balance rewritten for pH = 7.50

$$10^{-pH}+[Base] = \frac{[H_2CO_{3tot}]}{1+K_{CO2}+\dfrac{K_{H2CO3}}{10^{-pH}}+\dfrac{K_{HCO3}K_{H2CO3}}{10^{-2pH}}}\left(\frac{K_{H2CO3}}{10^{-pH}}+2\frac{K_{HCO3}K_{H2CO3}}{10^{-2pH}}\right)$$

$$+\frac{K_w}{10^{-pH}}$$

Which must be solved together with the balance for total carbonic acid.

We obtain:

$$[Base] = 1.14 \cdot 10^{-3} M$$

$$[H_2CO_{3tot}] = 1.22 \cdot 10^{-3} M$$

The daily amount of base to be added is given by

$$Q \cdot [Base] = 5.7 \frac{kmol}{day}$$

Example 4.21: Design of an activated sludge process
for carbon and nitrogen removal

It is required to design an activated sludge process for an influent flow rate of 20,000 m^3/day with an influent ammonia concentration of 60 mgN-NH_3/L. The chosen configuration is pre-denitrification followed by nitrification, with internal recycle. The flow rate of the recycle stream from the bottom of the settling tank to the pre-denitrification reactor is 10,000 m^3/day. It is estimated that the plant will produce 1000 kg biomass/day. It is desired that the effluent will have a reduction of at least 90% in the total soluble nitrogen compared to the influent. Calculate the required flow rate of the internal recycle. To simplify the calculations, assume that all the ammonia entering the aerobic reactor is converted to nitrate and that all the recycled nitrate is converted to molecular nitrogen in the anoxic reactor. Ignore any ammonia generation due to endogenous metabolism in the aerobic reactor.

Solution

This example shows a simplified and approximate method to estimate the internal recycle required for an anoxic–aerobic process for carbon and nitrogen removal. A more accurate calculation requires the solution of the mass balances reported in Section 4.2.

With the assumptions of this example, all the soluble nitrogen in the plant effluent will be present as nitrate. Therefore, the maximum allowable concentration of nitrate in the effluent will be 20% of 60 mgN/L, that is 12 mg N-NO_3/L. Since we assume that all the ammonia entering the aerobic reactor is converted to nitrate, the maximum allowable concentration of ammonia at the inlet of the aerobic reactor is 12 mgN-NH_3/L.

The influent ammonia concentration is 60 mg N-NH_3/L; however, part of this ammonia is converted to biomass. The ammonia converted to biomass is 12% of 1000 kg biomass/day, which is 120 kgN/day. The total influent ammonia is 20,000 m^3/day·0.06 kgN-NH_3/m^3 = 1200 kgN-NH_3/day. Therefore, 10% of the influent ammonia is converted to biomass, and the ammonia which is available for nitrification is 54 mgN-NH_3/L. The ammonia concentration is to be reduced from 54 to 12 mg N-NH_3/L due to dilution with the internal recycle and with the recycle from the bottom of the settling tank.

Therefore, the required flow rate of the internal recycle can be calculated from:

$$Q \cdot 54 = (Q + Q_R + Q_I) NH_{31} \quad \Rightarrow$$

$$Q_I = \frac{20{,}000 \cdot 54 - 30{,}000 \cdot 12}{12} = 60{,}000 \frac{m^3}{day}$$

This corresponds to an internal recycle $I = 3$.

Example 4.22: Activated sludge process for carbon and nitrogen removal

We have a wastewater with a COD concentration of 100 mg COD/L (which we assume totally biodegradable) and an ammonia concentration of 80 mgN/L. It is required to reduce the total nitrogen concentration in the effluent to below 10 mgN/L. Determine if this goal can be reached by using nitrification and denitrification. For simplicity, ignore the ammonia removal due to biomass growth and ignore biomass production in the denitrification process.

Solution

Ignoring biomass growth, this means that all the ammonia in the feed will be converted to nitrate by nitrification, so the concentration of produced nitrate will be 80 mgN/L. Nitrate can be removed via denitrification. Ignoring biomass production during denitrification means that all the COD is oxidised by nitrate. This assumption gives the maximum concentration of nitrate that can be removed via denitrification, in reality the nitrate removed will be somewhat less because part of the COD will be used for biomass growth.

The total oxidation of 100 mgCOD/L using nitrate will consume 100/2.86 mg N-NO3/L = 35 mg N-NO3/L. Since we generate 80 mg N-NO3/L and, at maximum, we can remove 35 mg N-NO3/L, it is not possible to achieve the desired concentration of total nitrogen of 10 mgN/L with a nitrification–denitrification process. The use of a nitrification–denitrification process will only be possible with the addition of an external source of biodegradable COD.

4.6 KEY POINTS

- In order to design the activated sludge process, we can write the mass balances for the main species involved. For the simplest activated sludge process (for carbon removal only), the mass balances will include the substrate and the microorganisms in the biological reactor and in the whole process. The mass balances, written for steady-state conditions, can be solved if values of the design parameters HRT, SRT (microorganisms' residence time) and R (recycle ratio) are chosen by the process designer;

- The SRT is the most important design parameters, because it determines, for a wastewater of given flow rate and composition, the effluent substrate concentration, the total mass of microorganisms in the reactor, the oxygen consumption and the sludge production. The choice of the HRT affects the volume of the reactor and the biomass concentration and the parameter R affects the biomass concentration in the recycle stream;

- The aeration system can be designed using mass balances for oxygen in the system on the basis of the calculated oxygen consumption by the microorganisms and of the required oxygen concentration in the reactor. The power consumption required for aeration can be calculated from the design data on the basis of mass transfer correlations;

- The required size of the settling tank can be calculated from the design results on the basis of a known expression for the settling velocity of the biomass. The main design parameter that affects the size of the settling tank is the HRT. If the HRT is too low, the biomass concentration becomes too high and the required diameter of the settling tank becomes too large;

- The pH in the activated sludge process can be calculated from the pH and alkalinity of the feed, using mass balances for the inorganic carbon. The mass balance for inorganic carbon includes the generation due to microbial activity and the loss due to carbon dioxide stripping. If the substrate is a neutral species, the pH tends to drop in the biological reactor because of the carbon dioxide generation. The pH in the reactor is buffered by the alkalinity of the feed and by the removal of carbon dioxide due to stripping;

- The activated sludge process for carbon and nitrogen removal can be designed with a procedure which is the same, in principle, as for the activated sludge for carbon removal only. For carbon and nitrogen removal, the mass balances for the substrate, microorganisms (hetero-trophs and autotrophs), ammonia and nitrate need to be written and solved. The design variables include, in addition to the total HRT, SRT and R, the HRT in one of the two reactors and the internal recycle R_I;

- The presence of gradients of substrate, as achievable, for example by having two or more smaller tanks in series rather than a single larger tank, is beneficial both in terms of substrate removal and of improving the settling properties of the microorganisms (selector for bulking control);

- If xenobiotics are present, usually at low concentration, in the feed of activated sludge processes, the effect of the design parameters on their removal is determined by which is the removal mechanism: if the xenobiotics are biodegradable, their removal is enhanced at high SRT, if they are adsorbed on the biomass surface, their removal is enhanced at short SRT, if they are removed by stripping, their removal is favoured by larger flow rates of the aeration gas

Questions and Problems

4.1 An activated sludge process can be described by the scheme below

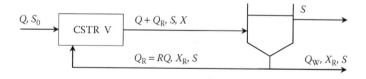

For this process, we have the following parameters:

$Q = 1000$ m³/day (influent flow rate)

$S_0 = 0.3$ kg COD/m³ (influent substrate concentration)

$Q_R = 800$ m³/day (recycle flow rate)

$X = 0.9$ kg biomass/m³ (biomass concentration in the biological reactor)

$X_R = 1.8$ kg biomass/m³ (biomass concentration in the recycle stream)

$S = 0.03$ kg COD/m³ (effluent substrate concentration)

$V = 1000$ m³ (volume of the biological reactor)

$k_L a = 10$ days⁻¹ (mass transfer coefficient for oxygen between the gas and the liquid phase)

Assume that biomass can be described by the formula $C_5H_7O_2N$ and that the molecular weights are the following: $C = 12$, $H = 1$, $O = 16$, $N = 14$.

Assume that the equilibrium relationship for oxygen between the gas and the liquid phase is given by:

$$C_{O2}^* = k^* p_{O2} \text{ with } k = 0.043 \text{ kg/m}^3/\text{atm}$$

where C_{O2}^* is the oxygen concentration in the liquid and p_{O2} is the oxygen partial pressure in the gas phase.

a. Determine the oxygen consumption rate by the biomass

b. Assuming aeration is provided by means of pure oxygen, calculate the oxygen concentration in the biological reactor. Assume atmospheric pressure in the reactor and neglect the oxygen content of the influent stream and of the recycle stream

4.2 A conventional activated sludge process operates with the following values of the process parameters:

$Q = 50,000$ m³/day (flow rate of the influent wastewater)

$Q_R = 70,000$ m³/day (recycle flow rate from the settling tank to the reaction tank)

$S_0 = 300$ mg COD/L (concentration of the influent wastewater)

$S = 30$ mg COD/L (concentration of the treated effluent stream)

$V = 60,000$ m³ (volume of the aeration tank)

$Q_w = 250$ m³/day (flow rate of the waste sludge stream)

$X = 2.3$ g/L (biomass concentration in the recycle and waste sludge streams)

$C_{O20} = 1$ mg/L (oxygen concentration in the influent wastewater)

$C_{O2} = 2.5$ mg/L (oxygen concentration in the aeration tank and in the recycle stream).

Oxygen is transferred by means of mechanical aerators. Assume that the solubility of oxygen in water under the process conditions is 9.2 mg/L.

Calculate the $k_L a$ in the aeration tank.

4.3 A conventional activated sludge process for carbon removal operates with the following values of the process parameters (assume no nitrification):

$Q = 30,000$ m³/day (flow rate of the influent wastewater)

$Q_w = 300$ m³/day (waste sludge flow rate)

$Q_R = 40,000$ m³/day (recycle flow rate)

$X = 3$ g/L (biomass concentration in the aeration tank)

$NH_{30} = 25$ mgN-NH₃/L (biomass concentration in the feed).

Calculate the concentration of ammonia in the effluent.

4.4 An activated sludge process has been designed, and the results of the design calculations are reported below:

$Q = 20,000$ m³/day (flow rate of the influent wastewater)

$Q_w = 100$ m³/day (waste sludge flow rate)

$Q_R = 30,000$ m³/day (recycle flow rate)

$S_0 = 500$ mg COD/L (concentration of the influent wastewater)

$S = 25$ mg COD/L (concentration of the treated effluent stream)

$V = 5000$ m³ (volume of the aeration tank)

$X_R = 4.5$ g/L (biomass concentration in the aeration tank)

The design assumes that there are no losses of microorganisms with the effluent from the top of the settling tank. A new process engineer is revising this design and is not convinced about the hypothesis of perfect settling. To be safe, the engineer wants to assume a solid loss with the clarified effluent of 5 mg/L. The engineer wants to maintain the same substrate concentration in the effluent (S), does not want to change the sludge waste flow rate and the reactor volume and only wants to change the recycle flow rate Q_R. Which value of Q_R has to be used in the revised design under the hypothesis of solid losses with the effluent?

4.5 An activated sludge process for carbon and nitrogen removal operates with an anoxic reactor followed by an aerobic one (pre-denitrification). The following values of the plant parameters are given:

$Q = 50,000$ m³/day (flow rate of the influent wastewater)

$Q_R = 50,000$ m³/day (recycle flow rate from the settling tank to the reaction tank)

$Q_I = 50,000$ m³/day (concentration of the influent wastewater)

$Q_W = 1250$ m³/day (sludge waste flow rate)

$NH_{30} = 50$ mg N/L (concentration of the ammonia in the influent stream)

$NH_{32} = 0.2$ mg N/L (concentration of the ammonia in the effluent stream)

$NO_{31} = 0.1$ mg N/L (concentration of nitrate in the anoxic reactor)

$X_{TOTR} = 0.9$ g/L (total concentration of microorganisms in the aerobic reactor)

What is the nitrate concentration in the effluent stream?

4.6 An activated sludge process for carbon and nitrogen removal operates with an anoxic reactor followed by an aerobic one (pre-denitrification). The final settling tank allows some microorganisms to escape with the effluent from the top.

The following values of the plant parameters are given:

$Q = 100,000$ m³/day (flow rate of the influent wastewater)

$Q_R = 200,000$ m³/day (recycle flow rate from the settling tank to the reaction tank)

$Q_I = 300,000$ m³/day (concentration of the influent wastewater)

$Q_W = 1500$ m³/day (sludge waste flow rate)

$NH_{30} = 50$ mg N/L (concentration of the ammonia in the influent stream)

$NH_{32} = 0.7$ mg N/L (concentration of the ammonia in the effluent stream)

$NO_{31} = 0.1$ mgN/L (concentration of nitrate in the anoxic reactor)

$X_{TOTR} = 1.3$ g/L (total concentration of microorganisms in the recycle stream)

$X_{TOT1} = 0.9$ g/L (total concentration of microorganisms in the anoxic reactor)

$X_{TOTEFF} = 10$ mg/L (total concentration of microorganisms in the plant effluent)

What is the ammonia concentration in the anoxic reactor?

4.7 A process for carbon and nitrogen removal operates with the performance described by the parameters below. Due to insufficient COD in the influent, the total nitrogen removal is not satisfactory. It is proposed to improve nitrogen removal by adding methanol to the anoxic reactor. It is estimated that methanol will be removed in the anoxic reactor with the consumption of 0.2 g N-NO3/g methanol. What is the amount of methanol (kg/day) that needs to be added to the system to achieve the maximum possible nitrogen removal in this system, without changing any of the flow rates of the various streams? Which could be a potential problem caused by the methanol addition?

$Q = 100,000$ m³/day (flow rate of the influent wastewater)

$Q_R = 50,000$ m³/day (recycle flow rate from the settling tank to the reaction tank)

$Q_I = 300,000$ m³/day (concentration of the influent wastewater)

$Q_W = 3500$ m³/day (sludge waste flow rate)

$NH_{30} = 50$ mg N/L (concentration of the ammonia in the influent stream)

$NH_{32} = 0.1$ mg N/L (concentration of the ammonia in the effluent stream)

$NO_{32} = 20$ mgN/L (concentration of nitrate in the aerobic reactor)

$X_{TOTR} = 2.0$ g/L (total concentration of microorganisms in the recycle stream)

CHAPTER 5

The Anaerobic Digestion Process

5.1 THE ANAEROBIC DIGESTER AS A CSTR WITHOUT RECYCLE

In the simplest approach, an anaerobic digester can be modelled as a continuous stirred tank reactor (CSTR). The influent is continuously fed to the reactor and the effluent is continuously withdrawn. A continuous flow rate of gas is also generated in the reactor. In a system like this the only design parameter, that is the only parameter that can be used to design the process, is the residence time (for a given composition and properties of the influent wastewater, and for given values of the temperature, pH and pressure of the reactor). In this case the residence time of the liquid and of the solids coincides, since there is no liquid–solid separation. The performance of the process as a function of the residence time can be simulated by writing and solving the appropriate mass balances.

In order to write the mass balances and simulate the performance of the process in a range of residence times, we will make the following assumptions:

- The only organic substrate in the feed is glucose;

- There are only three types of microorganisms in the reactor: the microorganisms that covert glucose to acetic acid and hydrogen (X_{GLU}), the ones that convert acetic acid to methane (X_{AC}) and the ones that convert hydrogen to methane (X_{H2});

- There are no microorganisms in the feed;

- The concentrations of the volatile species methane, hydrogen and carbon dioxide in the gas phase and in the liquid phase are in equilibrium;

- The gas phase is also saturated with water vapour.

A scheme of the anaerobic digester as a CSTR and of the reactions considered in this model is shown in Figure 5.1.

According to the stoichiometry and kinetics described in Chapter 2, the rate equations for the growth of the various microorganisms and for their endogenous metabolism can be written as follows.

Growth and endogenous metabolism of X_{GLU}:

$$r_{XGLU} = \mu_{maxGLU} \frac{GLU}{K_{SGLU} + GLU} X_{GLU} \quad r_{endXGLU} = -b_{GLU} X_{GLU}$$

Growth and endogenous metabolism of X_{AC}:

$$r_{XAc} = \mu_{maxAC} \frac{Ac}{K_{SAc} + Ac} X_{Ac} \quad r_{endXAc} = -b_{Ac} X_{Ac}$$

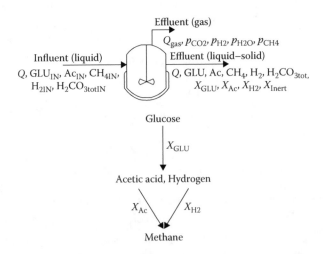

FIGURE 5.1 Scheme of the anaerobic digester modelled as a CSTR and of the main reactions used in the model.

Growth and endogenous metabolism of X_{H2}:

$$r_{XH2} = \mu_{maxH2} \frac{H_2}{K_{SH2} + H_2} X_{H2} \qquad r_{endXH2} = -b_{H2} X_{H2}$$

Once we have the expressions for the growth rates of the various micro-organisms, the production or consumption rates of the products and substrates can be expressed according to the method described in Chapter 2, introducing the growth yield of the microorganisms. They are reported below. All the rate Equations 5.1–5.5 written below are in the units of kg/m³.day (of course equivalent units can also be used).

Methane production rate:

$$r_{CH4} = r_{XAc} \left(\frac{0.267}{Y_{X/SGLU}} - 0.354 \right) + r_{XH2} \left(\frac{2}{Y_{X/SH2}} - 0.352 \right) \tag{5.1}$$

Acetic acid production rate:

$$r_{Ac} = r_{XGLU} \left(\frac{0.67}{Y_{X/SGLU}} - 0.88 \right) - \frac{r_{XAc}}{Y_{X/SAC}} \tag{5.2}$$

Hydrogen production rate:

$$r_{H2} = r_{XGLU} \left(\frac{0.044}{Y_{X/SGLU}} - 0.058 \right) - \frac{r_{XH2}}{Y_{X/SH2}} \tag{5.3}$$

Carbon dioxide production rate:

$$r_{CO2} = r_{XGLU} \left(\frac{0.49}{Y_{X/SGLU}} - 0.65 \right) + r_{XAc} \left(\frac{0.73}{Y_{X/SAc}} - 0.97 \right)$$

$$- r_{XH2} \left(\frac{5.50}{Y_{X/SH2}} + 0.097 \right) \tag{5.4}$$

Glucose 'production' rate (this is negative because glucose is consumed and not produced):

$$r_{GLU} = \frac{-r_{XGLU}}{Y_{X/SGLU}} \tag{5.5}$$

Once we have the expressions for the rate of production/consumption of the various components of the model, we can write the mass balances. From the solution of the mass balance the steady state of the digester will be calculated.

Mass balances can be written according to what was described in Chapter 3 and are written below (Equations 5.6 through 5.13), with reference to Figure 5.1. The variables GLU, Ac, H_2 and CH_4 represent the concentration of these species in the liquid phase (kg/m^3). p_{H2} and p_{CH4} are the partial pressures of the two gases in the gas phase (total pressure p_{tot}) and ρ_{H2} and ρ_{CH4} are the densities of these gases (kg/m^3) at the pressure and temperature of the system (of course, any alternative consistent sets of units can also be used).

Glucose:

$$Q \cdot GLU_{IN} + r_{GLU} \cdot V = Q \cdot GLU \tag{5.6}$$

Acetic acid:

$$Q \cdot AC_{IN} + r_{Ac} \cdot V = Q \cdot Ac \tag{5.7}$$

Hydrogen:

$$Q \cdot H_{2IN} + r_{H2} \cdot V = Q \cdot H_2 + Q_{gas} \frac{p_{H2}}{p_{tot}} \rho_{H2} \tag{5.8}$$

Methane:

$$Q \cdot CH_{4IN} + r_{CH4} \cdot V = Q \cdot CH_4 + Q_{gas} \frac{p_{CH4}}{p_{tot}} \rho_{CH4} \tag{5.9}$$

Microorganisms converting glucose to acetic acid (X_{GLU}):

$$\left(r_{XGLU} + r_{endGLU} \right) \cdot V = Q \cdot X_{GLU} \tag{5.10}$$

Microorganisms converting acetic acid to methane (X_{AC}):

$$\left(r_{XAc} + r_{endAC} \right) \cdot V = Q \cdot X_{Ac} \tag{5.11}$$

Microorganisms converting hydrogen to methane (X_{AC}):

$$\left(r_{XH2} + r_{endH2} \right) \cdot V = Q \cdot X_{H2} \tag{5.12}$$

Inerts (products of endogenous metabolism of microorganisms):

$$\left(-r_{endGLU} - r_{endH2} - r_{endAC} \right) \cdot V = Q \cdot X_{inert} \tag{5.13}$$

In addition to the mass balances above, the mass balance for carbon dioxide needs to be written. This balance is more complicated to write than the others because carbon dioxide reacts in water to form carbonic acid. Carbonic acid dissociates in water and this equilibrium is affected by pH. The easiest way to write the mass balance for carbon dioxide is to write the balance for the total concentration of carbonic acid. In this balance, differently from the previous balances, the concentrations will be expressed as mol/L, or kmol/m³, instead of kg/m³, because in the charge balance the concentrations need to be expressed in mol/L.

The mass balance for the total carbonic acid in the system is:

$$Q \cdot H_2CO_{3totin} + \frac{r_{CO2}}{44} V = Q \cdot H_2CO_{3tot} + Q_{gas} \frac{p_{CO2}}{p_{tot}} \frac{p_{CO2}}{44} \qquad (5.14)$$

In this balance the factor 44 (molecular weight of carbon dioxide) has been introduced to convert kg into kmol.

The balance for the total carbonic acid can be re-arranged to express H_2CO_{3tot} (the concentration of total carbonic acid in the liquid phase in the reactor) as a function of the concentration of dissolved carbon dioxide, assuming, as we always do, that the various species of carbonic acid in the liquid phase are in equilibrium. Using the equations for carbonic acid equilibrium developed in Chapter 2, we obtain:

$$Q \cdot H_2CO_{3totin} + \frac{r_{CO2}}{44} V = \frac{Q \cdot CO_2}{44} \cdot \frac{1 + K_{CO2} + \dfrac{K_{H2CO3}}{10^{-pH}} + \dfrac{K_{HCO3} K_{H2CO3}}{10^{-2pH}}}{K_{CO2}}$$

$$+ Q_{gas} \frac{p_{CO2}}{p_{tot}} \frac{p_{CO2}}{44} \qquad (5.15)$$

where the variable CO_2 represents the concentration of carbon dioxide in the liquid phase (as kg/m³; therefore, the factor 44 is required to convert it into kmol/m³). An additional equation is required to calculate the flow rate of the effluent gas Q_{gas}. We assume that the only species present in the gas phase are methane, carbon dioxide, hydrogen and water vapour. We assume that water vapour is present in the gas phase at the saturation partial pressure corresponding to the pressure and temperature of the system. Therefore, the flow rate of the effluent gas is given by the production rate of the volatile species minus the flow rate of these species which exit with the liquid effluent stream.

$$Q_{gas} = \frac{r_{CH4}V - Q(CH_4 - CH4_{IN})}{\rho_{CH4}} + \frac{r_{H2}V - Q(H_2 - H_{2IN})}{\rho_{H2}}$$

$$+ \frac{r_{CO2}V - Q(H_2CO_{3tot} - H_2CO_{3totIN})}{\rho_{CO2}} + Q_{H2O} \tag{5.16}$$

The assumption that the water content in the gas phase is at saturation means that:

$$Q_{H2O} = \frac{P_{SH2O}}{P_{tot}} Q_{gas} \tag{5.17}$$

where P_{SH2O} is the partial pressure of water in air at saturation, which is a function of temperature.

Re-arranging Equation 5.16 including Equation 5.17 we obtain:

$$Q_{gas} = \frac{\dfrac{r_{CH4}V - Q(CH_4 - CH4_{IN})}{\rho_{CH4}} + \dfrac{r_{H2}V - Q(H_2 - H_{2IN})}{\rho_{H2}} + \dfrac{r_{CO2}V - Q(H_2CO_{3tot} - H_2CO_{3totIN})}{\rho_{CO2}}}{1 - \dfrac{P_{SH2O}}{P_{tot}}}$$

$$\tag{5.18}$$

The mass balances written above can be re-arranged, dividing each term by the influent flow rate Q and introducing the hydraulic residence time defined as usual by:

$$HRT = \frac{V}{Q} \tag{5.19}$$

Also, since we are assuming that the liquid phase is in equilibrium with the gas phase, for the volatile species, methane, hydrogen and carbon dioxide, we can replace their concentration in the liquid phase as a function of their partial pressure in the liquid phase by using the following equilibrium relationships (where the k_{eq} factors are in kg/m³.atm):

$$H_2 = p_{H2} \cdot k_{eqH2} \tag{5.20}$$

$$CH_4 = p_{CH4} \cdot k_{eqCH4} \tag{5.21}$$

$$CO_2 = p_{CO2} \cdot k_{eqCO2} \tag{5.22}$$

With these re-arrangements the mass balances become:

$$GLU_{IN} + r_{GLU} \cdot HRT = GLU \tag{5.23}$$

$$AC_{IN} + r_{Ac} \cdot HRT = Ac \tag{5.24}$$

$$H_{2IN} + r_{H2} \cdot HRT = p_{H2} \cdot k_{eqH2} + \frac{Q_{gas}}{Q} \frac{p_{H2}}{p_{tot}} \rho_{H2} \tag{5.25}$$

$$CH_{4IN} + r_{CH4} \cdot HRT = p_{CH4} \cdot k_{eqCH4} + \frac{Q_{gas}}{Q} \frac{p_{CH4}}{p_{tot}} \rho_{CH4} \tag{5.26}$$

$$\left(r_{XGLU} + r_{endGLU} \right) \cdot HRT = X_{GLU} \tag{5.27}$$

$$\left(r_{XAc} + r_{endAc} \right) \cdot HRT = X_{Ac} \tag{5.28}$$

$$\left(r_{XH2} + r_{endH2} \right) \cdot HRT = X_{H2} \tag{5.29}$$

$$\left(-r_{endGLU} - r_{endH2} - r_{endAC} \right) \cdot HRT = X_{inert} \tag{5.30}$$

$$H_2CO_{3totin} + \frac{r_{CO2}}{44} HRT$$

$$= \frac{p_{CO2} \cdot k_{eqCO2}}{44} \cdot \frac{1 + K_{CO2} + \frac{K_{H2CO3}}{10^{-pH}} + \frac{K_{HCO3} K_{H2CO3}}{10^{-2pH}}}{K_{CO2}} + \frac{Q_{gas}}{Q} \frac{p_{CO2}}{p_{tot}} \frac{p_{CO2}}{44} \tag{5.31}$$

$$\frac{Q_{gas}}{Q} = \frac{\left(\dfrac{r_{CH4} \cdot HRT - \left(p_{CH_4} \cdot k_{eqCH4} - CH4_{IN} \right)}{p_{CH4}} + \dfrac{r_{H2} \cdot HRT - \left(p_{H_2} \cdot k_{eqH2} - H_{2IN} \right)}{p_{H2}} + \dfrac{r_{CO2} \cdot HRT - \left(H_2CO_{3tot} - H_2CO_{3totIN} \right)}{p_{CO2}} \right)}{1 - \dfrac{p_{SH2O}}{p_{tot}}} \tag{5.32}$$

Therefore, assuming that the inlet conditions and the temperature, pH and pressure of the reactor are known, assuming the hydraulic residence time (HRT) as a design parameter we have a system of 10 Equations, (5.23–5.32), in the following 10 unknowns: GLU, Q_{gas}, Ac, p_{H2}, p_{CH4}, p_{CO2}, X_{GLU}, X_{Ac}, X_{H2}, X_{inert}. The solution of the system of equations gives the values of all the variables that characterise the process.

Once the system of equations is solved, it is important to check the consistency of the solution by verifying the chemical oxygen demand (COD) and carbon balances. The equations for the COD and carbon balances are reported below, where for simplicity we have assumed that the feed only contains glucose and carbonic acid

COD balance:

$$\frac{COD_{out}}{COD_{in}}\left(\frac{kgCOD}{kgCOD}\right)=\frac{\left(\begin{array}{l}GLU\cdot1.067+Ac\cdot1.067+CH_4\cdot4+H_2\cdot8\\[4pt]+\left(X_{GLU}+X_{AC}+X_{H2}+X_{inert}\right)\cdot1.42+\\[4pt]\dfrac{Q_{gas}}{Q}\cdot\dfrac{1}{P_{tot}}\left(p_{H2}\cdot\rho_{H2}\cdot8+p_{CH4}\cdot\rho_{CH4}\cdot4\right)\end{array}\right)}{GLU_{IN}\cdot1.067} \qquad (5.33)$$

Carbon balance:

$$\frac{C_{out}}{C_{in}}\left(\frac{kmol}{kmol}\right)=\frac{\left(\begin{array}{l}GLU\cdot0.033+Ac\cdot0.033+CH_4\cdot0.0625+H_2CO_{3tot}\cdot1\\[4pt]+\left(X_{GLU}+X_{AC}+X_{H2}+X_{inert}\right)\cdot0.044+\\[4pt]\dfrac{Q_{gas}}{Q}\cdot\dfrac{1}{P_{tot}}\left(p_{CO2}\cdot\rho_{CO2}\cdot0.023+p_{CH4}\cdot\rho_{CH4}\cdot0.0625\right)\end{array}\right)}{GLU_{IN}\cdot0.033+H_2CO_{3tot}\cdot1} \qquad (5.34)$$

In Equation 5.34 it is worth remembering that the concentration of H_2CO_{3tot} is expressed in $kmol/m^3$, while all the other concentrations in the liquid phase are expressed in kg/m^3. The value of H_2CO_{3tot}, that is the concentration of total carbonic acid in the reactor, can be calculated from p_{CO2} (which is a calculated variable from Equations 5.23 through 5.32) from the equilibrium relationship of carbonic acid:

$$H_2CO_{3tot}=\frac{p_{CO2}\cdot k_{eqCO2}}{44}\cdot\frac{1+K_{CO2}+\dfrac{K_{H2CO3}}{10^{-pH}}+\dfrac{K_{HCO3}K_{H2CO3}}{10^{-2pH}}}{K_{CO2}}$$

5.1.1 Effect of the HRT on the Anaerobic Digestion Process

We have seen that, with the assumptions made in the previous section, the only design parameter for an anaerobic digester is the HRT. But how does the HRT affect the performance of the process?

There is a minimum residence time required for glucose removal. If the residence time is lower than this value, glucose is not consumed and its concentration in the reactor is equal to its concentration in the feed. The minimum residence time for glucose conversion to acetic acid can be calculated from the balance of X_{GLU}, Equation 5.27, which can be written, assuming $X_{GLUIN} = 0$:

$$\left(r_{GLU} + r_{endGLU} \right) \cdot HRT = X_{GLU} \tag{5.35}$$

Equation 5.35 can be re-arranged as, substituting the kinetic equations:

$$HRT = \frac{1}{\mu_{maxGLU} \dfrac{GLU}{K_{SGLU} + GLU} - b_{GLU}} \tag{5.36}$$

The minimum value of HRT for glucose removal can be obtained with the equation above by imposing that the glucose concentration in the reactor coincides with its concentration in the feed, that is $GLU = GLU_{IN}$:

$$HRT_{min,XGLU} = \frac{1}{\mu_{maxGLU} \dfrac{GLU_{IN}}{K_{SGLU} + GLU_{IN}} - b_{GLU}} \tag{5.37}$$

If the glucose concentration in the feed is much higher than K_{SGLU}, Equation 5.37 simplifies to:

$$HRT_{min} = \frac{1}{\mu_{maxGLU} - b_{GLU}} \tag{5.38}$$

Similarly, there is a minimum residence time for the growth of hydrogenotrophic and methanogenic microorganisms. This can be calculated by re-arranging the mass balances for X_{Ac} and X_{H2} (Equations 5.28 and 5.29):

$$HRT_{min,XH2} = \frac{1}{\mu_{maxH2} \dfrac{k_{eq} p_{H2}}{K_{SH2} + k_{eq} p_{H2}} - b_{H2}} \tag{5.39}$$

$$HRT_{min,XAc} = \frac{1}{\mu_{maxAc} \dfrac{Ac}{K_{SAc} + Ac} - b_{Ac}} \tag{5.40}$$

Equations 5.39 and 5.40 show that the minimum HRT for the growth of hydrogenotrophic and acetoclastic methanogens depends on the concentration of hydrogen and acetic acid in the liquid phase in the reactor, the hydrogen concentration being expressed as a function of the hydrogen partial pressure due to the assumption of gas–liquid equilibrium.

If

$$k_{eqH2}\, p_{H2} \gg K_{SH2} \qquad\qquad (5.41)$$

and

$$Ac \gg K_{SAC} \qquad\qquad (5.42)$$

the minimum residence time for the growth of hydrogenotrophic and acetoclastic methanogens becomes:

$$HRT_{min,XH2} = \frac{1}{\mu_{maxH2} - b_{H2}} \qquad\qquad (5.43)$$

$$HRT_{min,XAc} = \frac{1}{\mu_{maxAC} - b_{Ac}} \qquad\qquad (5.44)$$

In summary, under the stated hypotheses, assuming that the anaerobic digester is a CSTR without recycle (this means that the HRT and solids residence time [SRT] coincide), the performance of the anaerobic digester is determined by the HRT. Since an efficient conversion of glucose to methane is only possible with the coexistence of fermentative, acetoclastic and hydrogenotrophic microorganisms, the anaerobic digester has to be operated with an HRT which is high enough to avoid the washout of any of the three populations required. In other words, the minimum HRT for an anaerobic digester is given by:

$$HRT_{min} = max\left(HRT_{min,XGLU}, HRT_{min,XH2}, HRT_{min,XAc}\right) \qquad (5.45)$$

Of course, in practice the HRT in the anaerobic digester has to be considerably larger than HRT_{min}, if we want the maximum possible conversion of the feed into methane. This is shown numerically in Example 5.1.

From the mass balances it is also evident that the glucose concentration in the reactor does not depend on the influent feed concentration but only on the HRT. This is evident by re-arranging Equation 5.27 as:

$$\mu_{maxGLU} \frac{GLU}{K_{SGLU} + GLU} - b_{GLU} = \frac{1}{HRT} \qquad\qquad (5.46)$$

which shows that the glucose concentration in the reactor depends only on the HRT and on the kinetic parameters. The reason why the glucose concentration in the reactor does not depend on the influent concentration is that the higher the glucose concentration in the feed, the higher the concentration of microorganisms in the reactor, and therefore the higher the volumetric rate of glucose removal. This discussion is exactly the same done in Chapter 4 for the effluent substrate concentration of an activated sludge process, and Equation 5.44 is exactly analogous to Equation 4.15 (note that for the anaerobic digester without recycle considered in this section, HRT = SRT).

A similar discussion can be made for the concentration of acetic acid and hydrogen (or hydrogen partial pressure in the gas phase), which do not depend on the concentration of the feed but only on the HRT. This is evident by re-arranging Equations 5.28 and 5.29 as:

$$\mu_{maxAc} \frac{Ac}{K_{SGLU} + Ac} - b_{Ac} = \frac{1}{HRT} \qquad (5.47)$$

$$\mu_{maxH2} \frac{k_{eqH2} p_{H2}}{K_{SH2} + k_{eqH2} p_{H2}} - b_{H2} = \frac{1}{HRT} \qquad (5.48)$$

However, it is important to observe that Equations 5.47 and 5.48 are only valid if X_{AC} and X_{H2} are higher than 0; otherwise, if X_{AC} and X_{H2} are equal to 0, Equations 5.28 and 5.29 cannot be divided by X_{AC} and X_{H2}. If X_{AC} and X_{H2} are equal to 0, the concentration of acetic acid and hydrogen in the reactor will be dependent on the glucose concentration in the feed. If the HRT is large enough that acetoclastic and hydrogenotrophic microorganisms are present in the reactor, the concentration of acetic acid and hydrogen will decrease as the HRT increases. However, there is a minimum concentration of acetic acid and partial pressure of hydrogen which is possible to obtain even for very large values of the HRT and these are given by:

$$Ac_{min} = \frac{b_{Ac} K_{SAc}}{\mu_{maxAc} - b_{Ac}} \qquad (5.49)$$

$$p_{H2min} = \frac{1}{k_{eqH2}} \frac{b_{H2} K_{SH2}}{\mu_{maxH2} - b_{H2}} \qquad (5.50)$$

Note that Equations (5.46–5.48) correspond exactly to Equation 4.24 for the activated sludge process, considering that in a CSTR without recycle the HRT and the SRT coincide.

Example 5.1: Design of an anaerobic digester as a CSTR without recycle

a. Calculate the effect of the HRT on the performance of an anaerobic digester for the treatment of a wastewater with the parameters shown in Table 5.1;

b. Compare the results obtained in part a), with the case of reactor pH being equal to 6 and 7;

c. Compare the results obtained in part a) with the case of influent glucose concentration equal to 100 g/L.

TABLE 5.1   Summary of the Inlet Conditions, Reactor Conditions, Physical Properties and Kinetic Parameters

Inlet Conditions	
GLU_{IN}	$10 \ kg/m^3$
All the other inlet concentrations	$0 \ kg/m^3$
Reactor Conditions	
pH	8
p_{tot}	1 atm
Physical Properties	
k_{eqH2}	$0.014 \ kg/m^3/atm$
k_{eqCH4}	$0.017 \ kg/m^3/atm$
k_{eqCO2}	$1.2 \ kg/m^3/atm$
Kinetic and Stoichiometric Parameters	
μ_{maxGLU}	$4.2 \ day^{-1}$
K_{SGLU}	$0.02 \ kg/m^3$
$Y_{X/S \ GLU}$	$0.12 \ kg \ biomass/kg \ glucose$
b_{GLU}	$0.3 \ day^{-1}$
μ_{maxAc}	$0.36 \ day^{-1}$
K_{Sac}	$0.14 \ kg/m^3$
$Y_{X/S \ Ac}$	$0.04 \ kg \ biomass/kg \ acetic \ acid$
b_{Ac}	$0.2 \ day^{-1}$
μ_{maxH2}	$1.4 \ day^{-1}$
K_{SH2}	$0.000016 \ kg/m^3$
$Y_{X/S \ H2}$	$0.2 \ kg \ biomass/kg \ hydrogen$
b_{H2}	$0.3 \ day^{-1}$

Solution

a. The results of the calculations with the values of the parameters reported in Table 5.1 are shown in Figure 5.2. The main effects of the residence time on the performance of the digester is summarised below.

When HRT is higher than $HRT_{min,XGLU}$, the first process occurring is glucose conversion to acetic acid and hydrogen. This causes the concentration of acetic acid in the liquid medium and the partial pressure of hydrogen in the gas phase to increase. There is a narrow range of residence times in which hydrogen is the main component in the gas phase, the rest of the gases being carbon dioxide and water vapour. In this interval of residence times, the residence time in the reactor is too short to allow the growth of the microorganisms which consume hydrogen, X_{H2}. If the residence time is high enough to allow for the growth of X_{H2}, hydrogen is consumed and its partial pressure in the gas phase drops virtually to 0. The microorganisms X_{H2} produce methane while growing on hydrogen and carbon dioxide. That's why, when hydrogen concentration decreases methane concentration increases and carbon dioxide concentration decreases slightly. Acetic acid is not consumed until the residence time becomes high enough to allow the growth of X_{Ac}, the microorganisms that grow on acetic acid and produce methane and carbon dioxide. In this example, the minimum residence time for the growth of X_{Ac} is approximately 6 days. Acetate removal generates more gases, and the rate of gas production per unit volume of influent feed, Q_{gas}/Q, increases. The profiles for the various components of the biomass, X_{GLU}, X_{H2} and X_{AC} show that their concentration is 0 until the minimum residence time for their growth is reached. Then, as the residence time increases further, their concentration starts to decrease because of the endogenous metabolism with consequent increase in the inert solids. It is interesting to observe that the model predicts that as the residence time increases most of the solids will be inert, due to the endogenous metabolism.

In all the calculations, the COD and carbon balance, represented by Equations 5.33 and 5.34, are verified with an error of less than 1%, which can be attributed to rounding of the coefficients and to the inevitable numerical error in the solution of systems of equations.

b. In the previous simulations, the pH of the digester had been set to 8. The same simulations can be re-run with a different value of

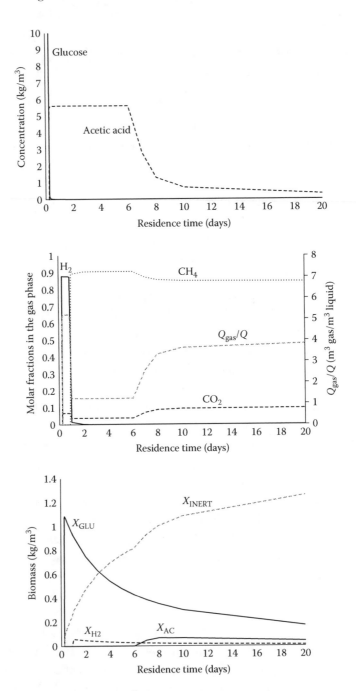

FIGURE 5.2   Effect of the residence time on the steady-state performance of the anaerobic digester. Parameter values in Table 5.1.

the pH, to see the effect of this parameter on the steady-state results. This is shown in Figure 5.3. The main effect of a decrease in pH is an increase in the fraction of carbon dioxide in the gas phase, with a consequent decrease of the molar fraction of methane. The reason for this is the effect of pH on carbonic acid equilibrium. As shown in Chapter 2, the fraction of the total carbonic acid that is present as carbon dioxide depends on pH according to the equation below:

$$\frac{[CO_2]}{[H_2CO_{3tot}]} = \frac{K_{CO2}}{1 + K_{CO2} + \dfrac{K_{H2CO3}}{10^{-pH}} + \dfrac{K_{HCO3}K_{H2CO3}}{10^{-2pH}}}$$

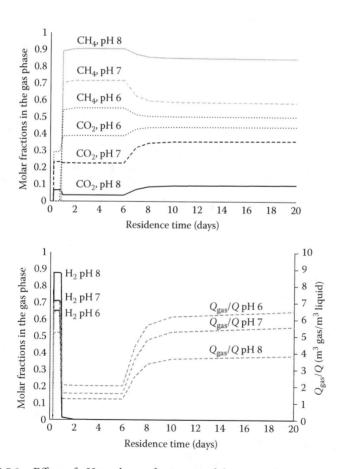

FIGURE 5.3   Effect of pH on the performance of the anaerobic digester.

If pH decreases, the concentration of carbon dioxide in the liquid phase increases, because the fraction of the total carbonic acid that is present as carbon dioxide increases. Since we assumed that the gas phase is in equilibrium with the liquid phase, then the partial pressure of carbon dioxide in the gas phase has to increase. The increase in the partial pressure of carbon dioxide causes a decrease in the partial pressure of hydrogen at low residence times and of methane at larger residence times. Also, the flow rate of produced gas increases, due to the larger volumes of carbon dioxide produced.

c. The effect of the concentration of the substrate in the feed is shown below in Figure 5.4. In this figure, the steady state of the digester for a glucose concentration in the feed of 10 and 100 g/L is compared. The minimum residence time which is required for glucose removal remains virtually unchanged. This is because the growth rate of the glucose-degrading microorganisms remains virtually the same for a glucose concentration of 10 and 100 g/L, because in both cases the glucose concentration is much higher than the K_{SGLU}. Similarly, the minimum residence time required for conversion of acetic acid to methane remains virtually unaffected by the feed concentration. Interestingly, increasing the substrate concentration in the feed causes a decrease in the partial pressure of methane and hydrogen in the gas phase. This is because the production rate of carbon dioxide increases, obviously, at higher substrate concentration in the feed. When the production rate of carbon dioxide is low, a larger fraction of this substance is absorbed by the liquid phase. However, when its production rate increases, the liquid phase cannot absorb most of the carbon dioxide produced which then transfers to the gas phase, causing its partial pressure to increase.

5.1.2 Calculation of pH in Anaerobic Digesters

In the simulations done so far, it was assumed that the pH in the digester was controlled and maintained at the desired value. In practice, however, the pH in anaerobic digesters is not controlled, so it is important to be able to calculate the pH as a function of the characteristics of the feed and of the residence time. The pH in the digester will depend on the various charged species that are present and that contribute to the charge balance. According to the simplified model used so far (i.e. glucose as only substrate, acetic acid, carbon dioxide and hydrogen, the only liquid

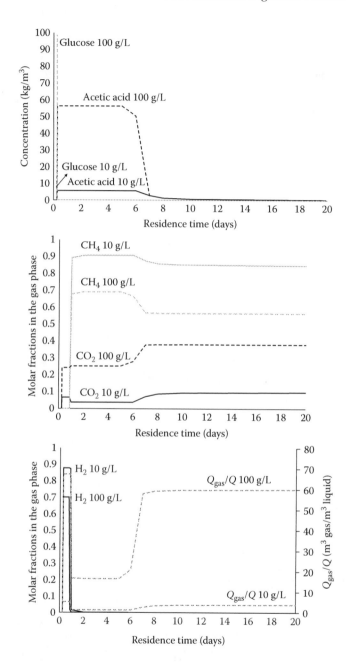

FIGURE 5.4   Effect of the substrate concentration in the feed on the performance of the anaerobic digester.

phase products of glucose fermentation), the species that contribute to the charge balance are acetic acid and carbonic acid.

The charge balance in the digester can be written therefore as follows:

$$\left[H_3O^+\right]+\left[\sum Cat\right]=\left[CH_3COO^-\right]+\left[HCO_3^-\right]+2\left[CO_3^{2-}\right]$$

$$+\left[\sum An\right]+\left[OH^-\right]$$

(5.51)

According to the equations derived in Chapter 2:

$$\frac{\left[CH_3COO^-\right]}{\left[CH_3COOH_{tot}\right]}=\frac{1}{1+\dfrac{10^{-pH}}{K_{CH3COOH}}}$$

where the concentration of total acetic acid $\left[CH_3COOH_{tot}\right]$ is the one calculated at the steady state of the digester using Equations 5.23–5.32, but expressed in mol/L, instead of kg/m³. Similarly the concentrations $\left[HCO_3^-\right]$ and $\left[CO_3^{2-}\right]$ can be expressed as a function of the concentration of carbon dioxide in the digester, also given by the solution of Equations 5.23–5.32. Using the equations derived in Chapter 2, and, according to the equilibrium conditions between the gas and liquid phases expressed by Equation 5.22, $\left[HCO_3^-\right]$ and $\left[CO_3^{2-}\right]$ can be expressed as:

$$\left[HCO_3^-\right]=\frac{K_{H2CO3}}{10^{-pH}}\frac{k_{eqCO2}\,p_{CO2}}{44\cdot K_{CO2}}$$

(5.52)

$$\left[CO_3^{2-}\right]=\frac{K_{H2CO3}K_{HCO3}}{10^{-2pH}}\frac{k_{eqCO2}\,p_{CO2}}{44\cdot K_{CO2}}$$

(5.53)

In the calculation of pH it is important to take into account the alkalinity and pH of the feed. As seen in Chapter 2, the alkalinity and pH of the feed determine the values of the concentrations $\left[\sum Cat-\sum An\right]$ and of the inlet total concentration of carbonic acid, $\left[H_2CO_{3tot,in}\right]$, which appears in Equation 5.31. The equations that give $\left[\sum Cat-\sum An\right]$ and $\left[H_2CO_{3tot,in}\right]$ as a function of the alkalinity and pH of the feed are reported below:

$$\left[\Sigma Cat-\Sigma An\right]_{feed}=\left[Alk_{mol}\right]_{feed}+\frac{K_w}{10^{-4.5}}-10^{-4.5}$$

$$[H_2CO_{3tot,in}] = \frac{10^{-pHfeed} - \dfrac{K_W}{10^{-pHfeed}} + [\Sigma Cat - \Sigma An]_{feed}}{\dfrac{K_{H2CO3}}{10^{-pHfeed}} + 2\dfrac{K_{HCO3}K_{H2CO3}}{10^{-2pHfeed}}}$$

$$\left(1 + K_{CO2} + \frac{K_{H2CO3}}{10^{-pHfeed}} + \frac{K_{HCO3}K_{H2CO3}}{10^{-2pHfeed}}\right)$$

Assuming a certain value of the alkalinity and pH of the feed, the concentration of total carbonic acid in the feed can be calculated and can be used in Equation 5.31.

Note that in the calculation of pH the charge balance has to be solved simultaneously to the mass balance Equations 5.23–5.32, because the value of the pH enters in Equation (5.31) and it affects carbon dioxide equilibrium.

The charge balance equation can therefore be re-written as:

$$10^{-pH} + \left[\sum Cat - \sum An\right]$$

$$= \frac{[CH_3COOH_{tot}]}{1 + \dfrac{10^{-pH}}{K_{CH3COOH}}} + \frac{K_{H2CO3}}{10^{-pH}} \frac{k_{eqCO2}p_{CO2}}{44 \cdot K_{CO2}} + 2\frac{K_{H2CO3}K_{HCO3}}{10^{-2pH}} \frac{k_{eqCO2}p_{CO2}}{44 \cdot K_{CO2}} + \frac{K_W}{10^{-pH}} \qquad (5.54)$$

where the term $\left[\sum Cat - \sum An\right]$ is calculated from the alkalinity of the feed. Equation 5.54, solved together with the mass balances (Equations 5.23 through 5.32), gives the pH of the digester.

Example 5.2: pH calculation in anaerobic digesters

a. Calculate the pH for the anaerobic digester of Example 5.1 (with glucose in the feed at 10 g/L) for a feed with pH 8 and an alkalinity in the range 0–2500 mgCaCO₃/L;

b. Assuming a feed of pH 8 and alkalinity 2500 mgCaCO₃/L, compare the pH in the digester calculated in part a) with the pH of an anaerobic digester where the glucose concentration in the feed is 100 g/L;

c. For a feed of alkalinity 0 mgCaCO₃/L and pH 8, calculate the concentration of strong base that needs to be added to the system to maintain its pH to 8, for feed concentration of 10 and 100 g/L of glucose.

Solution

a. Figure 5.5 shows the result of the pH calculations for various values of the alkalinity of the feed. It is evident that for low values of the alkalinity of the feed, the pH cannot be maintained in the right range under any residence time. Indeed, for methanogenic conditions, the pH should be at least equal to approximately 6.8. Instead with alkalinity of the feed up to 500 mgCaCO$_3$/L, the pH will be 6 or lower. This means that the digester will not be able to operate in these conditions in the absence of external pH control. Instead with values of the alkalinity of the feed of 1500 mgCaCO$_3$ or above, the pH of the digester can be maintained at approximately 6.8 or above under all the methanogenic conditions, that is when the residence time is 10 days or higher.

It is evident that all the pH profiles show a sharp drop, from a value of 8 to a value of 3–5, as soon as the minimum HRT for glucose fermentation is reached. This is because as soon as the HRT is large enough for glucose fermentation to start, acetic acid is the main product in the liquid phase, and this causes the pH to drop. The drop in pH is less for higher values of the alkalinity of the feed, because of the buffering effect of carbonic acid. When the HRT is large enough that acetic acid is removed from the system, the pH shows a sharp increase, but it still remains acidic because of the production of carbon dioxide and of the residual acetic acid.

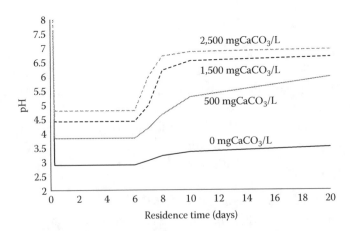

FIGURE 5.5 pH in the dige ster as a function of the residence time and of the alkalinity of the feed. Glucose concentration in the feed = 10 g/L.

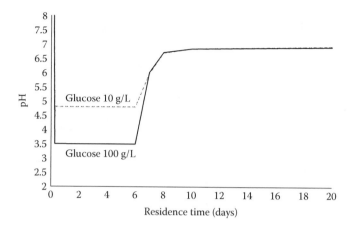

FIGURE 5.6 Calculated pH in the digester at two different concentrations of glucose in the feed. Feed alkalinity = 2500 mgCaCO₃/L.

b. Figure 5.6 shows the pH in the digester when the feed concentration is increased to 100 g/L. For the HRT values which give acetic acid as main product, the pH in the reactor is more acidic when the glucose concentration in the feed is larger. However, the concentration of the feed has no effect on the reactor pH when most of the acetic acid is removed from the reactor, because, as seen in Figure 5.4., the residual concentration of acetic acid at high values of the HRT does not depend on the feed concentration.

c. Another use of the charge balance is to calculate how much base needs to be added externally to maintain the pH at the desired value. This can be done by re-writing the charge balance as (assuming that the strong base used is, e.g. NaOH) shown in Equation 5.55, and solving for [Na⁺]. Figure 5.7 shows the concentration of strong base that needs to be added to the digester to maintain the pH to 8, for two values of the feed concentration.

$$10^{-pH} + \left[\sum Cat - \sum An\right] + \left[Na^+\right] = \frac{\left[CH_3COOH_{tot}\right]}{1 + \dfrac{10^{-pH}}{K_{CH3COOH}}} + \frac{K_{H2CO3}}{10^{-pH}} \frac{k_{eqCO2} p_{CO2}}{44 \cdot K_{CO2}}$$

$$+2 \frac{K_{H2CO3} K_{HCO3}}{10^{-2pH}} \frac{k_{eqCO2} p_{CO2}}{44 \cdot K_{CO2}} + \frac{K_w}{10^{-pH}}$$

(5.55)

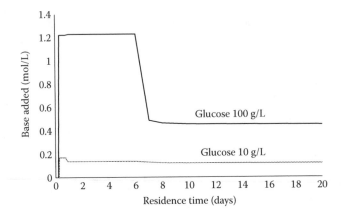

FIGURE 5.7   Concentration of strong base that needs to be added to the anaerobic digester to maintain the pH to 8. Feed alkalinity = 0 mgCaCO$_3$/L.

5.2 EXTENSION TO COMPLEX SUBSTRATES

So far we have hypothesised that glucose is the only organic species in the feed of the anaerobic digester. Usually, however, the feed of the digester is composed of many organic species, which, on anaerobic conditions, have a different metabolism depending on their chemical nature (see Chapters 1 and 2). Also, often at least some of the organic substrate in anaerobic digestion is insoluble and needs to be hydrolysed before its soluble monomeric building blocks can be metabolised. In this section we will adapt the model developed in Section 5.1.1 for a feed composed of a solid organic substrate, made of carbohydrates, proteins and lipids. For argument's sake, in the calculations done in this section we will assume that carbohydrates, proteins and lipids are equally abundant in the substrate, that is they each represent 33% by weight of the organic substrate. Of course, the model can be extended to any composition of the substrate, just by changing the coefficients. We will also assume that the only monomers which constitute carbohydrates and proteins are glucose and glycine, respectively. For lipids, we will assume that they are entirely composed of triglycerides of palmitic acid. Of course, the composition of the organic matter usually includes a variety of sugars, amino acids and fatty acids, and a more complex composition of the feed can also be modelled with the same approach, if enough information is available on the metabolism of the individual components. The anaerobic metabolism of glucose can be described by the reactions described in Chapter 2 and in Section 5.1.1,

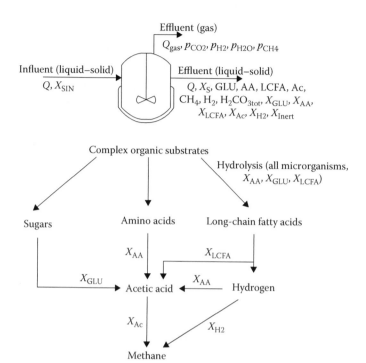

FIGURE 5.8 Scheme of the anaerobic digester fed with a complex organic substrate and of the main processes hypothesised in this section.

and, similarly, the reactions for the anaerobic metabolism of palmitic acid are reported in Chapter 2.

Figure 5.8 shows a scheme of the anaerobic digester and of the main processes hypothesised in this section.

In writing the rate equations, we hypothesise that all the active microorganisms hydrolyse the organic insoluble substrate, according to the hydrolysis kinetics described in Chapter 2 (and also used in Chapter 4, Section 4.1.6 for the hydrolysis of slowly biodegradable substrates, even though in that case the substrate was assumed to be soluble). We assume that all the microorganisms follow Monod kinetics. In addition to the notation used in Section 5.1.1, we call X_{AA} the microorganisms growing on amino acids (assumed to be glycine), we call X_{LCFA} the microorganisms growing on long-chain fatty acids (assumed to be palmitic acid), and we call AA and LCFA the concentrations of amino acids and LCFA in the liquid–solid phase. We will use the units of kg/m³ for all the concentrations in the liquid–solid phase, except for those relating to the carbonic acid equilibrium (which are in kmol/m^3, same as in Section 5.1.1). We will assume that the hydrolysis of the lipids (triglycerides) produces glycerol, alongside long-chain fatty acids, but we will assume that glycerol is not further metabolised (in reality glycerol is biodegradable under anaerobic conditions and it can be converted to organic acids, but for simplicity this is not taken into account here).

With the hypotheses described above, the rate equations for the hydrolysis of the organic substrate and for the growth of the various microbial populations are written below (all the rates have the units of kg/m^3.day).

Hydrolysis:

$$r_{hydr} = -k_h \frac{X_S \big/ \left(X_{GLU} + X_{AA} + X_{LCFA} \right)}{K_X + X_S \big/ \left(X_{GLU} + X_{AA} + X_{LCFA} \right)} \left(X_{GLU} + X_{AA} + X_{LCFA} \right) \quad (5.56)$$

Growth and endogenous metabolism of X_{GLU}:

$$r_{XGLU} = \mu_{maxGLU} \frac{GLU}{K_{SGLU} + GLU} X_{GLU} \quad r_{endXGLU} = -b_{GLU} X_{GLU}$$

Growth and endogenous metabolism of X_{AC}:

$$r_{XAc} = \mu_{maxAC} \frac{Ac}{K_{SAc} + Ac} X_{Ac} \quad r_{endXAc} = -b_{Ac} X_{Ac}$$

Growth and endogenous metabolism of X_{H2}:

$$r_{XH2} = \mu_{maxH2} \frac{H_2}{K_{SH2} + H_2} X_{H2} \quad r_{endXH2} = -b_{H2} X_{H2}$$

Growth and endogenous metabolism of X_{AA}:

$$r_{XAA} = \mu_{maxAA} \frac{AA}{K_{SAA} + AA} X_{AA} \quad r_{endXAA} = -b_{XAA} X_{AA}$$

Growth and endogenous metabolism of X_{LCFA}:

$$r_{XLCFA} = \mu_{maxLCFA} \frac{LCFA}{K_{SLCFA} + LCFA} X_{LCFA} \quad r_{endXLCFA} = -b_{XLCFA} X_{LCFA}$$

Once the rate equations for the growth of the microorganisms have been written, the rate equations for the production and consumption of the various components of the model can be written, using the reaction stoichiometries developed in Chapter 2. These rate equations are reported below, all of them in kg/m³.day. If the rate has a positive sign, this means the species is produced; if the sign is negative, the species is being consumed.

Production rates of the monomeric building blocks, sugars (assumed to be glucose), amino acids (assumed to be glycine), long-chain fatty acids (assumed to be palmitic acid), glycerol:

$$r_{GLU} = -0.33 r_{hydr} \cdot \frac{180}{162} - \frac{r_{XGLU}}{Y_{X/SGLU}} = -0.367 r_{hydr} - \frac{r_{XGLU}}{Y_{X/SGLU}} \tag{5.57}$$

$$r_{AA} = -0.33 r_{hydr} \cdot \frac{75}{57} - \frac{r_{XAA}}{Y_{X/SAA}} = -0.434 r_{hydr} - \frac{r_{XAA}}{Y_{X/SAA}} \tag{5.58}$$

$$r_{LCFA} = -0.33 r_{hydr} \cdot \frac{768}{806} - \frac{r_{XLCFA}}{Y_{X/SLCFA}} = -0.314 r_{hydr} - \frac{r_{XLCFA}}{Y_{X/SLCFA}} \tag{5.59}$$

$$r_{glycerol} = -0.33 r_{hydr} \cdot \frac{92}{806} = -0.038 r_{hydr} \tag{5.60}$$

Production rates of acetic acid, hydrogen, carbon dioxide and methane:

$$r_{Ac} = r_{XGLU}\left(\frac{0.67}{Y_{X/SGLU}} - 0.88\right) + r_{XAA}\left(\frac{0.805}{Y_{X/SAA}} - 1.77\right)$$
$$+ r_{XLCFA}\left(\frac{1.87}{Y_{X/SLCFA}} - 0.92\right) - \frac{r_{XAc}}{Y_{X/SAC}} \tag{5.61}$$

$$r_{H2} = r_{XGLU}\left(\frac{0.044}{Y_{X/SGLU}} - 0.058\right) - r_{XAA}\left(\frac{0.0268}{Y_{X/SAA}} - 0.059\right)$$
$$+ r_{XLCFA}\left(\frac{0.109}{Y_{X/SLCFA}} - 0.054\right) - \frac{r_{XH2}}{Y_{X/SH2}} \tag{5.62}$$

$$r_{CO2} = r_{XGLU}\left(\frac{0.49}{Y_{X/SGLU}} - 0.65\right) + r_{XAc}\left(\frac{0.73}{Y_{X/SAc}} - 0.97\right) + 0.65 r_{XAA}$$
$$- 0.59 r_{XLCFA} - r_{XH2}\left(\frac{5.50}{Y_{X/SH2}} + 0.097\right) \tag{5.63}$$

$$r_{CH4} = r_{XAc} \left(\frac{0.267}{Y_{X/SGLU}} - 0.354 \right) + r_{XH2} \left(\frac{2}{Y_{X/SH2}} - 0.352 \right) \tag{5.64}$$

Once the rate equations for the production/consumption of the various species have been written, we need to write the mass balances for all the components. In the mass balances below, the definition of HRT, Equation 5.19, is introduced.

Insoluble organic substrate:

$$X_{SIN} + r_{hydr} \cdot HRT = X_S \tag{5.65}$$

Sugars (glucose), amino acids (glycine), long-chain fatty acids (palmitic) and glycerol:

$$r_{GLU} \cdot HRT = GLU \tag{5.66}$$

$$r_{AA} \cdot HRT = AA \tag{5.67}$$

$$r_{LCFA} \cdot HRT = LCFA \tag{5.68}$$

$$r_{glycerol} \cdot HRT = GLYCEROL \tag{5.69}$$

Acetic acid, hydrogen and methane:

$$r_{Ac} \cdot HRT = Ac \tag{5.70}$$

$$r_{H2} \cdot HRT = p_{H2} \cdot k_{eq\,H2} + \frac{Q_{gas}}{Q} \frac{p_{H2}}{p_{tot}} p_{H2} \tag{5.71}$$

$$r_{CH4} \cdot HRT = p_{CH4} \cdot k_{eqCH4} + \frac{Q_{gas}}{Q} \frac{p_{CH4}}{p_{tot}} p_{CH4} \tag{5.72}$$

Microorganisms and inert products:

$$\left(r_{XGLU} + r_{endGLU} \right) \cdot HRT = X_{GLU} \tag{5.73}$$

$$\left(r_{XAA} + r_{endAA} \right) \cdot HRT = X_{AA} \tag{5.74}$$

$$\left(r_{\text{XLCFA}} + r_{\text{endLCFA}}\right) \cdot \text{HRT} = X_{\text{LCFA}} \tag{5.75}$$

$$\left(r_{\text{XAc}} + r_{\text{endAc}}\right) \cdot \text{HRT} = X_{\text{Ac}} \tag{5.76}$$

$$\left(r_{\text{XH2}} + r_{\text{endH2}}\right) \cdot \text{HRT} = X_{\text{H2}} \tag{5.77}$$

$$\left(-r_{\text{endGLU}} - r_{\text{endAA}} - r_{\text{endLCFA}} - r_{\text{endH2}} - r_{\text{endAC}}\right) \cdot \text{HRT} = X_{\text{inert}} \tag{5.78}$$

Total inorganic carbon (this is the only mass balance where the concentrations are in kmol/m^3, instead of kg/m^3) and is the same equation written in Section 5.1.1:

$$H_2CO_{3\text{totin}} + \frac{r_{CO2}}{44}\,\text{HRT} = \frac{p_{CO2} \cdot k_{eqCO2}}{44} \cdot \frac{1 + K_{CO2} + \dfrac{K_{H2CO3}}{10^{-pH}} + \dfrac{K_{HCO3}K_{H2CO3}}{10^{-2pH}}}{K_{CO2}}$$
$$+ \frac{Q_{gas}}{Q}\frac{p_{CO2}}{p_{tot}}\frac{\rho_{CO2}}{44} \tag{5.79}$$

The volumetric gas flow rate is expressed in the same way already seen in Section 5.1.1:

$$\frac{Q_{gas}}{Q} = \frac{\begin{pmatrix} \dfrac{r_{CH4} \cdot \text{HRT} - Q\left(p_{CH_4} \cdot k_{eqCH4} - CH4_{IN}\right)}{\rho_{CH4}} + \dfrac{r_{H2} \cdot \text{HRT} - Q\left(p_{H_2} \cdot k_{eqH2} - H_{2IN}\right)}{\rho_{H2}} \\[3mm] + \dfrac{r_{CO2} \cdot \text{HRT} - Q\left(H_2CO_{3tot} - H_2CO_{3totIN}\right)}{\rho_{CO2}} \end{pmatrix}}{1 - \dfrac{p_{SH2O}}{p_{tot}}} \tag{5.80}$$

In summary, Equations 5.65–5.80 constitute a system of 16 equations in the 16 unknowns X_S, GLU, AA, LCFA, GLYCEROL, Q_{gas}, Ac, p_{H2}, p_{CH4}, p_{CO2}, X_{GLU}, X_{AA}, X_{LCFA}, X_{Ac}, X_{H2}, and X_{inert}. By solving the system of equations as a function of the HRT, the concentration of all species can be determined and the value of the HRT that gives the desired performance can be found.

The COD and carbon balance need to be verified to check the consistency of the solution and they are shown below (adapted from Equations 5.33 and 5.34).

COD balance:

$$\frac{COD_{out}}{COD_{in}}\left(\frac{kgCOD}{kgCOD}\right)=\frac{\left(\begin{array}{c}X_S\cdot1.63+GLU\cdot1.067+AA\cdot0.64+LCFA\cdot2.875\\[4pt]+\,Glycerol\cdot1.217+Ac\cdot1.067+CH_4\cdot4+H_2\cdot8\\[4pt]+\left(X_{GLU}+X_{AC}+X_{H2}+X_{inert}\right)\cdot1.42\\[4pt]\dfrac{Q_{gas}}{Q}\cdot\dfrac{1}{p_{tot}}\left(p_{H2}\cdot\rho_{H2}\cdot8+p_{CH4}\cdot\rho_{CH4}\cdot4\right)\end{array}\right)}{X_{SIN}\cdot1.63} \tag{5.81}$$

Carbon balance:

$$\frac{C_{out}}{C_{in}}\left(\frac{kmol}{kmol}\right)=\frac{\left(\begin{array}{c}X_S\cdot0.0447+GLU\cdot0.033+AA\cdot0.027+LCFA\cdot0.021\\[4pt]+\,Ac\cdot0.033+CH_4\cdot0.0625+H_2CO_{3tot}\cdot1\\[4pt]+\left(X_{GLU}+X_{AC}+X_{H2}+X_{inert}\right)\cdot0.044\\[4pt]\dfrac{Q_{gas}}{Q}\cdot\dfrac{1}{p_{tot}}\left(p_{CO2}\cdot\rho_{CO2}\cdot0.023+p_{CH4}\cdot\rho_{CH4}\cdot0.0625\right)\end{array}\right)}{X_{SIN}\cdot0.0447+H_2CO_{3tot}\cdot1} \tag{5.82}$$

The COD factor of 1.63 for the substrate X_S is obtained considering its chemical composition, 33% carbohydrates (glucose), 33% proteins (glycine) and 33% triglycerides (palmitic acid and glycerol).

Example 5.3: Design of an anaerobic digester with complex substrates in the feed

Calculate the effect of the HRT on the performance of an anaerobic digester for the treatment of a wastewater with the parameters shown in Table 5.2. The feed is composed of a slowly biodegradable insoluble substrate composed of carbohydrates, proteins and triglycerides in equal proportions. Assume that the monomers constituting carbohydrates, proteins and triglycerides are, respectively, glucose, glycine and palmitic acid.

Solution

The solution of this example is obtained by simultaneously solving the system of mass balance Equations 5.65 through 5.80. The solution is shown in Figure 5.9.

TABLE 5.2   Summary of the Inlet Conditions, Reactor Conditions, Physical Properties and Kinetic Parameters. The Kinetic Parameters for X_{GLU}, X_{AA} and X_{LCFA} Have Been Assumed Equal for Simplicity

Inlet Conditions	
XS_{IN}	10 kg/m³
All the other inlet concentrations	0 kg/m³
Reactor Conditions	
pH	8
p_{tot}	1 atm
Physical Properties	
k_{eqH2}	0.014 kg/m³/atm
k_{eqCH4}	0.017 kg/m³/atm
k_{eqCO2}	1.2 kg/m³/atm
Kinetic and Stoichiometric Parameters	
k_h	5 kg substrate/kg biomass.day
K_X	1 kg substrate/kg biomass
μ_{maxGLU}	4.2 day⁻¹
K_{SGLU}	0.02 kg/m³
$Y_{X/S\ GLU}$	0.12 kg biomass/kg glucose
b_{GLU}	0.3 day⁻¹
μ_{maxAA}	4.2 day⁻¹
K_{SAA}	0.02 kg/m³
$Y_{X/S\ AA}$	0.12 kg biomass/kg glycine
B_{AA}	0.3 day⁻¹
$\mu_{maxLCFA}$	4.2 day⁻¹
K_{SLCFA}	0.02 kg/m³
$Y_{X/S\ LCFA}$	0.12 kg biomass/kg palmitic acid
B_{LCFA}	0.3 day⁻¹
μ_{maxAc}	0.36 day⁻¹
K_{Sac}	0.14 kg/m³
$Y_{X/S\ Ac}$	0.04 kg biomass/kg acetic acid
b_{Ac}	0.2 day⁻¹
μ_{maxH2}	1.4 day⁻¹
K_{SH2}	0.000016 kg/m³
$Y_{X/S\ H2}$	0.2 kg biomass/kg hydrogen
b_{H2}	0.3 day⁻¹

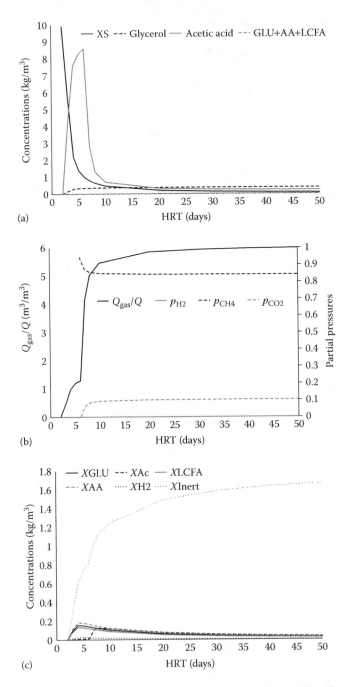

FIGURE 5.9 Effect of the HRT on the concentrations of insoluble substrate and monomers (a), gas production and composition (b) and microorganisms (c).

Figure 5.9 shows that the concentration of the substrate X_S increases rapidly if the HRT is too low, in this case lower than about five days. This is the minimum residence time required for an almost complete degradation of the substrate. If the HRT is slightly above this minimum value, acetate is the main product in the liquid phase and there is a modest gas production, almost entirely methane, due to the growth of hydrogenotrophic microorganisms on the hydrogen generated in the conversion of long-chain fatty acids and glucose. Note that in this example, since we have assumed for simplicity the same values of the kinetic parameters for X_{GLU}, X_{AA} and X_{LCFA}, the concentrations of their substrates GLU, AA and LCFA will be the same, as it is evident from Equations 5.70 through 5.72. Therefore, only the sum GLU+AA+LCFA is shown in Figure 5.9, showing that all these concentrations are very low, because they are consumed immediately after they are produced (in other words the hydrolysis of X_S is the rate limiting step). If the residence time is large enough to allow the growth of acetate-consuming microorganisms X_{AC}, acetate concentration decreases and gas production increases considerably. Note that the concentration of all microbial species increases when the HRT is above a certain minimum value corresponding to their wash-out, and then decreases at larger values of HRT due to endogenous metabolism. Since endogenous metabolism produces inert biomass, X_{inert} increases as the HRT increases. Also, note that, with the present model, glycerol concentration remains constant, because we are assuming that glycerol is not degraded. In reality, glycerol may be degradable under anaerobic conditions, in which case its concentration in the effluent would probably be very low.

The COD and carbon balances represented by Equations 5.81 and 5.82 are verified with an error of less than 1%.

5.3 ANAEROBIC DIGESTION WITH BIOMASS RECYCLE

So far we have modelled the anaerobic digestion process as a CSTR without recycle, which is a common configuration in many cases. However, anaerobic digestion can also be carried out by adding a settling tank after the digester and recycling the concentrated microorganisms back to the reactor. In this case the anaerobic digestion process is conceptually analogous to the activated sludge process and can be described, as we will see in this section, with mass balances derived using the same approach already seen in Chapter 4. The liquid–solid separation can also be carried out using membranes instead of a settling tank, and the mass balances will be the same.

In this section we will assume, for simplicity, that the only organic substrate in the feed is glucose (same assumption as in Section 5.1.1) and that separation of the microorganisms is perfect, that is there are no microorganisms in the liquid effluent stream after the liquid–solid separation. We will also assume that only the concentration of the microorganisms (insoluble species) change after the liquid–solid separation, while the concentration of the soluble species does not change (Figure 5.10).

The mass balances for the various species can be written using the same approach used in Section 5.1 and in Chapter 4. We need to write the balances for the soluble species in the reactor and for the microorganisms in both the reactor and the whole system. The mass balances for the soluble species in the reactor are the same as the equations used in Section 5.1.1 for the process without recycle.

Glucose in the reactor:

$$Q \cdot GLU_{IN} + r_{GLU} \cdot V = Q \cdot GLU \tag{5.6}$$

Acetic acid in the reactor:

$$Q \cdot AC_{IN} + r_{Ac} \cdot V = Q \cdot Ac \tag{5.7}$$

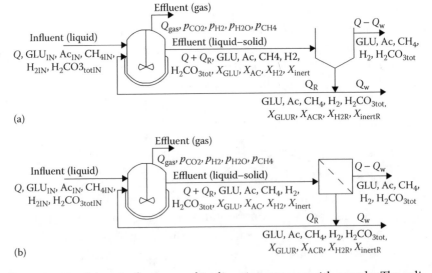

FIGURE 5.10   Scheme of an anaerobic digestion process with recycle. The soliliquid separation can be carried out using settling tanks (a) or membranes (b) or other methods.

Hydrogen in the reactor:

$$Q \cdot H_{2IN} + r_{H2} \cdot V = Q \cdot H_2 + Q_{gas} \frac{p_{H2}}{p_{tot}} \rho_{H2} \tag{5.8}$$

Methane in the reactor:

$$Q \cdot CH_{4IN} + r_{CH4} \cdot V = Q \cdot CH_4 + Q_{gas} \frac{p_{CH4}}{p_{tot}} \rho_{CH4} \tag{5.9}$$

X_{GLU} in the reactor:

$$\left(r_{XGLU} + r_{endGLU} \right) \cdot V + Q_R X_{GLUR} = \left(Q + Q_R \right) \cdot X_{GLU} \tag{5.83}$$

X_{AC} in the reactor:

$$\left(r_{XAc} + r_{endAC} \right) \cdot V + Q_R X_{ACR} = \left(Q + Q_R \right) \cdot X_{Ac} \tag{5.84}$$

X_{H2} in the reactor:

$$\left(r_{XH2} + r_{endH2} \right) \cdot V + Q_R X_{H2R} = \left(Q + Q_R \right) \cdot X_{H2} \tag{5.85}$$

Inerts in the reactor:

$$\left(-r_{endGLU} - r_{endH2} - r_{endAC} \right) \cdot V + Q_R X_{inertR} = \left(Q + Q_R \right) \cdot X_{inert} \tag{5.86}$$

Carbonic acid in the reactor:

$$Q \cdot H_2 CO_{3totin} + \frac{r_{CO2}}{44} V = Q \cdot H_2 CO_{3tot} + Q_{gas} \frac{p_{CO2}}{p_{tot}} \frac{\rho_{CO2}}{44} \tag{5.14}$$

The mass balances for microorganisms and inerts in the whole system are:

$$\left(r_{XGLU} + r_{endGLU} \right) \cdot V = Q_W \cdot X_{GLUR} \tag{5.87}$$

$$\left(r_{XAc} + r_{endAc} \right) \cdot V = Q_W \cdot X_{AcR} \tag{5.88}$$

$$\left(r_{XH2} + r_{endH2} \right) \cdot V = Q_W \cdot X_{H2R} \tag{5.89}$$

$$\left(-r_{endGLU} - r_{endH2} - r_{endAC} \right) \cdot V = Q_W \cdot X_{inertR} \tag{5.90}$$

The equation that gives the gas flow rate from the reactor is the same seen earlier:

$$Q_{gas} = \frac{r_{CH4}V - Q(CH_4 - CH4_{IN})}{\rho_{CH4}} + \frac{r_{H2}V - Q(H_2 - H_{2IN})}{\rho_{H2}}$$
$$+ \frac{r_{CO2}V - Q(H_2CO_{3tot} - H_2CO_{3totIN})}{\rho_{CO2}} + Q_{H2O} \tag{5.16}$$

We can define the hydraulic residence time, the solids residence time and the recycle ratio as usual:

$$HRT = \frac{V}{Q} \tag{5.19}$$

$$SRT = \frac{V(X_{GLU} + X_{AC} + X_{H2} + X_{inert})}{Q_W(X_{GLUR} + X_{ACR} + X_{H2R} + X_{inertR})} \tag{5.91}$$

$$R = \frac{Q_R}{Q} \tag{5.92}$$

With these definitions and introducing the equilibrium equations for hydrogen, carbon dioxide and methane, the mass balance equations can be re-written as follows:

$$GLU_{IN} + r_{GLU} \cdot HRT = GLU \tag{5.23}$$

$$AC_{IN} + r_{Ac} \cdot HRT = Ac \tag{5.24}$$

$$H_{2IN} + r_{H2} \cdot HRT = k_{eqH2} \cdot p_{H2} + \frac{Q_{gas}}{Q} \frac{p_{H2}}{p_{tot}} p_{H2} \tag{5.25}$$

$$CH_{4IN} + r_{CH4} \cdot HRT = k_{eqCH4} \cdot p_{CH4} + \frac{Q_{gas}}{Q} \frac{p_{CH4}}{p_{tot}} p_{CH4} \tag{5.26}$$

$$(r_{GLU} + r_{endGLU}) \cdot HRT + RX_{GLUR} = (1+R)X_{GLU} \tag{5.93}$$

$$(r_{XAc} + r_{endAC}) \cdot HRT + RX_{ACR} = (1+R)X_{Ac} \tag{5.94}$$

$$(r_{XH2} + r_{endH2}) \cdot HRT + RX_{H2R} = (1+R)X_{H2} \tag{5.95}$$

$$(-r_{endGLU} - r_{endH2} - r_{endAC}) \cdot HRT + RX_{inertR} = (1+R)X_{inert} \tag{5.96}$$

$$H_2CO_{3totin} + \frac{r_{CO2}}{44} HRT = \frac{p_{CO2} \cdot k_{eqCO2}}{44} \cdot \frac{\left(1 + K_{CO2} + \dfrac{K_{H2CO3}}{10^{-pH}} + \dfrac{K_{HCO3}K_{H2CO3}}{10^{-2pH}} \right)}{K_{CO2}}$$

$$+ \frac{Q_{gas}}{Q} \frac{p_{CO2}}{p_{tot}} \frac{p_{CO2}}{44} \tag{5.31}$$

$$\left(r_{GLU} + r_{endGLU} \right) = \frac{X_{GLUR}}{SRT} \frac{\left(X_{GLU} + X_{AC} + X_{H2} + X_{inert} \right)}{\left(X_{GLUR} + X_{ACR} + X_{H2R} + X_{inertR} \right)} \tag{5.97}$$

$$\left(r_{Ac} + r_{endAc} \right) = \frac{X_{AcR}}{SRT} \frac{\left(X_{GLU} + X_{AC} + X_{H2} + X_{inert} \right)}{\left(X_{GLUR} + X_{ACR} + X_{H2R} + X_{inertR} \right)} \tag{5.98}$$

$$\left(r_{H2} + r_{endH2} \right) = \frac{X_{H2R}}{SRT} \frac{\left(X_{GLU} + X_{AC} + X_{H2} + X_{inert} \right)}{\left(X_{GLUR} + X_{ACR} + X_{H2R} + X_{inertR} \right)} \tag{5.99}$$

$$\left(-r_{endGLU} - r_{endAc} - r_{endH2} \right) = \frac{X_{inertR}}{SRT} \frac{\left(X_{GLU} + X_{AC} + X_{H2} + X_{inert} \right)}{\left(X_{GLUR} + X_{ACR} + X_{H2R} + X_{inertR} \right)} \tag{5.100}$$

$$\frac{Q_{gas}}{Q} = \frac{\left(\dfrac{r_{CH4}HRT - \left(CH_4 - CH4_{IN} \right)}{p_{CH4}} + \dfrac{r_{H2}HRT - \left(H_2 - H_{2IN} \right)}{p_{H2}} + \dfrac{r_{CO2}HRT - \left(H_2CO_{3tot} - H_2CO_{3totIN} \right)}{p_{CO2}} \right)}{1 - \dfrac{p_{SH2O}}{P_{tot}}} \tag{5.32}$$

Equations 5.23–5.26, 5.31, 5.32, 5.93–5.100 represent a system of 14 equations in the 14 unknowns GLU, Q_{gas}, Ac, p_{H2}, p_{CH4}, p_{CO2}, X_{GLU}, X_{Ac}, X_{H2}, X_{inert}, X_{GLUR}, X_{AcR}, X_{H2R}, and X_{inertR}. For chosen design values of HRT, SRT and R, the solution of the system of equations gives the values of all the variables that characterise the process. Note that once the solution of the equations is calculated, we can immediately calculate from Equations 5.19, 5.91 and 5.92 the required values of the volume of the reactor, recycle flow rate and sludge waste flow rate Q_W.

5.3.1 Effect of the Choice of the Design Parameters on the Design Results

In order to choose appropriate values of the HRT, SRT and R, the same considerations done in Chapter 4, Section 4.1.1, can be done here. The most important consideration is that the effluent concentrations of all the species is only dependent on the SRT, and does not depend on the HRT or R. Similarly to what was seen in Section 4.1.1, this is evident by re-writing, example Equation 5.97 as:

$$\mu_{maxGLU} \frac{GLU}{K_{SGLU} + GLU} - b_{GLU} = \frac{1}{SRT} \tag{5.101}$$

where we have used:

$$\frac{\left(X_{GLU} + X_{AC} + X_{H2} + X_{inert} \right)}{\left(X_{GLUR} + X_{ACR} + X_{H2R} + X_{inertR} \right)} = \frac{X_{GLU}}{X_{GLUR}} \tag{5.102}$$

which comes from the fact that all the species settle (or in general are separated) together and the relative concentrations of all the species do not change with the liquid–solid separation.

Equations analogous to 5.101 can also be written for the other types of microorganisms, X_{Ac}, X_{H2}, and so we conclude that the effluent concentration of all the species, GLU, Ac, H_2, in the liquid phase is only determined by the SRT. Since the SRT determines the H_2 concentration in the liquid phase, due to the equilibrium condition it also determines p_{H2} in the gas phase.

Again in complete analogy with the activated sludge process, the SRT also determines the total mass of each type of microorganisms in the reactor, for a given value of GLU_{IN} and for a given influent flow rate. This is evident, for example by re-writing Equation 5.23 as:

$$GLU_{IN} - GLU = \frac{r_{GLU} \cdot X_{GLU} V}{Q} \tag{5.103}$$

and, since both GLU and r_{GLU} are only dependent on the SRT, we can conclude that $X_{GLU}V$, that is the total mass of the biomass that grows on glucose, depends only on the SRT. The same conclusion can be drawn for all the other species of microorganisms in the reactor.

With similar considerations as the ones done in Section 4.1.1, we can also conclude that the HRT determines the value of the biomass concentration in the reactor and that the recycle ratio R determines the concentration of the biomass in the recycle stream.

From these considerations it is evident that a process with biomass recycle gives the significant advantage over a process without recycle (Section 5.1), of achieving the same COD concentration in the effluent and the same gas production with a potentially much lower volume of the digester. Indeed, we can have a high value of the SRT, with consequently low effluent COD concentration and high gas production, with a low value of the HRT, that is a small reactor volume. However, in a process without recycle (Section 5.1), the HRT and the SRT coincide, and therefore, in order to achieve low COD in the effluent and high gas production, the HRT ($\equiv$SRT) needs to be large, and consequently the reactor volume needs also to be large.

Example 5.4: Design of an anaerobic digestion process with biomass recycle

Calculate the effect of the design parameters SRT, HRT and R for an anaerobic digestion process with biomass recycle. Use the parameters of Example 5.1 (Table 5.1).

Solution

As we have seen in Section 5.3.1 the concentration of the soluble species and the gas production depends only on the SRT. Since the influent glucose concentration and the kinetic parameters are the same as in Example 5.1, the concentration of all the soluble species, the gas production and the composition of the gas will be the same, as a function of the SRT, as in Example 5.1 (remember that in Example 5.1, the SRT and the HRT coincide).

What is different from Example 5.1 is the biomass concentration in the reactor, which depends on both the SRT and the HRT and is lower for higher values of the HRT. The total biomass concentration (i.e. $X_{GLU} + X_{H2} + X_{Ac} + X_{Inert}$) as a function of the SRT and HRT is shown in Figure 5.11 (the relative ratios of the various types of biomass is the same as in Example 5.1). Figure 5.11 also shows the total biomass concentration in the recycle stream.

The HRT also affects the value of the required sludge waste flow rate Q_W (Figure 5.12). The value of R affects the value of the biomass concentration in the recycle stream and of the required sludge waste flow rate (Figure 5.13).

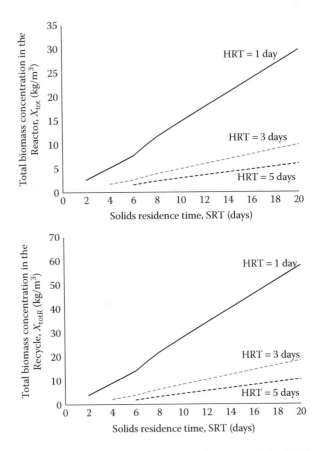

FIGURE 5.11 Total biomass concentration as a function of the HRT and SRT.

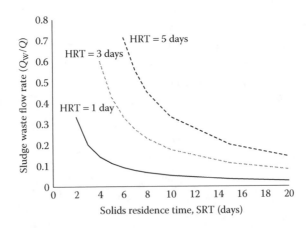

FIGURE 5.12 Effect of the SRT and HRT on the required sludge waste flow rate ($R = 1$).

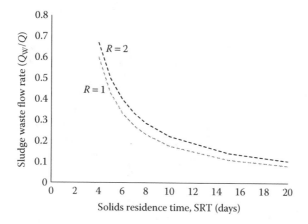

FIGURE 5.13 Effect of the SRT and R on the biomass concentration in the recycle stream and on the required sludge waste flow rate (HRT = 3 days).

5.4 TEMPERATURE CALCULATION IN ANAEROBIC DIGESTION

So far we have assumed that the digester works at constant temperature. However, the temperature in the reactor depends on the rate of heat generation due to the chemical reactions occurring and on the heat of heat supply or removal via the external heating or cooling medium (if provided). Once an anaerobic digestion process has been designed and the composition of all the inlet and outlet streams is known, the temperature in the digester can be calculated using the enthalpy balances for continuous processes developed in Chapter 3. In this section we will write the enthalpy balances for the digester without recycle, fed with glucose as only carbon source, described in Section 5.1 (Figure 5.1). We assume that the composition of all the streams is known, and also that the temperature of the feed is known. In analogy with the notation used in Chapters 2 and 3, we will call $H_i(T)$ the specific enthalpy of species i at the temperature T of the stream. The units of H_i are J/kg for all the species except for carbonic acid, where they are in J/kmol (because the concentration of H_2CO_{3tot} is expressed in kmol/m³ in our notation). We will call T_{feed} the temperature of the feed and T the temperature inside the reactor, which coincides with the outlet temperature of the liquid–solid and of the gas streams.

With this notation, considering an adiabatic digester, the enthalpy balance is:

$$
Q\begin{pmatrix} GLU_{IN}H_{GLU}(T_{feed})+ \\ NH_{3IN}H_{NH3}(T_{feed})+ \\ H_2O_{IN}H_{H2O}(T_{feed})+ \\ H_2CO_{3totIN}H_{H2CO3}(T_{feed}) \end{pmatrix} = Q\begin{pmatrix} GLU\,H_{GLU}(T)+ \\ NH_3H_{NH3}(T)+ \\ X_{tot}H_X(T)+ \\ H_2OH_{H2Oliq}(T)+ \\ H_2CO_{3tot}H_{H2CO3}(T) \end{pmatrix} + Q_{gas}\begin{pmatrix} \dfrac{pH_2}{p_{tot}}\rho_{H2}H_{H2}(T)+ \\ \dfrac{pCH_4}{p_{tot}}\rho_{CH4}H_{CH4}(T)+ \\ \dfrac{pCO_2}{p_{tot}}\rho_{CO2}H_{CO2}(T)+ \\ \dfrac{p_{H_2O}}{p_{tot}}\rho_{H2O}H_{H2Ovap}(T) \end{pmatrix}
\tag{5.104}
$$

Equation 5.104 can immediately be re-arranged as:

$$
GLU_{IN}H_{GLU}(T_{feed})+NH_{3IN}H_{NH3}(T_{feed})
$$
$$
+H_2O_{IN}H_{H2O}(T_{feed})+H_2CO_{3totIN}H_{H2CO3}(T_{feed})
$$
$$
= GLU\cdot H_{GLU}(T)+NH_3H_{NH3}(T)+X_{tot}H_X(T)
$$
$$
+H_2OH_{H2Oliq}(T)+H_2CO_{3tot}H_{H2CO3}(T)
\tag{5.105}
$$
$$
+\frac{Q_{gas}}{Q}\begin{pmatrix} \dfrac{pH_2}{p_{tot}}\rho_{H2}H_{H2}(T)+\dfrac{pCH_4}{p_{tot}}\rho_{CH4}H_{CH4}(T) \\[2mm] +\dfrac{pCO_2}{p_{tot}}\rho_{CO2}H_{CO2}(T)+\dfrac{p_{H_2O}}{p_{tot}}\rho_{H2O}H_{H2Ovap}(T) \end{pmatrix}
$$

If the concentrations of all the species are known and the heat of formations and specific heats of the various species are known, Equation 5.105 has the digester temperature T as only unknown. Solving Equation 5.105, therefore, gives the temperature in the digester for an adiabatic process.

If the digester is not adiabatic, the heat transfer rate with the fluid in the jacket needs to be considered, as described in Chapter 3. In this case the enthalpy balance follows immediately from a simple modification of Equation 5.104:

$$
Q\big(GLU_{IN}H_{GLU}(T_{feed})+NH_{3IN}H_{NH3}(T_{feed})+H_2O_{IN}H_{H2O}(T_{feed})+H_2CO_{3totIN}H_{H2CO3}(T_{feed})\big)
$$
$$
+UA(T_j-T)=Q\big(GLU\cdot H_{GLU}(T)+NH_3H_{NH3}(T)+X_{tot}H_X(T)+H_2OH_{H2Oliq}(T)
$$
$$
+H_2CO_{3tot}H_{H2CO3}(T)\big)+Q_{gas}\bigg(\frac{pH_2}{p_{tot}}\rho_{H2}H_{H2}(T)+\frac{pCH_4}{p_{tot}}\rho_{CH4}H_{CH4}(T)+\frac{pCO_2}{p_{tot}}\rho_{CO2}H_{CO2}(T)
$$
$$
+\frac{p_{H_2O}}{p_{tot}}\rho_{H2O}H_{H2Ovap}(T)\bigg)
\tag{5.106}
$$

which can be immediately re-arranged as:

$$GLU_{IN}H_{GLU}(T_{feed})+NH_{3IN}H_{NH3}(T_{feed})+H_2O_{IN}H_{H2O}(T_{feed})$$

$$+H_2CO_{3totIN}H_{H2CO3}(T_{feed})+\frac{UA(T_J-T)HRT}{V}=GLU\cdot H_{GLU}(T)$$

$$+NH_3H_{NH3}(T)+X_{tot}H_X(T)+H_2OH_{H2Oliq}(T)+H_2CO_{3tot}H_{H2CO3}(T)$$

$$+\frac{Q_{gas}}{Q}\left(\begin{array}{l}\dfrac{pH_2}{P_{tot}}\rho_{H2}H_{H2}(T)+\dfrac{pCH_4}{P_{tot}}\rho_{CH4}H_{CH4}(T)+\\[3mm]\dfrac{pCO_2}{P_{tot}}\rho_{CO2}H_{CO2}(T)+\dfrac{pH_2O}{P_{tot}}\rho_{H2O}H_{H2Ovap}(T)\end{array}\right)$$

(5.107)

Assuming that the heat transfer coefficient U and the heat transfer area A are known, Equation 5.107 has two unknowns, T and T_J; therefore, it needs to be coupled with the enthalpy balance for the jacket fluid:

$$w_J c_{pJ}(T_{JIN}-T_J)=UA(T_J-T)$$ (5.108)

where w_J and T_{JIN} are the mass flow rate and temperature of the jacket fluid (known quantities). Equations 5.107 and 5.108 constitute a system of two equations in the two unknowns T and T_J. The solution of the equations gives the desired temperature in the reactor. Another use of Equations 5.107 and 5.108 is the calculation of the required flow rate of the jacket fluid to maintain a desired temperature in the digester.

Example 5.5: Enthalpy balances in anaerobic digesters

Consider the anaerobic digester of Example 5.2, with alkalinity of the feed equal to 2500 mgCaCO$_3$/L. Assume the feed is at a temperature of 15 C and the HRT is 20 days.

a. Assuming adiabatic conditions, calculate the temperature of the digester for a glucose feed concentration of 10 and 100 g/L;
b. With the feed concentration of 10 g/L, calculate the flow rate of the heating fluid which is required to maintain the reactor at 35°C. Assume that the heating fluid is available at 50°C, the heat transfer coefficient U is 100 W/m².0C, reactor volume is 1000 m³ and the area available for heat transfer is 600 m².

Solution

a. The steady-state values of the reactor conditions at HRT = 10 days, obtained by solving the mass balances as described in Sections 5.1 and 5.2, are summarised below.

Parameter	Feed 10 g/L	Feed 100 g/L
	Feed	
Glucose	$10 \ kg/m^3$	$100 \ kg/m^3$
H_2CO_{3tot}	$5.09 \cdot 10^{-2} \ kmol/m^3$	$5.09 \cdot 10^{-2} \ kmol/m^3$
	Reactor	
Glucose	$0.017 \ kg/m^3$	$0.017 \ kg/m^3$
Acetic acid	$0.26 \ kg/m^3$	$0.32 \ kg/m^3$
X_{tot}	$1.49 \ kg/m^3$	$14.9 \ kg/m^3$
pH_2	0.000357 atm	0.00038 atm
pCO_2	0.462 atm	0.473 atm
pCH_4	0.482 atm	0.470 atm
pH_2O	0.055 atm	0.055 atm
Q_{gas}/Q	$6.8 \ m^3/m^3$	$71.8 \ m^3/m^3$
pH	6.93	6.91

First of all we need to calculate the other concentrations required for the enthalpy balance.

For ammonia, we assume that the inlet concentration of ammonia is the minimum required to support the growth of the microorganisms, that is:

$$NH3_{IN}\left(\frac{kgNH_3}{m^3}\right) = X_{tot} \cdot 0.12 \cdot \frac{17}{14}$$

This gives $NH3_{IN} = 0.18 \left(kg/m^3\right)$ and $NH3_{IN} = 1.8 \left(kg/m^3\right)$ for a feed concentration of 10 and 100 kg/m³ respectively. The assumption that the ammonia is the minimum required makes the ammonia concentration in the reactor (and in the effluent) equal to 0.

The concentration of total inorganic carbon in the reactor, H_2CO_{3tot}, can be calculated immediately from the partial pressure of carbon dioxide in the reactor, assuming, as usual, equilibrium conditions:

$$\left[H_2CO_{3tot}\right]\left(\frac{kmol}{m^3}\right) = \left(1 + K_{CO2} + \frac{K_{H2CO3}}{10^{-pH}} + \frac{K_{HCO3}K_{H2CO3}}{10^{-2pH}}\right)\frac{k_{eqCO2}\,p_{CO2}}{K_{CO2}}\frac{1}{44}$$

and, by substituting the values of p_{CO2} and of pH for feed concentration of 10 and 100 kg/m³, we obtain:

$$[H_2CO_{3tot}] = 5.82 \times 10^{-2} \text{ and } [H_2CO_{3tot}] = 5.76 \times 10^{-2} \text{ kmol/m}^3 \text{ for}$$
feed concentrations of 10 and 100 kg/m³, respectively.

Regarding water, the inlet concentration (H_2O_{IN}) can be taken equal to 990 and 900 kg/m³, for the low and high feed concentrations, respectively, where for simplicity we have considered only glucose as other component in the feed. In the reactor some water is generated by the reaction and, although this has been neglected in all the mass balances, it has to be considered in the enthalpy balance in order to obtain results with reasonable accuracy.

The water formed due to the reaction can be calculated using an overall mass balance on the reaction, that is:

$$GLU_{IN} + H_2CO_{3totIN} \cdot 61 + NH_{3IN}$$

$$= GLU + Ac + H_2CO_{3tot} \cdot 61 + \frac{Q_{gas}}{Q} \frac{pH_2}{p_{tot}} \rho_{H2}$$

$$+ \frac{Q_{gas}}{Q} \frac{pCH_4}{p_{tot}} \rho_{CH4} + \frac{Q_{gas}}{Q} \frac{pCO_2}{p_{tot}} \rho_{CO2} + H_2O_{formed}$$

From this equation we obtain values for H_2O_{formed} of 1.97 and 5.83 kg/m³ for feed concentrations of 10 and 100 kg/m³, respectively.

The concentration of the water in the reactor is equal to the concentration in the feed, plus the water formed in the reaction minus the water evaporated.

$$H_2O = H_2O_{IN} + H_2O_{formed} - \frac{Q_{gas}}{Q} \frac{pH_{2O}}{p_{tot}} \rho_{H2O}$$

This equation gives a concentration of water of 991.7 and 903.0 kg/m³ for the low and high feed concentrations, respectively.

Once all the concentrations in and out have been calculated, we need to find expression for the enthalpies as a function of the temperature. This can be done as shown in Chapters 2 and 3, and the values for the various species are reported below.

For glucose:

$$H_{GLU}(T)\left(\frac{J}{kg}\right) = \Delta H_{fGLU}(25°C) + \lambda_{dissolutionGLU}(25°C) + c_{PH2O}(T-25)$$

$$= -7.1 \times 10^6 + 6.1 \times 10^4 + 4186(T-25)$$

$$= -7.04 \times 10^6 + 4186(T-25)$$

For ammonia:

$$H_{NH3}(T)\left(\frac{J}{kg}\right) = \Delta H_{fNH3}(25°C) + c_{PH2O}(T-25)$$

$$= -4.8 \times 10^6 + 4186(T-25)$$

For biomass:

$$H_X(T)\left(\frac{J}{kg}\right) = \Delta H_{fX}(25°C) + c_{PH2O}(T-25)$$

$$= -6.8 \times 10^6 + 4186(T-25)$$

For water:

$$H_{H2Oliq}(T)\left(\frac{J}{kg}\right) = \Delta H_{fH2O}(25°C) + c_{PH2O}(T-25)$$

$$= -1.6 \cdot 10^7 + 4186(T-25)$$

For water vapour:

$$H_{H2Ovap}(T)\left(\frac{J}{kg}\right) = \Delta H_{fH2O}(25°C) + \lambda_{evap}(25°C) + c_{PH2Ovap}(T-25)$$

$$= -1.6 \cdot 10^7 + 2.44 \cdot 10^5 + 1.86 \cdot 10^3 (T-25)$$

$$= -1.58 \cdot 10^6 + 1.86 \cdot 10^3 (T-25)$$

For carbon dioxide:

$$H_{CO2}(T)\left(\frac{J}{kg}\right) = \Delta H_{fCO2}(25°C) + c_{PCO2gas}(T-25)$$

$$= -8.9 \times 10^6 + 910(T-25)$$

For carbonic acid:

$$H_{H2CO3}(T)\left(\frac{J}{kmol}\right) = \Delta H_{fH2CO3}(25°C) + c_{PH2CO3}(T-25)$$

$$= -6.9\times10^8 + 2.5\times10^5(T-25)$$

For hydrogen:

$$H_{H2}(T)\left(\frac{J}{kg}\right) = \Delta H_{fH2}(25°C) + c_{H2PCO2gas}(T-25)$$

$$= 0 + 14,300(T-25)$$

For methane:

$$H_{CH4}(T)\left(\frac{J}{kg}\right) = \Delta H_{fCH4}(25°C) + c_{CH4gas}(T-25)$$

$$= -4.7\times10^6 + 2200(T-25)$$

Therefore, the enthalpy balance can be written as follows for the 10 kg/m³ feed:

$$10\left[-7.04\times10^6 + 4186(15-25)\right] + 0.18\left[-4.8\times10^6 + 4186(15-25)\right]$$

$$+ 990\left[-1.6\times10^7 + 4186(15-25)\right] + 0.0509\left[-6.9\times10^8 + 2.5\times10^5(15-25)\right]$$

$$= 0.0017\times\left[-7.04\times10^6 + 4186(T-25)\right] + 1.49\left[-6.8\times10^6 + 4186(T-25)\right]$$

$$+ 991.7\left[-1.6\times10^7 + 4186(T-25)\right] + 0.0582\left[-6.9\times10^8 + 2.5\times10^5(T-25)\right]$$

$$+ 6.8\times0.00036\times0.0079\times\left[14,300(T-25)\right]$$

$$+ 6.8\times0.482\times0.63\left[-4.7\times10^6 + 2200(T-25)\right]$$

$$+ 6.8\times0.462\times1.74\left[-8.9\times10^6 + 910(T-25)\right]$$

$$+ 6.8\times0.055\times0.711\left[-1.58\times10^6 + 1.86\times10^3(T-25)\right]$$

The solution of this equation gives $T = 23.5°C$ and this is the temperature in the digester under adiabatic conditions.

The calculation can be repeated for the case of feed concentration equal to 100 kg/m³ and in this case we obtain a temperature of $T = 44.6°C$. As expected the temperature increase is higher for the more concentrated feed, because the heat generated is proportional to the mass of reagents which have been processed.

b. If heat is provided via a jacket, or a generic heat surface, we need to add the term of enthalpy addition in the heat balance for the fluid in the digester, and we need to couple this enthalpy balance with the enthalpy balance for the fluid in the jacket.

$$10\left[-7.04\cdot10^6 +4186(15-25)\right]+0.18\left[-4.8\cdot10^6 +4186(15-25)\right]$$

$$+990\left[-1.6\cdot10^7 +4186(15-25)\right]+0.0509\left[-6.9\cdot10^8 +2.5\cdot10^5(15-25)\right]$$

$$=0.0017\cdot\left[-7.04\cdot10^6 +4186(T-25)\right]+1.49\left[-6.8\cdot10^6 +4186(T-25)\right]$$

$$+991.7\left[-1.6\cdot10^7+4186(T-25)\right]+0.0582\left[-6.9\cdot10^8+2.5\cdot10^5(T-25)\right]$$

$$+6.8\cdot0.00036\cdot0.0079\cdot\left[14{,}300(T-25)\right]$$

$$+6.8\cdot0.482\cdot0.63\left[-4.7\cdot10^6 +2200(T-25)\right]$$

$$+6.8\cdot0.462\cdot1.74\left[-8.9\cdot10^6 +910(T-25)\right]$$

$$+6.8\cdot0.055\cdot0.711\left[-1.58\cdot10^6 +1.86\cdot10^3(T-25)\right]$$

$$+\frac{100\cdot86{,}400\cdot600\cdot(T-T_\text{J})\cdot20}{1{,}000}$$

$$w_\text{J}4186(50-T_J)=100\cdot86{,}400\cdot600(T_\text{J}-T)$$

The solution of these two equations gives the temperature in the reactor and in the jacket. The results are plotted in Figure 5.14 as a function of the flow rate of the heating fluid in the jacket. From the figure it is evident that a jacket flow rate of approximately 40,000 kg/day is required to maintain the digester at a temperature of 35°C.

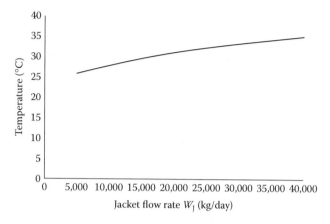

FIGURE 5.14   Effect of the jacket flow rate on the temperature in the anaerobic digester.

5.5 KEY POINTS

- The design of anaerobic digestion processes can be made with the same approach used for the activated sludge process, that is writing the mass balances for all the species of interest, choosing the values of the design parameters and solving the system of equations. Since in anaerobic digestion multiple populations of microorganisms need to coexist, mass balances need to be written for all the populations in the system. The rate of production and consumption of all the species need to be written on the basis of the stoichiometry of the respective reactions;

- The design approach is the same, in principle, for a feed composed of readily biodegradable substrates (glucose was used as an example) and for a feed composed of more complex substrates (a hypothetical substrate composed of carbohydrates, proteins and fats has been used as an example). For complex substrates, the hydrolysis process needs to be accounted for in the model, with a suitable choice of the kinetic model and of the stoichiometry;

- In the simplest process configuration, anaerobic digester as CSTR without recycle, the only design parameter is the HRT, which coincides with the SRT. By increasing the HRT (and therefore the SRT) the degradation of the fed substrate improves. For short values of the SRT, the digester mainly converts the substrate (glucose was used as an example) to acetic acid and hydrogen, while for longer values of the SRT acetoclastic and hydrogenotrophic methanogens are able to grow and convert acetic acid and hydrogen to methane. Therefore, at long SRT methane is the main digestion product;

- pH is an important parameter in anaerobic digestion and needs to be maintained in the right range for the successful performance of the plant. The pH of the digester can be calculated from the design results using the charge balance and the mass balance for inorganic carbon. The pH in the digester tends to be more acidic than the feed due to the generation of organic acids (e.g. acetic acid) and carbon dioxide. The pH in the digester is a function of the SRT and of the alkalinity of the feed. The higher the alkalinity of the feed, the easier it is to maintain the digester at the desired pH;

- In anaerobic digestion processes with recycle, there are three design parameters, SRT, HRT and R, analogously as for the activated sludge process. The SRT is the main design parameter, because it determines the effluent substrate concentration, the total mass of biomass in the reactor, the gas production and the gas composition. The effect of HRT and R is analogous to their effect on the activated sludge process;

- Once the process has been designed, and the values of all the variables that characterise the process are known, enthalpy balances can be carried out to calculate the temperature of the digester, with or without an external heating or cooling fluid. Anaerobic digestion typically generates heat and the increase in the reactor temperature is higher for higher concentrations of the feed. Based on heat balances, the required temperature and/or flow rate of the heating or cooling fluid can be calculated.

Questions and Problems

5.1 An anaerobic digester operating as a CSTR at steady state is fed with a wastewater at a concentration of 10 gCOD/L. The flow rate of the wastewater is 100 m³/day. The effluent of the reactor contains 0.5 g/L of microorganisms, 1 g/L of acetic acid and 0.3 g/L of butyric acid. Calculate the methane production rate in Nm³/day.

5.2 An anaerobic digester is fed with an influent stream of 20 gCOD/L, at a flow rate of 1000 m³/day. The liquid–solid effluent has a total COD of 1.5 gCOD/L. What is the methane production rate (Nm³/day)?

5.3 It is desired to use anaerobic digestion to generate the electricity required by a small community. The electricity energy required by the community is 200,000 kWh/year. The anaerobic digester is to be fed with an organic waste which is 20% solids (20 g of solid per 100 g of total waste). The COD of the solids is 1.5 gCOD/g solid. Assume that the influent solids are converted to methane with a yield of 0.9 g methane (as COD)/g waste (as COD). Assume that the biogas obtained from a digester is converted into electricity with an efficiency of 50%. Which is the daily amount of waste which needs to be processed in the digester? Assume that the digester operates throughout the year continuously.

The molecular weight of methane is 16 and its heat of combustion is 890 kJ/mol.

5.4 An anaerobic digester is operated with a feed flow rate of 5 tonne/hr. The feed contains 5% of organic matter. This organic matter has a carbon content of 40% (by weight). The carbon in the feed is converted 45% into methane, 45% into CO_2 and 10% into microorganisms. The methane produced is converted into electricity at the site with an efficiency of 60%. The AD plant is operated for 8000 hr/year. The microorganisms produced are used as fertiliser. Calculate the following:

a. volume of methane produced in Nm³/hr;

b. the electricity produced per year (kWh/year);

c. the amount of solid fertiliser produced per year.

The calorific value of methane is 890 kJ/mol.

5.5 An anaerobic digester processes a feed with the following characteristics:

Feedstock	Organic Matter Content (%, w/w)	Flow Rate (m³/day)
Cattle manure	5	10
Food waste	10	10
Whisky wastewater	3	20

The residence time in the digester is 30 days and the biogas produced is 1000 Nm³/day. The composition of the biogas is 60% CH_4, 40% CO_2 in volume. Calculate the following:

a. the volume of the digester;

b. the fraction of the organic matter that is converted to methane (g methane/g solids).

CHAPTER 6

The Sequencing Batch Reactor[*]

6.1 THE SEQUENCING BATCH REACTOR FOR CARBON REMOVAL

As we have seen in Chapter 1, the sequencing batch reactor (SBR) is an activated sludge process that is operated as a sequence of phases, with reaction and settling taking place in the same vessel. For carbon removal, the typical sequence of phases is fill, reaction, sludge withdrawal, settling and effluent withdrawal, followed by a new feed that starts a new cycle. Sludge withdrawal is typically done either from the completely mixed reactor after the reaction phase or from the settled sludge at the end of the settling phase. In developing the model in this section, we will assume that sludge withdrawal is made from the completely mixed reactor after the reaction phase. Figure 6.1 shows the sequence of phases for an SBR for carbon removal assumed in this section, with the corresponding nomenclature.

Figure 6.2 shows typical profiles of substrate and biomass during an SBR cycle. The substrate concentration increases during the fill phase, because of the addition of the feed, and then it is removed completely

---

[*] This chapter has been written together with Mr. Adamu Abubakar Rasheed, a PhD candidate in Chemical Engineering, University of Aberdeen.

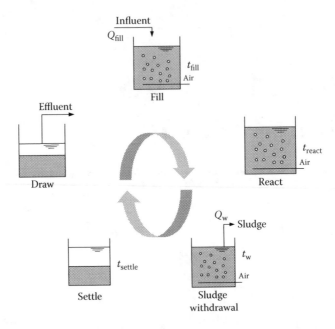

FIGURE 6.1 Sequence of phases for an SBR for carbon removal assumed in this section.

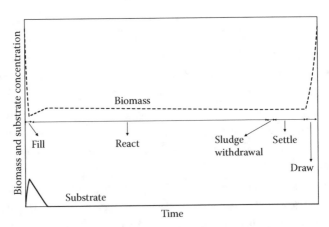

FIGURE 6.2 Typical profiles of biomass and substrate (arbitrary units) during a cycle of the SBR.

during the react phase. The biomass concentration decreases during the fill phase, because of the dilution with the feed, and then during the react phase, it initially increases slightly because the substrate is removed, and then, after the substrate is removed completely, the biomass concentration decreases slightly due to the endogenous metabolism. The biomass concentration increases during the draw (effluent removal) phase, because the clarified effluent is removed and therefore the same amount of biomass occupies a lower volume. From Figure 6.2, it is evident that an important difference between SBR processes and the continuous-flow-activated sludge process is that in the SBR, the substrate concentration changes during the cycle and is removed completely (depending on the design parameters) at the end of the cycle. On the other hand, we have seen in Chapter 4 that in the activated-sludge process, the influent substrate cannot be removed completely and the effluent substrate concentration will always be higher than zero (even though it is usually very low).

Assuming that we have to treat wastewater with an average daily flow rate Q and with a chemical oxygen demand (COD) concentration S_0, designing an SBR means finding the values of its operating parameters that give the required removal of the COD. For an SBR for carbon removal, the operating parameters that need to be defined are the following:

Number of cycles per day: No cycles

Length of the filling phase and flow rate during this phase: t_{fill}, Q_{fill}

Length of the reaction phase: t_{react}

Length of the sludge withdrawal phase and flow rate during this phase: t_W, Q_W

Length of the settling phase: t_{settle}

Length of the effluent withdrawal phase and flow rate during this phase: t_{eff}, Q_{eff}

Volume of the reactor: V_{full}

So, in order to define an SBR for carbon removal, we need to set the values of 10 operating parameters. However, these parameters are not all independent, because there are certain basic consistency equations that link some of them. These equations are as follows:

- The volume fed per day must be equal to the influent flow rate:

$$Q = Q_{fill} \cdot t_{fill} \cdot \text{No cycles} \qquad (6.1)$$

- The sum of the lengths of the phases must be equal to the length of a cycle:

$$t_{fill} + t_{react} + t_W + t_{settle} + t_{eff} = \frac{24\,h}{No\ cycles} \qquad (6.2)$$

- The overall volume fed in a cycle must be equal to the volume withdrawn in the same cycle:

$$Q_{eff} t_{eff} + Q_W t_W = Q_{fill} t_{fill} \qquad (6.3)$$

Therefore, designing the SBR means determining the values of $10 - 3 = 7$ operating parameters. For example, an SBR is fully specified if we set the values of the volume V_{full}, the number of cycles No cycles, the sludge withdrawal flow rate Q_W and the length of all the phases minus one (e.g. t_{fill}, t_{react}, t_W, t_{settle}).

However, how can we ensure that our choice of the operating parameters satisfies the main requirement of the process, that is the desired extent of COD removal?

We need the mass balances of substrate and biomass in the system. In the case that the substrate is composed only of readily biodegradable substrates, the mass balances for biomass and substrate can be written as follows:

$$\frac{d(VS)}{dt} = -\frac{r_X}{Y_{X/S}} V\Big|_{Fill,React,W} + Q_{fill} S_{feed}\Big|_{Fill} - Q_{eff} S\Big|_{Eff} - Q_W S\Big|_W \qquad (6.4)$$

$$\frac{d(VX)}{dt} = \left(r_X + r_{end}\right) V\Big|_{Fill,React,W} - Q_W X\Big|_W \qquad (6.5)$$

In these equations, and throughout this chapter, the notation $f(x)\big|_{Name\,of\,the\,phase(s)}$ means that the function f(x) has to be calculated only during the specified phases(s) and has to be taken equal to zero during the rest of the cycle. So, for example, $\left(r_X + r_{end}\right) V\big|_{Fill,React,W}$ means that the function $\left(r_X + r_{end}\right) V$ has to be calculated during the fill, react and sludge withdrawal phases, and has to be taken equal to zero during the rest of the cycle. Similarly, for example, the term $Q_W X\big|_W$ has to be calculated only during the sludge withdrawal phase and has to be taken equal to zero during the rest of the cycle.

In these equations, the volume V is the volume of the reactor during the cycle, which is variable, and equal to V_{full} only when the reactor is completely full, that is at the end of the feed and for all the length of the reaction phase. Therefore, we also need an equation that expresses the variable reactor volume, which is:

$$\frac{dV}{dt} = Q_{fill}\big|_{Fill} - Q_{eff}\big|_{Eff} - Q_w\big|_W \tag{6.6}$$

In order to rearrange these equations, it is useful to define the solids residence time (SRT) and the hydraulic residence time (HRT) in an analogous way, as it was done for continuous-flow reactors (Chapter 4). The HRT and the SRT are defined as the nominal average residence time of the liquid and of the microorganisms, respectively, in the reactor, and, therefore, in the notation used so far, they can be expressed as follows:

$$HRT = \frac{V_{full}}{Q_{fill}\, t_{fill}\, No\ cycles} \tag{6.7}$$

$$SRT = \frac{V_{full}\, X}{Q_w\, X t_w\, No\ cycles} = \frac{V_{full}}{Q_w\, t_w\, No\ cycles} \tag{6.8}$$

where the biomass concentration that appears in the definition of SRT is taken as its value during the sludge withdrawal phase and can, therefore, be cancelled out.

Introducing the HRT and SRT definitions, and rearranging, we obtain:

$$\frac{dS}{dt} = -\frac{r_X}{Y_{X/S}} V\bigg|_{Fill,React,W} + \frac{1}{HRT \cdot t_{fill} \cdot No\ cycles} \frac{(S_{feed} - S)}{V\big/V_{full}}\bigg|_{Fill} \tag{6.9}$$

$$\frac{dX}{dt} = (r_X + r_{end})\big|_{Fill,React,W} - \frac{X}{V\big/V_{full}} \frac{1}{HRT \cdot t_{fill} No\ cycles}\bigg|_{Fill} \tag{6.10}$$

$$+ \frac{X}{V\big/V_{full}} \frac{1}{No\ cycles \cdot t_{eff}}\left(\frac{1}{HRT} - \frac{1}{SRT}\right)\bigg|_{Eff}$$

$$\frac{d\left(V\big/V_{full}\right)}{dt} = \frac{1}{HRT \cdot t_{fill} No\ cycles}\bigg|_{Fill} - \frac{1}{No\ cycles \times t_{eff}}\left(\frac{1}{HRT} - \frac{1}{SRT}\right)\bigg|_{Eff} \tag{6.11}$$

$$- \frac{1}{SRT \cdot t_w \cdot No\ cycles}\bigg|_W$$

These equations give the profiles of substrate, biomass and volume during the operation of the SBR and can be used to calculate the time profiles of these variables as a function of time from the startup of the reactor. After the startup phase, the SBR will eventually reach the 'periodic steady-state', that is a condition where the time profiles of substrate and biomass in consecutive cycles do not change. When the reactor reaches the periodic steady state, the substrate and biomass concentrations at any given time of the cycle, for example, at the end of the feed or at the end of the reaction phase, are the same in each cycle.

In the design of the SBR, we are usually interested only in the periodic steady state and not in the startup phase. Therefore, we need to be able to calculate the periodic steady state of the SBR from Equations 6.9–6.11. This can be done by imposing the condition that at the periodic steady state, the biomass and substrate concentrations do not change over a complete cycle, that is:

$$\int_{cycle} dS = 0 \tag{6.12}$$

$$\int_{cycle} dX = 0 \tag{6.13}$$

Imposing these conditions means that at a given time of the cycle, the concentrations of substrate and biomass do not change for successive cycles, that is we are at the periodic steady state. With the conditions given by Equations 6.12 and 6.13, the time profiles of substrate and biomass at the periodic steady-state cycle can be obtained from Equations 6.9 and 6.10. At the periodic steady state, these equations become, after rearrangements:

$$\int_{cycle} \left[\left. \left(r_X + r_{end} \right) \right|_{Fill,React,W} - \left. \frac{X}{V/V_{full}} \frac{1}{HRT \cdot t_{fill} No\ cycles} \right|_{Fill} + \left. \frac{X}{V/V_{full}} \frac{1}{No\ cycles \cdot t_{eff}} \left(\frac{1}{HRT} - \frac{1}{SRT} \right) \right|_{W} \right] dt = 0 \tag{6.14}$$

$$\int_{cycle} \left[-\left. \frac{r_X}{Y_{X/S}} \right|_{Fill,React,W} + \left. \frac{1}{HRT \cdot t_{fill} \cdot No\ cycles} \frac{(S_{feed} - S)}{V/V_{full}} \right|_{Fill} \right] dt = 0 \tag{6.15}$$

Equations 6.14 and 6.15 can be solved, together with the volume Equation 6.11, to calculate the periodic steady-state of the SBR.

In order to solve Equations 6.14, 6.15 and 6.11, the values of HRT, SRT, No cycles and the length of all the phases minus one, for example, t_{fill}, t_{react}, t_W, t_{settle}, need to be set. Once these operating parameters are set, Equations 6.14, 6.15 and 6.11 will give the periodic steady state of the SBR, and the values of the operating parameters Q_{fill}, Q_W and V_{full} will be immediately calculated by the definitions of HRT and SRT:

$$Q_{fill} = \frac{Q}{t_{fill} \cdot \text{No cycles}} \tag{6.16}$$

$$V_{full} = HRT \cdot Q_{fill} \, t_{fill} \, \text{No cycles} \tag{6.17}$$

$$Q_W = \frac{V_{full}}{SRT \cdot t_W \cdot \text{No cycles}} \tag{6.18}$$

$$Q_{eff} = \frac{Q_{fill} \, t_{fill} - Q_W \, t_W}{t_{eff}} \tag{6.19}$$

In summary, the SBR design procedure can be summarised as follows:

1. Set the values for HRT, SRT, number of cycles and the length of all the phases minus one;

2. Calculate the length of the remaining phase and solve Equations 6.11, 6.14 and 6.15 simultaneously. This will give the profiles of substrate and biomass (and also of the filling ratio) as a function of the chosen operating parameters;

3. Using Equations 6.16, 6.17, 6.18 and 6.19, calculate the values of the remaining operating parameters that still need to be determined, that is Q_{fill}, Q_W, Q_{eff} and V_{full};

Of course, the obtained design is satisfactory only if the substrate concentration in the discharged effluent is lower than the maximum possible value. If this is not the case, other values of the design parameters need to be chosen and the procedure needs to be repeated.

Once the SBR is designed, the average daily sludge production can be calculated immediately:

$$P_X \left(\frac{\text{kg biomass}}{\text{day}} \right) = Q_W \, t_W \, \text{No cycles} \cdot X = \frac{V_{full}}{SRT} X \tag{6.20}$$

where, in this case, X is the biomass concentration during the sludge withdrawal phase.

The daily average oxygen consumption by the microorganisms can be calculated as usual from the overall COD balance:

$$Q_{O2biomass}\left(\frac{\text{kg oxygen}}{\text{day}}\right) = Q(S_0 - S) - 1.42 P_X = Q(S_0 - S) - 1.42 \frac{V_{full}}{SRT} X \quad (6.21)$$

where S is the residual substrate concentration in the effluent.

It is important to highlight again that Equations 6.20 and 6.21 give the average daily sludge production and oxygen consumption, and not their instantaneous values, which will be variable during the cycle. Sludge production will be zero during all phases except in the sludge withdrawal phase and oxygen consumption will be variable due to the variable concentrations of substrate and biomass during the cycle. The oxygen profile during a cycle will be discussed further in Section 6.1.2.

6.1.1 Effect of the Choice of the Design Parameters on the Performance of the SBR

Compared to a conventional activated sludge process for carbon removal (Chapter 4), the SBR has more degrees of freedom. Indeed, to fully specify an SBR for carbon removal, we need to specify the values of seven parameters—HRT, SRT, the number of cycles per day and the length of the four phases. On the other hand, to design a conventional activated sludge process, we only need to specify three variables—HRT, SRT and the recycle ratio. The effect of the choice of the values of the various design parameters for the SBR can be summarised as follows:

- Similarly as for the conventional activated sludge process, SRT is the main parameter that determines the extent of substrate removal. As for the conventional activated sludge process, higher values of the SRT give higher substrate removal, higher biomass concentration, higher oxygen consumption and lower sludge production;

- For a given SRT, the HRT determines the concentration of biomass in the reactor. Higher values of HRT give higher volumes of the SBR and lower biomass concentrations, while lower values of HRT give lower reactor volumes and higher biomass concentrations in the reactor. The effect of HRT is analogous to the effect of this parameter in the conventional activated sludge process;

- The number of cycles per day has no equivalent in the conventional activated sludge process and, coupled with HRT, has an important effect on the optimisation of the reactor volume. Indeed, increasing the number of cycles per day allows operating at shorter HRT, giving therefore a smaller reactor volume. Indeed, the daily flow rate Q can be expressed as a function of the flow rate during the fill phase, the length of the fill phase and the number of cycles as follows: $Q = Q_{fill} \cdot t_{fill} \cdot$ No cycles. In order to minimise the volume of the SBR required to treat this daily flow rate, the value of the variable $(Q / V_{full}) = (1/HRT)$ needs to be maximised. Therefore, in order to minimise the required volume of the SBR, the HRT has to be set to its minimum possible value. There are several constraints on the minimum value of the HRT that is possible to achieve. The first constraint comes from the basic fact that the volume of feed that can be added per cycle must be lower than the total volume of the reactor, because some volume has to be left in the reactor for the settled biomass. If we assume the limit case of $Q_{fill} \cdot t_{fill} = V_{full}$, we obtain the condition for the minimum possible HRT, $HRT_{min} = (1/\text{No cycles})$, which gives the maximum daily flow rate per unit of reactor volume $(Q/V_{full})_{max} = (1/HRT_{min}) = $ No cycles. This shows that increasing the number of cycles allows, in principle, to work at lower HRT and, therefore, allows a smaller reactor volume. However, it has to be observed that increasing the number of cycles means shorter lengths for all the phases and this can be a problem, in particular for the settling phase, for which a minimum length needs to be guaranteed in order to allow complete separation of the microorganisms from the effluent;

- The length of the various phases has no equivalent in the conventional activated sludge process. The length of the settling phase represents an inactive part of the cycle, where no biological reactions occur, and therefore, should be set to the minimum value that ensures good separation of the biomass from the liquid effluent and minimal loss of the biomass with the effluent. However, since the settling performance of the biomass can be difficult to predict and subject to a good degree of variability, it is advisable to add some safety margin to the length of the settling phase. Similarly, the length of the effluent withdrawal phase should be minimised because it is an inactive phase of the cycle. However, in practice, care should be taken to avoid turbulence during efflu-ent withdrawal, in order to avoid destabilising the settled solids. The length of the feed phase has an effect on the substrate profile within a cycle. A short feed gives a higher substrate concentration at the start of the reaction phase than a slow feed, because with a short feed phase substrate removal during the feed is minimised. Since both the biomass growth rate and the substrate removal rate depend on the substrate concentration, it is in general a kinetic advantage to use a short feed. However, for the same reason, if the influent substrate is inhibiting at high concentrations, a long feed gives advantage because the lower substrate concentration will decrease the inhibition effect. The length of the reaction phase should be maximised, because it affects substrate removal. If sludge withdrawal is done immediately after the reaction phase from the totally mixed reactor, the length of the sludge withdrawal phase should be kept to the minimum. This ensures that the length of the other phases, and in particular of the reaction phase, is maximised, therefore minimising the substrate concentration in the sludge withdrawal stream.

Example 6.1: SBR for carbon removal

An SBR for carbon removal treats 5000 m^3/day of wastewater. The HRT is 0.5 day, there are four cycles per day, and the reactor produces a waste sludge stream with an average daily flow rate of 50 m^3/day. The length of the phases is as follows: feed 30 min, sludge withdrawal 10 min, settling 45 min and effluent withdrawal 15 min. Calculate the volume of the reactor, the length of the reaction phase, SRT and the flow rates during the individual phases.

Solution

The overall daily flow rate is $Q = Q_{fill} \cdot t_{fill} \cdot \text{No cycles} = 5000 \ m^3/\text{day}$ and, from the definition of HRT, Equation 6.7, it follows that the reactor volume is:

$$V_{full} = 2500 \ m^3$$

Since there are four cycles per day (No cycles = 4), the total length of the phases is six hours, and since we have:

$$t_{\text{fill}} = 30 \text{ min} = 0.0208 \text{ day}$$

$$t_{\text{W}} = 10 \text{ min} = 0.0069 \text{ day}$$

$$t_{\text{settle}} = 45 \text{ min} = 0.03125 \text{ day}$$

$$t_{\text{eff}} = 15 \text{ min} = 0.0104 \text{ day}$$

The length of the reaction phase will be:

$$t_{\text{react}} = (0.25 - 0.0208 - 0.069 - 0.03125 - 0.0104)\text{day}$$

$$= 0.18065 \text{ day} = 260 \text{ min}$$

The total average daily flow rate of the waste sludge stream is:

$$Q_{\text{W}} \cdot t_{\text{W}} \cdot \text{No cycles} = 50 \text{ m}^3/\text{day}$$

and, since $t_{\text{W}} = 0.0069$ day and No cycles = 4, the flow rate during the sludge withdrawal phase is:

$$Q_{\text{W}} = 1812 \frac{\text{m}^3}{\text{day}} = 1.258 \frac{\text{m}^3}{\text{min}}$$

With similar considerations, we calculate the feed flow rate during the fill (or feed) phase:

$$Q_{\text{fill}} = 60,096 \frac{\text{m}^3}{\text{day}} = 41.73 \frac{\text{m}^3}{\text{min}}$$

According to Equation 6.8, SRT is equal to 50 days.

Example 6.2: Design of an SBR for carbon removal

Calculate the effect of the design parameters for an SBR for carbon removal. Assume the kinetic parameters of Example 4.1 and the same influent substrate concentration of 0.5 gCOD/L.

Also, choose a suitable design for an influent flow rate of 10,000 m³/day, and calculate the corresponding oxygen consumption by the microorganisms and sludge production.

Solution

The effect of the design parameters can be calculated by solving the system of equations that describe the periodic steady state of SBR for a readily biodegradable feed (Equations 6.11, 6.14 and 6.15). Figure 6.3 shows the effect of SRT on carbon removal in SBR. In reality, changing the SRT means changing the sludge withdrawal rate, Q_W, and as expected, the biomass concentration increases when the SRT is increased, with an asymptotic value reached at longer SRTs. On the basis of the kinetic parameters adopted in this example, there is complete removal of substrates even at a very short SRT (e.g. 0.5 days). Note that all the values of biomass concentration reported in this and in other examples refer to the concentration at the end of the sludge withdrawal phase.

Figure 6.4 shows the effect of HRT on carbon removal in SBR. In practice, changing the HRT means changing the feed flow rate and as expected, the biomass concentration decreases as the HRT increases.

The effect of the length of the reaction phase is shown in Figure 6.5. It can be observed that the substrate is removed completely even at very short reaction lengths, while the biomass concentration increases when the length is shortened. This is due to the fact that the reduction in the length of the reaction phase has been

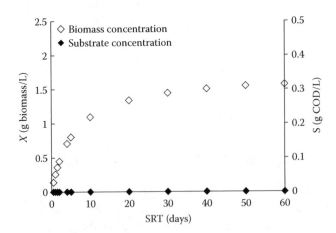

FIGURE 6.3   Effect of SRT on the biomass and substrate concentrations at steady state. The HRT is equal to 0.5 day in a pattern of 4 cycles per day. The length of the phases is: fill 5 min, react 290 min, sludge withdrawal 5 min, settle 45 min, draw 15 min.

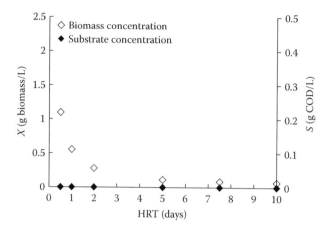

FIGURE 6.4 Effect of HRT on the biomass and substrate concentrations at steady state. SRT is equal to 10 days and all the other parameters are the same as in Figure 6.3.

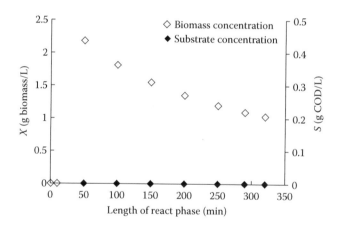

FIGURE 6.5 Effect of the length of the react phase on the biomass and substrate concentrations at steady state. HRT and SRT are equal to 0.5 and 10 days, respectively, in a pattern of 4 cycles per day. The shorter length of the react phase corresponds to longer lengths of the settling phase, and all other phases have the lengths used for Figure 6.3.

obtained by increasing the length of the settling phase, where biomass is inactive. Therefore, for longer lengths of the reaction phase, endogenous metabolism becomes more important and the biomass concentration decreases. The effect of the number of cycles per day was also calculated and shown in Figure 6.6. It turns out to have very

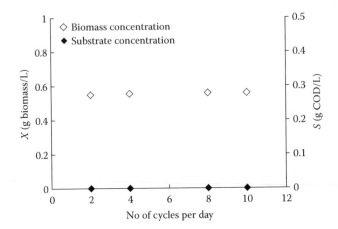

FIGURE 6.6 Effect of number of cycles per day on the biomass and substrate concentrations at steady state. HRT and SRT are equal to 0.5 and 10 days, respectively, and all other parameters are the same as in Figure 6.3.

little effect on the biomass concentration since the overall amount of substrate fed per day is the same.

The second part of this example is about calculating all the variables that characterise the SBR for a certain flow rate of wastewater. The chosen design parameters and the calculated values of all the variables are reported below. Note that the flow rates Q_{eff}, Q_{fill} and Q_W are the instantaneous flow rates during the effluent withdrawal, fill and sludge withdrawal phases, respectively, while Q is the average daily flow rate of the influent.

V_{full}	5000	m³
No Cycle	4	day⁻¹
SRT	2	day
HRT	0.5	day
X	0.45	kg/m³
Q	10,000	m³/day
Q_W	2.08	m³/s
Q_{eff}	2.08	m³/s
Q_{fill}	8.3	m³/s
t_{fill}	5	min
t_{react}	290	min
t_w	5	min
t_{settle}	45	min
t_{eff}	15	min

Sludge production is calculated from Equation 6.20, and we obtain $P_X = 1125$ kg biomass/day.

Oxygen consumption is calculated from Equation 6.21, where the effluent substrate concentration is zero. We obtain 3402 kg oxygen/day.

6.1.2 Calculation of the Oxygen Profile during the SBR Cycle

We have seen that Equations 6.11, 6.14 and 6.15 allow the design of the SBR and the calculation of the profiles of substrate, biomass and volume during the cycle at the periodic steady state. We have also seen that from the effluent data of biomass and substrate concentrations, we can calculate the average daily oxygen consumption rate by the microorganisms (Equation 6.21). However, in reality, the rate of oxygen consumption will be variable during the SBR cycle, and it will be higher in the initial part of the cycle when the substrate is present and lower in the remaining part of the cycle when the substrate has been removed completely and the oxygen consumption is only due to endogenous metabolism.

During the phases when we have assumed biological activity takes place (i.e. all the phases except the settling and effluent withdrawal phases), the mass balance for oxygen is given by:

$$\frac{dC_{O_2}}{dt}\left(\frac{\text{kg oxygen}}{\text{m}^3 \cdot \text{day}}\right) = -\frac{\mu_{max} S}{K_S + S} \times \left(\frac{1}{Y_{X/S}} - 1.42\right) \cdot X$$

$$-1.42 \cdot b \cdot X + k_L a\left(C_{O2}^* - C_{O2}\right)$$

Similarly to what we have done for the substrate and biomass concentrations, the dissolved oxygen profile during a cycle at the periodic steady state is obtained by imposing that:

$$\int_{\text{cycle}} dC_{O2} = 0 \tag{6.22}$$

which is:

$$\int_{\text{cycle}} \left[\begin{array}{l} -\dfrac{\mu_{max} S}{K_S + S} \cdot \left(\dfrac{1}{Y_{X/S}} - 1.42\right) \cdot X \\[2mm] -1.42 \cdot b \cdot X + k_L a\left(C_{O2}^* - C_{O2}\right) \end{array} \right]_{\text{Fill,React,W}} dt = 0 \tag{6.23}$$

With the values of S and X already obtained from the solution of the design Equations (6.11, 6.14 and 6.15), the solution of Equation 6.23 gives the oxygen profile during the SBR cycle at the periodic steady state.

Note that the function inside the integral in Equation 6.23 needs to be calculated only during the fill, reaction and sludge withdrawal phases, that is during the phases when we have assumed there is biological activity. We have assumed there is no biological activity during the settling and effluent withdrawal phases. The consequence of this assumption is that the oxygen concentration at the start of the cycle, that is at the start of the fill phase, will be equal to the value at the end of the sludge withdrawal phase. In practice, this assumption is questionable because during the settling and effluent withdrawal phases, there is no aeration, but the oxygen concentration will drop due to endogenous metabolism. However, in the actual practice of the SBR operation, an aerated idle phase (not considered in this model) is usually added before the new feed and this allows the oxygen concentration to rise close to the saturation values. Therefore, in practice, the assumption that the oxygen concentration at the start of the cycle is equal to its value before the start of the settling phase is probably very close to the industrial practice.

Typically, the oxygen profile obtained from the solution of Equation 6.23 shows that oxygen concentration drops during the fill phase and until there is substrate present, and then it rapidly increases up to the initial value. Indeed, when the substrate has been totally removed, the only contribution to oxygen consumption is the endogenous metabolism of the microorganisms, which proceeds at a much lower rate. An important use of Equation 6.23 is the calculation of the minimum value of the $k_L a$ that is required to maintain the oxygen concentration above the minimum value, which allows unrestricted microbial growth. If the $k_L a$ is not high enough, the oxygen concentration during the initial part of the cycle when the substrate is present will drop to zero (or close to it), and microorganism growth will be limited by oxygen supply, which is undesirable. Using Equation 6.23 for various values of $k_L a$ will allow the calculation of the minimum $k_L a$ required to maintain the oxygen concentration above a desired minimum value.

Example 6.3: Oxygen profiles in SBR cycles

Calculate the oxygen profiles for the SBR designed in Example 6.2 for various values of $k_L a$. Use as design parameters: SRT = 2 days,

HRT = 1 day, number of cycles 4 per day and the following lengths of the phases: fill 30 min, react 265 min, sludge withdrawal 5 min, settle 45 min, effluent withdrawal 15 min.

Solution

The first step is the calculation of the substrate and biomass profiles during a steady state cycle, using Equations (6.11, 6.14 and 6.15), and obtaining results similar to the ones reported in Example 6.2. Once the substrate and biomass profiles are calculated, the oxygen profile during the cycle can be obtained from Equation 6.23 for given values of $k_L a$. The oxygen profiles have been reported in Figure 6.7. From the simulations, oxygen decreases during feeding until all there is complete substrate removal, and then rapidly increases. From this example, it is evident that the minimum $k_L a$ required to maintain an oxygen concentration of at least 1 mg/L in the reactor is 300 days⁻¹.

6.1.3 Extension to Slowly Biodegradable Substrates

If the feed of the reactor also contains slowly biodegradable substrates (X_S), we need to add the mass balance for these substrates, which can be written as follows:

$$\frac{d(VX_S)}{dt} = r_{hydr} V\Big|_{Fill,React,W} + Q_{fill} X_{S\,feed}\Big|_{Fill} - Q_{eff} X_S\Big|_{Eff} - Q_W X_S\Big|_W \quad (6.24)$$

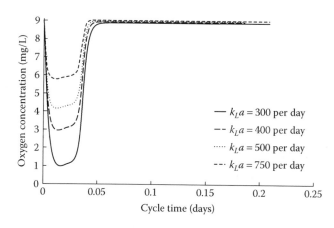

FIGURE 6.7 Oxygen concentration profiles during the cycle for various values of $k_L a$.

And the mass balance for readily biodegradable substrates becomes:

$$\frac{d(VS)}{dt} = -r_{hydr} V\big|_{Fill,React,W} + r_S V\big|_{Fill,React,W}$$

$$+ Q_{fill} S_{feed}\big|_{Fill} - Q_{eff} S\big|_{Eff} - Q_W S\big|_W \qquad (6.25)$$

The rate equation for the hydrolysis processes, r_{hydr}, has been defined in Chapter 2 and used in Chapter 4 for the case of feed with slowly biodegradable substrates. The mass balance for the biomass is still given by Equation 6.5, since biomass growth occurs only on readily biodegradable substrates.

By introducing the definitions of HRT and SRT, rearranging and imposing the periodic steady-state conditions:

$$\int_{cycle} dS = 0 \qquad (6.12)$$

$$\int_{cycle} dX = 0 \qquad (6.13)$$

$$\int_{cycle} dX_S = 0 \qquad (6.26)$$

We obtain the following equations, which can be solved to obtain the periodic steady state of an SBR with slowly biodegradable substrates in the feed. Biomass:

$$\int_{cycle} \left[\left(r_X + r_{end} \right)\big|_{Fill+React+W} - \frac{X}{V/V_{full}} \frac{1}{HRT \cdot t_{fill} \, No \, cycles}\bigg|_{Fill} \right.$$

$$\left. + \frac{X}{V/V_{full}} \frac{1}{No \, cycles \cdot t_{eff}} \left(\frac{1}{HRT} - \frac{1}{SRT} \right)\bigg|_{Eff} \right] dt = 0 \qquad (6.14)$$

Slowly biodegradable substrate:

$$\int_{cycle} \left[r_{hydr}\big|_{Fill+React+W} + \frac{1}{HRT \cdot t_{fill} \cdot No \, cycles} \frac{(X_{Sfeed} - X_S)}{V/V_{full}}\bigg|_{Fill} \right] dt = 0 \qquad (6.27)$$

Readily biodegradable substrate:

$$\int_{cycle}\left[\left(r_S-r_{hydr}\right)\Big|_{Fill+React+W}-\frac{1}{HRT\cdot t_{fill}\cdot No\ cycles}\frac{\left(S_{feed}-S\right)}{V\big/V_{full}}\Big|_{Fill}\right]dt=0 \quad (6.28)$$

The design of the SBR for slowly biodegradable substrates in the feed is obtained by simultaneously solving Equations 6.11, 6.14, 6.27 and 6.28, and it is conceptually the same as for readily biodegradable substrates, with the observation that the lower biodegradation rate might mean the need for higher values of the SRT and for a different choice of the other design parameters.

Example 6.4: Design of an SBR for carbon removal with a feed composed of slowly biodegradable substrates

Show the effect of design parameters for an SBR for carbon removal with a feed composed exclusively of slowly biodegradable substrates.

Assume the kinetic parameters used in Example 4.10 and the same concentration of slowly biodegradable substrate in the feed, 0.5 gCOD/L. Calculate the SBR size, oxygen consumption and sludge production for an influent flow rate of 10,000 m³/day.

Solution

The effect of the design parameters for a feed composed only of slowly biodegradable feed can be calculated by solving the system of Equations 6.11, 6.14, 6.27 and 6.28. Figure 6.8 shows the effect of SRT on carbon removal in SBR for a slowly biodegradable feed. The results are similar to the ones obtained for a readily biodegradable feed in Example 6.2. However, in this case, at shorter SRT values (<2 days), the SRT is not long enough to ensure complete biodegradation of the slowly biodegradable substrates, resulting in biomass washout and incomplete degradation of the substrate.

Figure 6.9 shows the same trend for the effect of HRT on carbon removal in SBR as in Example 6.2.

The second part of this example is about calculating the sludge produced and oxygen consumed for the design conditions below. The parameters were calculated for HRT and SRT of 0.5 and 5 days, respectively, for a slowly biodegradable feed.

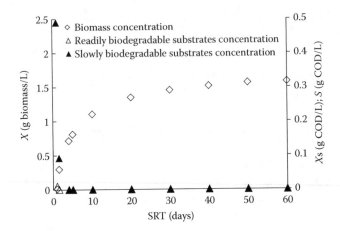

FIGURE 6.8 Effect of SRT on the biomass and substrate concentrations at steady state for a feed composed of slowly biodegradable substrates. The HRT is equal to 0.5 days in a pattern of 4 cycles per day. The length of the phases is: fill 5 min, react 290 min, sludge withdrawal 5 min, settle 45 min, draw 15 min.

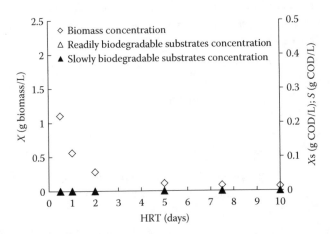

FIGURE 6.9 Effect of HRT on the biomass and substrate concentrations at steady state for a feed composed of slowly biodegradable substrates. The SRT is equal to 10 days and all other parameters are the same as in Figure 6.8.

V_{full}	5000	m^3
No Cycle	4	day^{-1}
SRT	5	day
HRT	0.5	day
X_W	0.81	kg/m^3
Q	10,000	m^3/day
		(*Continued*)

Q_W	0.83	m³/s
Q_{eff}	2.49	m³/s
Q_{fil}	8.3	m³/s
t_{fill}	5	min
t_{react}	290	min
t_w	5	min
t_{settle}	45	min
t_{eff}	15	min

Sludge production is calculated from Equation 6.20, and we obtain $P_X = 810$ kg biomass/day.

Oxygen consumption is calculated from Equation 6.21, with the observation that in this case, the substrate is slowly biodegradable, where the effluent substrate concentration is zero. We obtain 3849 kg oxygen/day.

6.2 SBR FOR CARBON AND NITROGEN REMOVAL

The SBR for nitrogen removal was described in Chapter 1 and a typical sequence of phases, including sludge withdrawal from the completely mixed reactor after the end of the reaction phase, is shown below. In terms of the number of phases, the only difference between an SBR for carbon and nitrogen removal vs nitrogen removal is that the reaction phase is split into two phases—anoxic and aerobic. This corresponds to a scheme with predenitrification, when the feed is fed during the anoxic phase, so that the nitrate generated during the aerobic phase of the previous cycle is used as electron acceptor during the feed and anoxic reaction phases. After the anoxic reaction phase, an aerobic reaction phase is used to nitrify the ammonia to nitrate.

In the SBR for carbon and nitrogen removal (Figure 6.10), we have six phases, instead of five, so we have one additional degree of freedom. In total, therefore, we have eight degrees of freedom, that is we need to specify the values of eight variables in order to design the process. In writing the mass balances, we have used the usual kinetic equations for the growth rate and endogenous metabolism of the microorganisms.

Growth rate and endogenous metabolism rate of heterotrophic microorganisms during the anoxic (not aerated) phase:

$$r_{XAnox} = \mu_{max} \frac{S}{K_S + S} \frac{NO_3}{K_{SNO3} + NO_3} X \qquad r_{endAnox} = -b \frac{NO_3}{K_{SNO3} + NO_3} X$$

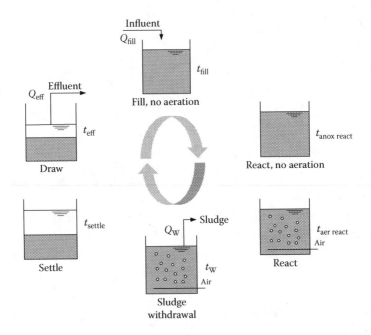

FIGURE 6.10 Sequence of phases for an SBR for carbon and nitrogen removal.

Growth rate and endogenous metabolism of heterotrophic microorganisms during the aerobic (aerated) phase:

$$r_{XAer} = \mu_{max} \frac{S}{K_S + S} X \quad r_{endAer} = -bX$$

Growth rate and endogenous metabolism of autotrophic microorganisms during the aerobic phase (like in previous chapters, we assume that during the anoxic phase autotrophic microorganisms are inactive):

$$r_{XA} = \mu_{maxA} \frac{NH_3}{K_{SNH3} + NH_3} X_A \quad r_{endA} = -b_A X_A$$

The mass balances for the various variables in the system have been reported below and have been derived with the same approach previously used only for carbon removal.

Substrate:

$$\frac{d(VS)}{dt} = -\frac{r_{XAnox}}{Y_{X/S}} V \bigg|_{Anox\,Fill,Anox\,React} - \frac{r_{XAer}}{Y_{X/S}} V \bigg|_{Aer\,React,W} + Q_{fill} S_{feed} \big|_{Fill}$$

$$- Q_{eff} S \big|_{Eff} - Q_W S \big|_W$$

(6.29)

Heterotrophic biomass:

$$\frac{d(VX)}{dt} = \left(r_X + r_{end}\right)_{Anox} V\bigg|_{Anox\ Fill,Anox\ React}$$

$$+\left(r_X + r_{end}\right)_{Aer} V\bigg|_{Aer\ React,W} - Q_W X\big|_W \qquad (6.30)$$

Autotrophic biomass:

$$\frac{d(VX_A)}{dt} = \left(r_{XA} + r_{endA}\right)V\bigg|_{Aer\ React} - Q_W X_A\big|_W \qquad (6.31)$$

Ammonia:

$$\frac{d(VNH_3)}{dt} = -\left(r_X + r_{end}\right)_{Anox} 0.12 \cdot V\bigg|_{Anox\ Fill,Anox\ React}$$

$$-\left(r_X + r_{end}\right)0.12 \cdot V\bigg|_{Aer\ React} - \left(r_{XA}\left(0.12 + \frac{1}{Y_{XA/NO3}}\right) + r_{endA}\right)V\bigg|_{Aer\ Reac} \qquad (6.32)$$

$$+ Q_{fill}NH_{3\,feed}\big|_{Fill} - Q_{eff}NH_3\big|_{Eff} - Q_W S\big|_W$$

Nitrate:

$$\frac{d(VNO_3)}{dt} = \left[-r_X\left(\frac{1}{Y_{X/S}} - 1.42\right) + r_{end}\right] \times \frac{V}{2.86}\bigg|_{Anox\ Fill,Anox\ React}$$

$$+\frac{r_{XA}}{Y_{XA/NO3}} V\bigg|_{Aer\ React} - Q_{eff}NO_3\big|_{Eff} - Q_W NO_3\big|_W \qquad (6.33)$$

These mass balances can be rearranged, as done for the SBR for carbon removal (Section 6.1) using the same definitions of HRT and SRT, Equations 6.7 and 6.8, and imposing the following conditions that define the periodic steady state:

$$\int_{cycle} dS = 0 \qquad (6.12)$$

$$\int_{cycle} dX = 0 \qquad (6.13)$$

$$\int_{\text{cycle}} dX_A = 0 \tag{6.34}$$

$$\int_{\text{cycle}} dNH_3 = 0 \tag{6.35}$$

$$\int_{\text{cycle}} dNO_3 = 0 \tag{6.36}$$

The periodic steady-state conditions are, therefore, the ones reported below.
Substrate:

$$\int_{\text{cycle}} \left[-\frac{r_{\text{XAnox}}}{Y_{X/S}}\bigg|_{\text{Anox Fill,Anox React}} -\frac{r_{\text{XAer}}}{Y_{X/S}}\bigg|_{\text{Aer React,W}} + \frac{1}{\text{HRT} \cdot t_{\text{fill}} \cdot \text{Nocycles}} \frac{(S_{\text{feed}} - S)}{V/V_{\text{full}}}\bigg|_{\text{Anox Fill}} \right] dt = 0 \tag{6.37}$$

Heterotrophic biomass:

$$\int_{\text{cycle}} \left[\begin{array}{l} \left(r_X + r_{\text{end}}\right)_{\text{Anox}}\bigg|_{\text{Anox Fill,Anox React}} + \left(r_X + r_{\text{end}}\right)\bigg|_{\text{Aer React,W}} \\ -\frac{X}{V/V_{\text{full}}} \frac{1}{\text{HRT} \cdot t_{\text{fill}} \text{Nocycles}}\bigg|_{\text{Fill}} \\ +\frac{X}{V/V_{\text{full}}} \frac{1}{\text{Nocycles} \cdot t_{\text{eff}}} \left(\frac{1}{\text{HRT}} - \frac{1}{\text{SRT}}\right)\bigg|_{\text{Eff}} \end{array} \right] dt = 0 \tag{6.38}$$

Autotrophic biomass:

$$\int_{\text{cycle}} \left[\begin{array}{l} \left(r_{\text{XA}} + r_{\text{endA}}\right)\bigg|_{\text{Aer React,W}} - \frac{X_A}{V/V_{\text{full}}} \frac{1}{\text{HRT} \cdot t_{\text{fill}} \text{Nocycles}}\bigg|_{\text{Fill}} \\ +\frac{X_A}{V/V_{\text{full}}} \frac{1}{\text{Nocycles} \cdot t_{\text{eff}}} \left(\frac{1}{\text{HRT}} - \frac{1}{\text{SRT}}\right)\bigg|_{\text{Eff}} \end{array} \right] dt = 0 \tag{6.39}$$

Ammonia:

$$
\int_{cycle} \left[
\begin{array}{c}
-\left(r_X + r_{end}\right)_{Anox} 0.12 \cdot \Big|_{Anox\,Fill,Anox\,React} \\[2mm]
-\left(r_X + r_{end}\right)_{Aer} 0.12 \cdot \Big|_{Aer\,React,W} \\[2mm]
-\left(r_{XA}\left(0.12 + \dfrac{1}{Y_{XA/NO3}} \right) + r_{endA} \right)\Bigg|_{Aer\,Reac} \\[4mm]
+ \dfrac{1}{HRT \cdot t_{fill} \cdot Nocycles} \dfrac{\left(NH_{3feed} - NH_3\right)}{V/V_{full}}\Bigg|_{Anox\,Fill}
\end{array}
\right] dt = 0 \quad (6.40)
$$

Nitrate:

$$
\int_{cycle} \left[
\begin{array}{c}
\left[-r_X\left(\dfrac{1}{Y_{X/S}} - 1.42 \right) + r_{end} \right] \cdot \dfrac{1}{2.86}\Bigg|_{Anox\,Fill,Anox\,React} \\[4mm]
+ \dfrac{r_{XA}}{Y_{XA/NO3}}\Bigg|_{Aer\,Reac} + \dfrac{1}{HRT \cdot t_{fill} \cdot Nocycles} \dfrac{-NO_3}{V/V_{full}}\Bigg|_{Anox\,Fill}
\end{array}
\right] dt = 0 \quad (6.41)
$$

In summary, the periodic steady state for an SBR for carbon and nitrogen removal can be obtained by solving Equations 6.37 through 6.41, coupled with the volume Equation 6.11. In order to solve Equations (6.37 through 6.41 and 6.11), we need to set some values for HRT, SRT, the number of cycles and for the length of all phases minus one, that is we need to specify the values of eight parameters, as discussed above. The solution of these equations will give the profiles of substrate, heterotrophic biomass, autotrophic biomass, ammonia and nitrate during a cycle at the periodic steady state.

6.2.1 Effect of the Choice of the Operating Parameters on the Design of the SBR for Carbon and Nitrogen Removal

The choice of the operating parameters for an SBR for carbon and nitrogen removal can be done using criteria similar to the ones used for the SBR only for carbon removal. However, there are some parameters that have a distinct effect on the efficiency of nitrogen removal. In summary, the main criteria for the choice of the operating parameters have been reported below:

- The SRT should be chosen at a value that is high enough for the growth for heterotrophic and autotrophic microorganisms. Since autotrophic microorganisms usually grow slower than heterotrophic ones, the SRT for an SBR for carbon and nitrogen removal is typically longer than that for an SBR only for carbon removal;

- The HRT has the same meaning as for the SBR only for carbon removal (and for the conventional activated sludge process). High values of HRT correspond to low biomass concentration, while low values of HRT give high biomass concentration;

- The number of cycles coupled to the HRT is particularly important. In general, nitrogen removal efficiency increases with increase in the number of cycles. This is because assuming the other design parameters are chosen so to ensure that all or most of the ammonia is converted to nitrate within a cycle, having a large number of cycles means that the amount of ammonia fed per cycle is lower and so the maximum concentration of nitrate at the end of the cycle is also lower, thereby minimising the concentration of total nitrogen in the effluent;

- The length of the various phases can be chosen with criteria similar to the ones used for SBR only for carbon removal. It is important to ensure that the respective lengths of the anoxic and aerobic reaction phases are enough to ensure full nitrate removal (anoxic phase) and full ammonia conversion to nitrate (aerobic phase)

6.3 ANAEROBIC SBR

The typical sequence of phases for anaerobic SBR is essentially the same as that for the aerobic SBR, the only difference being the absence of aeration (Figure 6.11).

In order to develop the mass balances for the anaerobic SBR, we will assume here that the substrate is composed entirely of glucose, and we will assume the kinetic model already used in Chapter 5. The rate equations are the ones used in Chapter 5 (Equations 5.1 through 5.5).

The mass balances for the various components of the model have been written below.

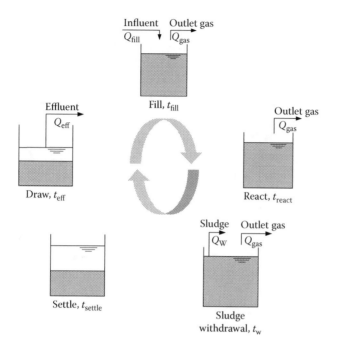

FIGURE 6.11   Typical sequence of phases for an anaerobic SBR.

Glucose:

$$\frac{d(V\text{GLU})}{dt} = r_{\text{GLU}}V\big|_{\text{Fill,React,W}} + Q_{\text{fill}}\text{GLU}_{\text{IN}}\big|_{\text{Fill}}$$
$$- Q_{\text{eff}}\text{GLU}\big|_{\text{Eff}} - Q_{\text{W}}\text{GLU}\big|_{\text{W}}$$

(6.42)

Acetate:

$$\frac{d(V\text{Ac})}{dt} = r_{\text{Ac}}V\big|_{\text{Fill,React,W}} - Q_{\text{eff}}\text{Ac}\big|_{\text{Eff}} - Q_{\text{W}}\text{Ac}\big|_{\text{W}}$$

(6.43)

Hydrogen:

$$\frac{d(V\text{H}_2)}{dt} = r_{\text{H2}}V\big|_{\text{Fill,React,W}} - Q_{\text{eff}}\text{H}_2\big|_{\text{Eff}} - Q_{\text{W}}\text{H}_2\big|_{\text{W}} - Q_{\text{gas}}\frac{p_{\text{H2}}}{p_{\text{tot}}}\rho_{\text{H2}}\bigg|_{\text{Fill,React,W}}$$

(6.44)

Methane:

$$\frac{d(V\text{CH}_4)}{dt} = r_{\text{CH4}}V\big|_{\text{Fill,React,W}} - Q_{\text{eff}}\text{CH}_4\big|_{\text{Eff}} - Q_{\text{W}}\text{CH}_4\big|_{\text{W}}$$
$$- Q_{\text{gas}}\frac{p_{\text{CH4}}}{p_{\text{tot}}}\rho_{\text{CH4}}\bigg|_{\text{Fill,React,W}}$$

(6.45)

Total carbonic acid:

$$\frac{d(VH_2CO_{3tot})}{dt} = \frac{r_{CO2}}{44} V\Big|_{Fill,React,W} + Q_{fill}H_2CO_{3totIN}\Big|_{Fill} - Q_{eff}H_2CO_{3tot}\Big|_{Eff}$$
$$- Q_W H_2CO_{3tot}\Big|_W \tag{6.46}$$

The gas flow rate is variable during the cycle, because the rate of the reactions is variable and because we assume there is no gas flow rate during the settling and effluent withdrawal phases:

$$Q_{gas} = \frac{\left(r_{CH4}V - \frac{d(VCH_4)}{dt}\right)\Big|_{Fill,React,W}}{\rho_{CH4}} + \frac{\left(r_{H2}V - \frac{d(VH_2)}{dt}\right)\Big|_{Fill,React,W}}{\rho_{H2}}$$
$$+ \frac{\left(r_{CO2}V - \frac{d(VH_2CO_{3tot})}{dt}\right)\Big|_{Fill,React,W}}{\rho_{H2}} + Q_{H2O} \tag{6.47}$$

With the usual rearrangements, imposing the conditions of periodic steady state and taking the definitions of HRT and SRT into account, we obtained the following equations that describe the periodic steady state of the SBR:

$$\int_{cycle} dGLU = \int_{cycle} \left[r_{GLU}\Big|_{Fill,React,W} + \frac{1}{HRT \cdot t_{fill} \cdot Nocycles} \frac{(GLU_{IN} - GLU)}{V/V_{full}}\Big|_{Fill} \right] dt = 0 \tag{6.48}$$

$$\int_{cycle} dAc = \int_{cycle} \left[r_{Ac}\Big|_{Fill,React,W} + \frac{1}{HRT \cdot t_{fill} \cdot Nocycles} \frac{-Ac}{V/V_{full}}\Big|_{Fill} \right] dt = 0 \tag{6.49}$$

$$\int_{cycle} dH_2 = \int_{cycle} \left[\begin{array}{c} \left. r_{H2} \right|_{Fill,React,W} - \left. \dfrac{Q_{gas}}{V} \dfrac{p_{H2}}{p_{tot}} \rho_{H2} \right|_{Fill,React,W} \\[2em] + \dfrac{1}{HRT \cdot t_{fill} \cdot Nocycles} \left. \dfrac{-H_2}{V/V_{full}} \right|_{Fill} \end{array} \right] dt = 0 \quad (6.50)$$

$$\int_{cycle} dCH_4 = \int_{cycle} \left[\begin{array}{c} \left. r_{CH4} \right|_{Fill,React,W} - \left. \dfrac{Q_{gas}}{V} \dfrac{p_{CH4}}{p_{tot}} \rho_{CH4} \right|_{Fill,React,W} \\[2em] + \dfrac{1}{HRT \cdot t_{fill} \cdot Nocycles} \left. \dfrac{-CH_4}{V/V_{full}} \right|_{Fill} \end{array} \right] dt = 0 \quad (6.51)$$

$$\int_{cycle} dCO_2 = \int_{cycle} \left[\begin{array}{c} \left. \dfrac{r_{CO2}}{44} \right|_{Fill,React,W} - \left. \dfrac{Q_{gas}}{V} \dfrac{p_{CO2}}{p_{tot}} \dfrac{p_{CO2}}{44} \right|_{Fill,React,W} \\[2em] + \dfrac{1}{HRT \cdot t_{fill} \cdot Nocycles} \\[1em] \left. \dfrac{\left(H_2CO_{3totIN} - \dfrac{p_{CO2} \cdot k_{eqCO2}}{44} \cdot \dfrac{1 + K_{CO2} + \dfrac{K_{H2CO3}}{10^{-pH}} + \dfrac{K_{HCO3} K_{H2CO3}}{10^{-2pH}}}{K_{CO2}} \right)}{V/V_{full}} \right|_{Fill} \end{array} \right] dt = 0 \quad (6.52)$$

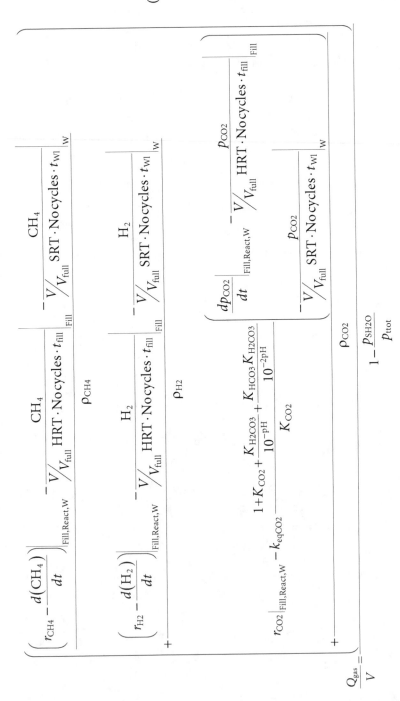

$$
\frac{Q_{gas}}{V} = \left\{ \left[r_{CH4} - \frac{d(CH_4)}{dt} \Big|_{Fill,React,W} \right] - \frac{V}{V_{full}} \Big|_{\frac{CH_4}{HRT \cdot Nocycles \cdot t_{fill}}} \Big|_{Fill} \rho_{CH4} - \frac{V}{V_{full}} \Big|_{\frac{CH_4}{SRT \cdot Nocycles \cdot t_{W1}}} \Big|_W \right.
$$
$$
+ \left[r_{H2} - \frac{d(H_2)}{dt} \Big|_{Fill,React,W} \right] - \frac{V}{V_{full}} \Big|_{\frac{H_2}{HRT \cdot Nocycles \cdot t_{fill}}} \Big|_{Fill} \rho_{H2} - \frac{V}{V_{full}} \Big|_{\frac{H_2}{SRT \cdot Nocycles \cdot t_{W1}}} \Big|_W
$$
$$
+ \left[r_{CO2} \Big|_{Fill,React,W} - k_{eqCO2} \frac{1 + K_{CO2} + \frac{K_{H2CO3}}{10^{-pH}} + \frac{K_{HCO3}K_{H2CO3}}{10^{-2pH}}}{K_{CO2}} \left(\frac{dp_{CO2}}{dt} \Big|_{Fill,React,W} - \frac{V}{V_{full}} \Big|_{\frac{p_{CO2}}{HRT \cdot Nocycles \cdot t_{fill}}} \Big|_{Fill} \right. \right.
$$
$$
\left. \left. \rho_{CO2} - \frac{V}{V_{full}} \Big|_{\frac{p_{CO2}}{SRT \cdot Nocycles \cdot t_{W1}}} \Big|_W \right) \right] \rho_{CO2} \right\} \frac{1}{1 - \frac{p_{SH2O}}{p_{ttot}}}
$$

(6.53)

The solution of the system of Equations 6.48 through 6.53 gives the profiles of glucose and acetic acid in the liquid phase, of the partial pressure of methane, hydrogen and carbon dioxide in the gas phase and of the gas flow rate during a cycle at the periodic steady state. Similar to what we discussed for the SBR only for carbon removal and for carbon and nitrogen removal, in order to solve Equations 6.48 through 6.53, we need to set the values of HRT, SRT, the number of cycles and the length of all the phases minus one (the length of the remaining phases can then be immediately calculated from the overall length of the cycle).

In order to set the values of the design parameters, considerations similar to the ones done for the anaerobic digestion process with biomass recycle (Section 5.3) can be done. The performance of the process, in terms of substrate removal, gas production and gas composition, is essentially determined by the SRT, which therefore needs to be chosen at a high-enough value that ensures the desired treatment efficiency. The HRT determines the concentration of the microorganisms in the reactor. Therefore, as for other types of wastewater treatment processes, the choice of the HRT is a compromise between the need of having a small reactor (low HRT) and the need for avoiding a too high biomass concentration, which would give a low settling velocity and therefore would make the process unfeasible. As far as the length of the phases is concerned, in general, it may be advantageous to have a short fill phase, in order to have high substrate concentration at the end of the feed, which gives higher substrate removal rates. However, if the feed contains substrates that are inhibitory at high concentrations, the fill phase should be long enough to limit their buildup. The length of the settling phase should be long enough to allow the complete settling of the microorganisms.

6.4 FURTHER EXAMPLES ON THE SBR PROCESS

Example 6.5

An SBR for carbon removal is designed with a cycle composed of fill, react, sludge withdrawal, settle and effluent withdrawal phases. The design value of the HRT is 0.25 day, the SRT is 20 days, the number of cycles is 6 per day and the calculated biomass concentration at the end of the sludge withdrawal phase is 2.5 kg/m³. The estimated SVI is 150 ml/g. Is the process well designed, that is will all the biomass be able to settle with no losses with the effluent?

Solution

Since we have an HRT of 0.25 day, this means that every day, the volume of the feed to the reactor is four times the volume of the reactor. Since we have six cycles per day, in each cycle, we need to feed a volume equal to 4/6 = 67% of the reactor volume. Therefore, at the end of each cycle, we will be left with 33% of the reactor volume occupied by the settled microorganisms. The concentration of micro-organisms in the full reactor is 2.5 g/L. When this biomass is settled, it will occupy a volume of:

$$2.5\frac{g}{l} \cdot 0.15\frac{l}{g} = 0.375\frac{l}{l}$$

which is 37.5% of the full reactor volume.

Therefore, the process is not well designed, because the volume occupied by the settled biomass will be larger than the available volume for settling. This will cause the loss of some biomass during the effluent withdrawal phase, with negative consequences on the process.

The design of the process needs to be modified, and a possible option is to increase the number of cycles per day, leaving the HRT and the SRT unchanged. This will not change the biomass concentration at the end of the sludge withdrawal phase. Using, for example, eight cycles per day, in each cycle we will need to feed 4/8 = 50% of the reactor volume, and 50% of the volume will be left after the effluent removal phase. This means that 50% of the full reactor volume will be available for the microorganisms to settle. Since we estimate that the settled biomass will need 37.5% of the reactor volume, the design of a process with eight cycles per day is satisfactory and meets the requirements for biomass settling. Of course, it needs to be checked that the length of the phases will still be enough for complete removal of the substrate and in particular that the length of the settling phase will be enough to allow the settling of the microorganisms.

Example 6.6

An SBR process operates with a substrate concentration in the feed equal to 700 mgCOD/L, and the effluent substrate concentration is 30 mgCOD/L. The length of the fill phase is 40 min, during which time the flow rate of the feed pump is 7 m³/min. The length of the sludge withdrawal phase is 10 min, during which the sludge withdrawal pump operates at a flow rate of 0.5 m³/min. The biomass concentration during the sludge withdrawal phase is 2 g/L. There are four cycles per day. Calculate the average daily oxygen consumption by the microorganisms.

Solution

The COD removed per day is:

$$Q_{fill}\, t_{fill} \text{No cycles}(S_0 - S) = 7\frac{m^3}{min}\, 40 \text{ min} \cdot \frac{4}{day}(700 - 30)\frac{gCOD}{m^3}$$

$$= 187.6\frac{kgCOD}{day}$$

The biomass produced per day is:

$$Q_W\, t_W \text{No cycles} \cdot X = 0.5\frac{m^3}{min}\, 10 \text{ min} \cdot \frac{4}{day}\, 2\frac{kg}{m^3} = 40\frac{kg\, biomass}{day}$$

which, converted into COD using the usual factor of 1.42 gCOD/gbiomass, gives 56.8 kgCOD/day.

Therefore, the average daily oxygen consumption by the microorganisms is:

$$Q_{O2biomass} = 187.6 - 56.8 = 130.8\frac{kgO_2}{day}$$

Example 6.7

A wastewater has an ammonia concentration of 60 mgN/L. It is desired to remove ammonia in an SBR using a sequence of anoxic and aerobic phases. The desired effluent nitrogen concentration is 8 mgN/L. The chosen design value for the HRT is 0.5 day. Ignoring the contribution of ammonia removal due to biomass growth and assuming that the process conditions are such that all the ammonia is converted to nitrate in the aerobic phase and all the nitrate is removed during the anoxic phase, what is the minimum required number of cycles per day? If this number of cycles is not satisfactory, which design parameter needs to be changed to obtain a more reasonable number of cycles?

Solution

An HRT value of 0.5 day means that the volume of wastewater fed per day is equal to twice the full volume of the reactor. If, for example, we have four cycles per day, we will have that in each cycle the volume of feed is half of volume of the reactor. Therefore, in this case, the initial ammonia concentration at the end of the feed will be $60/2 = 30$ mgN/L, which will correspond to 30 mg N-NO$_3$/L at the end of the cycle and in the effluent. Note that here we have assumed that all the ammonia is converted to nitrate in the aerobic phase and that all the nitrate is removed in the anoxic phase.

These considerations show that four cycles per day are not enough to achieve the desired nitrogen removal. With similar considerations, we can calculate that 16 cycles per day are required to produce an effluent nitrogen concentration equal to 7.5 mgN/L and compliant to the specification. Indeed, with 16 cycles per day, 1/8th of the reactor volume needs to be fed per day, and therefore, the initial ammonia concentration and the final nitrate concentration are both equal to $60/8 = 7.5$ mgN/L.

Having 16 cycles per day may be impractical, because the length of each cycle will be only 1.5 hours, leaving not much time for the settling phase. In order to achieve the desired nitrogen removal with fewer cycles, it is necessary to increase the HRT. For example, with an HRT of one day and eight cycles per day, the volume fed per cycle will be equal to 1/8th of the total reactor volume, obtaining an effluent nitrogen concentration of 7.5 mgN/L, in agreement with the specification.

Example 6.8

An anaerobic SBR is fed with wastewater having a substrate concentration of 8 kgCOD/m^3. The volume of the reactor is 1000 m^3, the HRT is two days and the SRT is 30 days. The concentration of microorganisms in the sludge withdrawal stream is 15 kg/m^3. The average gas composition of the produced is 50% CH$_4$/50% CO$_2$ by volume. Assuming that all the COD that is not converted to microorganisms is converted to methane, calculate the average daily gas flow rate produced by the reactor.

Solution

With an HRT of two days and a reactor volume of 1000 m^3, the volume of wastewater fed per day is:

$$Q = \frac{V}{HRT} = 500 \frac{m^3}{day}$$

This means that the total mass flow rate of substrate is:

$$QS_0 = 8\frac{kgCOD}{m^3} \cdot 500\frac{m^3}{day} = 4000\frac{kgCOD}{day}$$

The average daily flow rate of the sludge withdrawal stream is:

$$Q_W t_W \text{No cycles} = \frac{V_{full}}{SRT} = \frac{1000\,m^3}{30\,day} = 33.3\frac{m^3}{day}$$

Therefore, the average daily sludge production is:

$$P_X = 33.3\frac{m^3}{day} \cdot 15\frac{kg}{m^3} = 500\frac{kg}{day}$$

With the assumption that all the COD that is not converted to micro-organisms is converted to methane, the amount of methane produced is equal to $4000 - 1.42 \cdot 500 = 3290$ kgCOD/day. This corresponds to $3290/4 = 822.5$ kg methane/day (where 4 is the conversion factor of methane into COD). This is equal to $822.5/16 = 51.4$ kmol methane/day. From the ideal gas law, for $0\,°C$ and 1 atm (in order to obtain Nm³):

$$V = \frac{nRT}{P} = \frac{51.4 \cdot 0.0821 \cdot 273.15}{1} = \frac{1153\,Nm^3\,methane}{day}$$

Since the produced biogas is 50% methane by volume, the daily gas production is 2306 Nm³/day.

6.5 KEY POINTS

- In SBRs, wastewater treatment is obtained as a temporal sequence of phases and cycles in one single vessel, rather than as a spatial sequence of vessels, like in conventional activated sludge processes. In particular, reaction and settling take place in the same vessel rather than in separate vessels;

- Substrate and biomass concentrations change during the cycle due to dilution with the feed, biomass metabolism on the substrate, settling and effluent withdrawal. However, at the periodic steady state, the profiles of biomass and substrate during the cycle are the same for any cycle. The profiles of biomass and substrate (and of any impor-tant variable) at the periodic steady state can be calculated using the appropriate mass balances under dynamic conditions and imposing the condition that the integral of all the differential variables over the length of the cycle is equal to zero. This ensures that the values of the variables at any given time of the cycle are the same between consecutive cycles;

- The SBR has more degrees of freedom than the continuous flow activated sludge process, because in addition to the SRT and the HRT, we also have the length of the various phases. This means that for the SBR, there are more parameters that the process designer needs to choose before calculating the process design. However, the most important design parameter is still the SRT, which determines the effluent concentration of the substrate and the total mass of biomass in the system;

- Once the substrate and biomass profiles at the periodic steady state have been calculated, the oxygen profile during a cycle at the periodic steady state can also be calculated with the same approach, that is by writing the mass balance for oxygen under dynamic conditions and imposing the condition that the integral of the differential oxygen concentration over the length of the cycle is equal to zero. The oxygen profile depends on the mass transfer coefficient $k_L a$, and it can be used to determine the minimum value of the $k_L a$ required to maintain the oxygen concentration above the desired minimum level. It is important to observe that due to the variable biomass and substrate concentrations during the cycle, the oxygen concentration is also variable during the SBR cycle and the required $k_L a$ may be larger than for conventional activated sludge processes due to the faster rate of reaction, especially for shorter feed lengths (which correspond to higher substrate concentrations in the initial part of the cycle);

- The dynamic nature of the SBR, characterised by substrate gradients during the cycle, gives distinct advantages of continuous flow activated sludge processes: the influent biodegradable substrate can be removed completely and filamentous bulking can be controlled better, because the high substrate concentrations in the initial part of the cycle favour the growth of floc-formers over filaments. However, the extent of this effect depends on the chosen values of the design parameters. In particular, the length of the feed is the key parameter that determines the presence and extent of substrate gradients during the cycle.

CHAPTER 7

Attached Growth Processes

S O FAR WE HAVE only considered processes with dispersed growth, that is, where the microorganisms are suspended in the liquid medium. Another category of biological processes are the so-called attached growth processes, where the microorganisms grow attached to a solid support medium. The main advantage of attached growth processes is that higher biomass concentrations can be obtained than in suspended growth processes; therefore, an attached growth process can obtain the same treatment efficiency with a lower reactor volume. In an ideal attached growth process where all the microorganisms are attached to the support and there are no suspended cells, there would be no need of solid–liquid separation after the biological reactor. However, usually a small fraction of the microorganisms detach from the support and are present in the effluent and therefore need to be separated before discharge of the wastewater.

In this section, we will see two types of attached growth processes, packed bed processes and rotating biological reactors (also called rotating biological contactors or rotating disc reactors). In attached growth processes, microorganisms usually grow as a thick biofilm attached to the support medium. Under these conditions, it is possible that significant mass transfer resistances develop, causing the substrate and/or oxyvgen concentration available to microorganisms inside the biofilm to be lower than in the bulk liquid phase. Under these conditions, the rate of microorganisms growth, and therefore of substrate removal, may be limited by the mass transfer of substrate and/or oxygen rather than by the metabolic processes. In this chapter however, we will ignore mass transfer limitations and we will assume that the substrate (and oxygen and any other nutrients) concentration inside the biofilm is the same as in the bulk liquid phase. We will make this assumption in order to simplify our study and to develop simple design equations. Also, for the same reason, we will always make the assumption that the liquid phase is perfectly mixed, even though this assumption may be questionable, especially for packed bed reactors.

7.1 PACKED BED PROCESSES

7.1.1 Aerobic Packed Bed Processes

Consider an aerobic packed bed reactor, where the influent wastewater has a flow rate Q and a substrate concentration S_0 (Figure 7.1).

The total volume of the reactor, that is, the total volume of the packed bed, is V and the overall biomass concentration, referred to the total volume of the reactor, is X. X_{eff} is the biomass concentration in the effluent and we call δ the ratio between the biomass concentration in the effluent and in the reactor, that is,

$$\delta = \frac{X_{eff}}{X} \tag{7.1}$$

We assume that δ is constant, that is, that the ratio between the biomass concentration in the effluent and in the packed bed is the same under all conditions. The hydraulic retention time is defined as usual as

$$HRT = \frac{V}{Q} \tag{4.11}$$

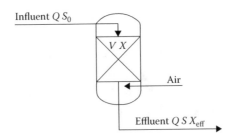

Influent $Q\,S_0$

$V\,X$

Air

Effluent $Q\,S\,X_{eff}$

FIGURE 7.1 Conceptual scheme of an aerobic packed bed reactor.

and the solids residence time (SRT) is given by

$$SRT = \frac{VX}{QX_{eff}} = \frac{HRT}{\delta} \qquad (7.2)$$

Note that with the model used here the SRT is a function of the HRT; therefore, we have only one design parameter. Designing an aerobic packed bed reactor means essentially deciding the value of its volume as a function of the desired concentration of substrate in the effluent. The design of the packed bed reactor can be made by using the mass balances for substrate and biomass. Under the assumptions explained above, that is, perfect mixing of the liquid phase and absence of mass transfer resistances, the mass balances are particularly straightforward.

Biomass:

$$\left(r_X + r_{end}\right)V = QX_{eff} \Rightarrow \left(\mu - b\right) = \frac{QX_{eff}}{XV} = \frac{1}{SRT} = \frac{\delta}{HRT} \qquad (7.3)$$

By solving for S we obtain

$$S = \frac{\delta K_S + bK_S HRT}{\left(\mu_{max} - b\right)HRT - \delta} = f(HRT) \qquad (7.4)$$

Equation (7.4) is the same as the equation for the effluent substrate concentration in the activated sludge process, as we can observe by replacing δ using Equation (7.2):

$$S = \frac{bK_S SRT + K_S}{\left(\mu_{max} - b\right)SRT - 1} \qquad (4.22)$$

The substrate balance is

$$Q(S_0 - S) = \frac{r_X V}{Y_{X/S}} = \frac{\mu XV}{Y_{X/S}} \Rightarrow \frac{XV}{Q} = \frac{(S_0 - S)Y_{X/S} HRT}{\delta + bHRT} = f(HRT) \qquad (7.5)$$

Therefore, it is evident is that the choice of the HRT determines both the substrate concentration in the effluent and the total mass of biomass in the reactor per unit of influent flow rate. It is an obvious consequence of Equation 7.5 that the choice of the HRT also determines the value of the biomass concentration in the reactor, that is,

$$X = \frac{(S_0 - S)Y_{X/S}}{\delta + b\text{HRT}} \qquad (7.6)$$

For a given packing material, characterised by a bulk density ρ_{pack}, the choice of the HRT also determines the mass of packing material M_{pack} (kg) that is required to fill the reactor (per unit of influent flow rate):

$$\frac{V}{Q} = \text{HRT} = \frac{M_{pack}}{Q\rho_{pack}} \quad \Rightarrow \quad M_{pack} = Q\rho_{pack}\text{HRT} \qquad (7.7)$$

It is also a consequence of the previous equations that the choice of the HRT determines the concentration of biomass per unit mass of the packing materials, X_{pack} (kg biomass/kg support). Indeed, the total mass of biomass in the reactor has to be equal to the total mass of biomass on the support material:

$$XV = X_{pack}M_{pack} \qquad (7.8)$$

Equation 7.8, combined with Equation 7.6, gives

$$X_{pack} = \frac{(S_0 - S)Y_{X/S}}{\delta + b\text{HRT}}\frac{1}{\rho_{pack}} \qquad (7.9)$$

In summary, the design of an aerobic packed bed reactor requires the choice of the HRT and the calculation of the substrate concentration from Equation 7.4 and of the biomass concentration from Equation 7.6. From the choice of the HRT, the reactor volume can be immediately calculated. The required mass of the packing material can be calculated from its bulk density using Equation 7.7. Table 7.1 reports some literature values for some typical support materials.

TABLE 7.1   Values of the Bulk Density and of the Biomass Concentration on the Support for Some Packing Materials

Packing Material	Bulk Density (kg/m³)	Biomass Concentration (kg Biomass/kg Support)
Bone char	850	0.087
Glass beads	1470	0.044
Glass wool	62	0.05
Clay brick	790	0.093
Activated carbon	493	0.032
Expanded clay	610	0.007

7.1.1.1 Effect of the Choice of the Design Parameters

We have seen that designing an aerobic packed bed reactor means essentially deciding the value of the HRT. Having set a value for the HRT, with known kinetic parameters of the microorganisms and with a known value of the parameter δ which gives the solid losses with the effluent, the effluent substrate concentration, the total mass of biomass in the reactor (per unit of influent flow rate) and the biomass concentration in the reactor can be calculated. Once a packing material is chosen, with its value of the bulk density, the required mass of packing material and the biomass concentration on the support (kg biomass/kg support) can also be calculated.

Looking at Equation 7.4, increasing the HRT gives a lower value of the effluent substrate concentration. However this is only true if we have biomass losses with the effluent, that is, $X_{eff} \neq 0, \delta \neq 0$. If there are no biomass losses with the effluent ($X_{eff} = 0, \delta = 0$), the effluent substrate concentration becomes independent of the HRT. This is because in reality what determines the effluent substrate concentration is the SRT and not the HRT. If $\delta = 0$, the SRT becomes infinite for any value of the HRT and the effluent substrate concentration becomes, as we have already seen in Chapter 4,

$$S = \frac{bK_S}{\mu_{max} - b} \tag{4.24}$$

Similar considerations can be made for the total mass of biomass in the reactor which, if there are biomass losses with the effluent, depends only on the HRT, and increases with the HRT. If there are no biomass losses with the effluent ($\delta = 0$), the total mass of biomass in the reactor per unit of influent flow rate becomes

$$\frac{XV}{Q} = \frac{(S_0 - S)Y_{X/S}}{b} \tag{7.10}$$

that is, it is independent of the HRT. The explanation of this is as discussed previously, that is, it is the SRT and not the HRT that actually matters, and with $\delta = 0$, the SRT becomes infinite for any value of the HRT.

However, the choice of the HRT is important because it determines, for any value of δ, the value of the biomass concentration in the reactor and especially the value of the biomass concentration on the support, X_{pack}. X_{pack} increases with increasing the HRT, and if the HRT is too low, X_{pack} will become too high and the support material will not be able to handle the biomass anymore, causing high losses with the effluent and therefore compromising the performance of the reactor. Therefore, the HRT should be chosen so to avoid exceeding the maximum concentration of biomass on the support. The values of biomass concentration on various types of support reported in Table 7.1 can be used as a guidance to see whether the chosen HRT value is acceptable.

Example 7.1

It is desired to treat 5000 m³/day of a wastewater having a substrate concentration of 5 kgCOD/m³. It is desired to obtain an effluent substrate concentration of less than 1 mgCOD/L. It is chosen to use an aerobic packed bed process, using activated carbon as support material. Choose an appropriate value of the HRT and calculate the effluent substrate concentration, the biomass concentration in the reactor and in the effluent, the required mass of activated carbon. Assume that the ratio between biomass concentration in the effluent and in the reactor (parameter δ) is equal to 0.02. Assume the kinetic parameters of Example 4.1. Assume that the support material has a bulk density of 500 kg/m³ and that the maximum biomass concentration that this support can hold is 0.04 kg biomass/kg support.

Solution

The substrate and biomass concentration in the reactor can be calculated from Equations 7.4 and 7.6 as a function of the HRT. These profiles are reported in Figure 7.2. The profiles are analogous to the profiles of Example 4.1 as a function of the SRT. Indeed, with the assumptions used in this section, the HRT and SRT are proportional by the constant factor δ. From Figure 7.2, it is evident that the required effluent quality is obtained even for the smallest HRT values used in the simulations. Therefore, purely based on the removal of the substrate, even very small values of the HRT could be chosen. However, the HRT also affects the biomass concentration on the support, X_{pack}, which is obtained from Equation (7.9). Figure 7.3 reports the effect of the HRT on X_{pack} and on the mass of packing materials required, M_{pack} (Equation 7.7). From Figure 7.3, it is evident that HRT values below approximately 0.3 day would give an unacceptably high biomass concentration on the support, above the maximum value of 0.04 kg biomass/kg support. Therefore, a value of the HRT higher than 0.3 day needs to be chosen. For example if we

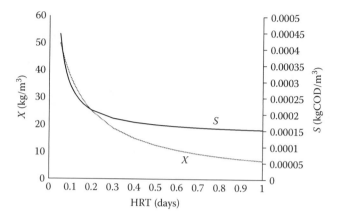

FIGURE 7.2   Example 7.1. Effect of the HRT on the effluent substrate concentration and on the biomass concentration inside the reactor.

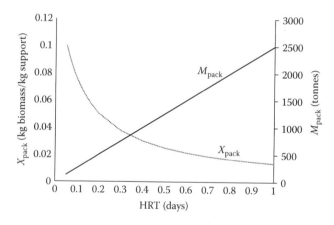

FIGURE 7.3   Example 7.1. Effect of the HRT on the biomass concentration on the support and on the required mass of packing material.

choose an HRT of 0.4 day (which corresponds to an SRT of 20 days), that gives a reactor volume of 2000 m³, a mass of packing material of 1000 tonnes and a X_{pack} value of 0.03 kg biomass/kg support. Note that the biomass concentration referred to the whole reactor volume is in this case equal to 15 kg/m³. It would be probably not practicable to use a suspended biomass process like activated sludge with a so high biomass concentration, because the settling velocity would be very low and therefore the required settling area would be unreasonably high. If we consider an activated sludge process and we assume, that the maximum biomass concentration in the biological reactor is, for example, 4 kg/m³, with a wastewater like the one considered here (5 kg COD/m³, 5000 m³/day) and with an SRT of 20 days, we would need a minimum HRT of 1.5 day, with a corresponding minimum volume of 7500 m³. Therefore, this example shows the benefits of attached growth processes over suspended growth processes for wastewaters of high COD loading: attached growth processes allow to have higher biomass concentrations in the reactor, therefore requiring lower reactor volumes.

7.1.2 Anaerobic Packed Bed Reactors

The design of the anaerobic packed bed reactor is in principle similar to the design of the aerobic reactor. Let us consider for example an anaerobic packed bed reactor fed with a wastewater containing glucose as the only carbon source. As described in Chapter 5, we assume that there are three population of microorganisms, X_{GLU}, X_{AC} and X_{H2}, plus the inert biomass which is a product of the endogenous metabolism. The effluent of the reactor will have biomass concentrations equal to X_{GLUeff}, X_{Aceff}, X_{H2eff}, $X_{inerteff}$ and a gas flow rate is generated composed of carbon dioxide, hydrogen, methane and water vapour. The conceptual scheme of an anaerobic packed bed reactor is shown in Figure 7.4.

The ratio δ is defined, in the same way as for the aerobic packed bed reactor, as the ratio between the biomass concentration in the effluent and in the reactor:

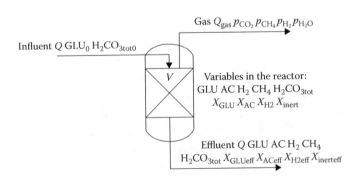

FIGURE 7.4 Conceptual scheme of an anaerobic packed bed reactor.

$$\delta = \frac{X_{GLUeff}}{X_{GLU}} = \frac{X_{ACeff}}{X_{AC}} = \frac{X_{H2eff}}{X_{H2}} = \frac{X_{INERTeff}}{X_{INERT}}$$

$$= \frac{\left(X_{GLUeff} + X_{ACeff} + X_{H2eff} + X_{INERTeff}\right)}{X_{GLU} + X_{AC} + X_{H2} + X_{INERT}} \tag{7.11}$$

And we define the HRT and the SRT as done previously:

$$HRT = \frac{V}{Q} \tag{4.11}$$

$$SRT = \frac{V\left(X_{GLU} + X_{AC} + X_{H2} + X_{INERT}\right)}{Q\left(X_{GLUeff} + X_{ACeff} + X_{H2eff} + X_{INERTeff}\right)} = \frac{HRT}{\delta} \tag{7.12}$$

Analogously to what was done for the aerobic packed bed reactor, in order to design the anaerobic packed bed reactor we need to choose a value of the HRT, and then all the variables can be calculated using mass balances. The mass balances for the various components can be written with the usual procedure and are very similar, although not identical, to the ones for the CSTR anaerobic digester in Section 5.1.

$$GLU_0 + r_{GLU} \cdot HRT = GLU \tag{7.13}$$

$$r_{Ac} \cdot HRT = Ac \tag{7.14}$$

$$r_{H2} \cdot HRT = p_{H2} \cdot k_{eqH2} + \frac{Q_{gas}}{Q} \frac{p_{H2}}{p_{tot}} p_{H2} \tag{7.15}$$

$$r_{CH4} \cdot HRT = p_{CH4} \cdot k_{eqCH4} + \frac{Q_{gas}}{Q} \frac{p_{CH4}}{p_{tot}} p_{CH4} \tag{7.16}$$

$$\left(r_{XGLU} + r_{endGLU}\right) \cdot HRT = \delta \tag{7.17}$$

$$\left(r_{XAC} + r_{endAC}\right) \cdot HRT = \delta \tag{7.18}$$

$$\left(r_{XH2} + r_{endH2}\right) \cdot HRT = \delta \tag{7.19}$$

$$\left(-r_{endGLU} - r_{endAC} - r_{endH2}\right) \cdot HRT = X_{INERTeff} \tag{7.20}$$

$$H_2CO_{3totin} + \frac{r_{CO2}}{44} HRT = \frac{p_{CO2} \cdot k_{eqCO2}}{44}$$

$$\frac{1 + K_{CO2} + \dfrac{K_{H2CO3}}{10^{-pH}} + \dfrac{K_{HCO3} K_{H2CO3}}{10^{-2pH}}}{K_{CO2}} + \frac{Q_{gas}}{Q} \frac{p_{CO2}}{p_{tot}} \frac{p_{CO2}}{44} \tag{5.31}$$

$$\frac{Q_{gas}}{Q} = \frac{\left(\dfrac{r_{CH4} \cdot HRT - \left(p_{CH_4} \cdot k_{eqCH4} - CH4_{IN} \right)}{p_{CH4}} + \dfrac{r_{H2} \cdot HRT - \left(p_{H_2} \cdot k_{eqH2} - H_{2IN} \right)}{p_{H2}} \\ + \dfrac{r_{CO2} \cdot HRT - \left(H_2CO_{3tot} - H_2CO_{3totIN} \right)}{p_{CO2}} \right)}{1 - \dfrac{p_{SH2O}}{p_{tot}}} \tag{5.32}$$

Equations 7.13–7.20, 5.31 and 5.32 are a system of 10 equations with 10 unknowns and, by giving a value to the parameter HRT, the values of all the unknowns, GLU, AC, X_{GLU}, X_{H2}, X_{AC}, X_{INERT}, Q_{gas}/Q, p_{CO2}, p_{H2} and p_{CH4}, can be calculated. Note that the anaerobic packed bed reactor is a system in which the HRT is different from the SRT and therefore the design Equations 7.13–7.20, 5.31 and 5.32 are essentially the same as the one for the anaerobic digester with biomass recycle (Section 5.3), with the only difference that here we do not have the recycle ratio R, and therefore the biomass balances in the reactor and in the whole system coincide (while they were different equations for the anaerobic digester with biomass recycle). Also, in the anaerobic packed bed reactor as modelled here the HRT and SRT are different but proportional to each other, while they are independent parameters in the anaerobic digester with biomass recycle.

The required mass of packing material can be calculated with the same approach used for the aerobic packed bed reactor:

$$\frac{V}{Q} = HRT = \frac{M_{pack}}{Q\rho_{pack}} \implies M_{pack} = Q\rho_{pack}HRT \tag{7.7}$$

The total concentration of biomass referred to the support material, $X_{packtot} = X_{GLUpack} + X_{ACpack} + X_{H2pack} + X_{INERTpack}$, can be calculated from

$$X_{tot}V = X_{packtot} M_{pack} \tag{7.21}$$

where

$$X_{tot} = X_{GLU} + X_{AC} + X_{H2} + X_{INERT} \qquad (7.22)$$

which gives

$$X_{packtot} = \frac{X_{tot}}{\rho_{pack}} \qquad (7.23)$$

In summary, from the chosen value of the HRT all the variables that characterise the process, the required mass of packing material and the biomass concentration on the packing material can be calculated.

7.1.2.1 Effect of the Choice of Design Parameters

The only design parameter is the HRT. Since, with the assumptions made in this chapter, the HRT and the SRT are proportional to each other, the effect of the HRT on the anaerobic packed bed reactor is the same as the effect of the SRT discussed in Chapter 5 on the anaerobic digesters suspended biomass. High values of the HRT (which correspond in this case to high values of the SRT) give higher conversion to methane, while at shorter HRT (i.e. shorter SRT) values methanogenic microorganisms may be washed out and acetic acid may be the main product in the liquid phase.

An important difference between the anaerobic and the aerobic packed bed reactor is the effect of the HRT on the biomass concentration per unit of support material, $X_{packtot}$. Under aerobic conditions, according to the model used here, biomass decay due to endogenous metabolism gives a reduction in biomass concentration. Therefore, as we have seen in Chapter 4, biomass concentration increases less than linearly, while the SRT increases and eventually plateaus out. Since increasing the HRT causes a decrease in biomass concentration, the overall effect of increasing the HRT in an aerobic packed bed reactor is a decrease in the biomass concentration X and in the biomass concentration on the packing material X_{pack} (as we have seen in Example 7.1). By contrast, under anaerobic conditions with the modelling assumptions done here, increasing the SRT gives an approximately linear increase in the total biomass concentration (with a decrease in the active biomass concentration and an increase of the inert biomass). Therefore, in an anaerobic packed bed reactor, where with our assumptions the HRT and the SRT are linked and directly proportional, by increasing the HRT (and therefore the SRT) the effects of the HRT and the SRT on the biomass concentration approximately cancel out.

Therefore, under anaerobic conditions the total biomass concentration in the reactor (X) and on the support (X_{pack}) stays approximately constant as the HRT is changed, as long as the SRT is long enough to ensure that most of the inlet substrate is converted to methane. This means that for the anaerobic packed bed reactor the main criterion for the choice of the HRT is only that the HRT, and therefore the SRT, should be long enough to ensure the desired removal of the substrate and the desired gas production. Different from the aerobic packed bed reactor, there is no advantage increasing the HRT above the minimum value which ensures the desired performance.

Example 7.2

Using the kinetic parameters of Example 5.1, design an anaerobic packed bed reactor for a feed composed of glucose at 10 g/L. The flow rate of the feed is 1000 m^3/day. Assume that the value of the parameter δ (ratio biomass concentration in the effluent/biomass concentration in the reactor) is 0.04. Assume that the packing material has a bulk density 600 kg/m^3 and with a maximum allowable biomass concentration of X_{pack} 0.08 kg biomass/kg support. Assume that the pH in the reactor is controlled at the value of 8 and that there is no H_2CO_{3tot} in the feed.

Solution

The solution of this problem consists in solving Equations 7.13–7.20, 5.31 and 5.32 for various values of the HRT (which give corresponding values of the SRT). Note that the solution of these equations is the same as the solution of the equations for the anaerobic digester with biomass recycle, that is, Equations 5.23–5.26, 5.31, 5.32, 5.93–5.100, for the same values of the HRT and the SRT (note that the value of the parameter R, present in Equations 5.93–5.96, has no effect on the substrate and biomass concentration in the reactor, so its value is irrelevant in this case).

Since the HRT and the SRT are linked, in the choice of the HRT what matters is that the retention time for the microorganisms (i.e. the SRT) is long enough to ensure that conversion of the substrate to methane is as complete as possible. The profiles of the substrate concentrations and gas production will be the same as for Example 5.1 (see Figure 5.2). As discussed previously, assuming that most of the substrate is converted to methane, with the model used here the HRT will have very little impact on the biomass concentration on the support X_{pack}.

If we choose HRT = 1 day, which corresponds in this case to SRT = 1/0.04 = 25 days, we calculate the following values of the process variables:

GLU = 0.0018 kg/m³, Ac = 0.28 kg/m³, p_{H2} = 0.0004 atm, p_{CO2} = 0.098 atm, p_{CH4} = 0.846 atm, X_{GLU} = 3.53 kg/m³, X_{Ac} = 0.89 kg/m³, X_{H2} = 0.22 kg/m³, X_{inert} = 32.6 kg/m³, Q_{gas} = 3851 m³/day, M_{pack} = 500 tonnes, X_{tot} = 37.2 kg/m³, X_{pack} = 0.062 kg biomass/kg support.

These values give a satisfactory performance. However, as long as the SRT is long enough to ensure that most of the substrate is converted to methane (Figure 5.2 can be used as a guidance), lower values of the HRT would be good as well, because the biomass concentration on the support would stay relatively unaffected. Lower values of the HRT would give a reduced size of the reactor and a reduced requirement for the support material.

7.2 ROTATING BIOLOGICAL REACTORS

The design of Rotating Biological Reactor (RBR) can be done, conceptually, with the same approach and the same mass balances of the aerobic packed bed reactor. Consider a rotating biological contactor (RBC), where the influent wastewater has a flow rate Q and a substrate concentration S_0 (see Figure 7.5).

The submerged volume of the packed bed is V:

$$V = f_{subm} \frac{\pi D^2}{4} L \qquad (7.24)$$

where D and L are the diameter and length of the cylinder, respectively.

The overall biomass concentration, referred to the submerged volume of the reactor, is X. X_{eff} is the biomass concentration in the effluent and,

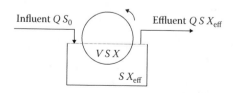

Influent Q S_0 Effluent Q S X_{eff}

$V S X$

$S X_{eff}$

FIGURE 7.5   Scheme of an RBR.

similarly as what we have done for the packed bed reactor, we call δ the ratio between the biomass concentration in the effluent and in the reactor, that is,

$$\delta = \frac{X_{\text{eff}}}{X} \tag{7.1}$$

And we assume that this ratio is a property of the system and does not depend on the HRT or SRT of the process. The hydraulic retention time is defined as usual as

$$\text{HRT} = \frac{V}{Q} \tag{4.11}$$

and the SRT is given by

$$\text{SRT} = \frac{VX}{QX_{\text{eff}}} = \frac{\text{HRT}}{\delta} \tag{7.2}$$

Designing an RBR essentially means deciding the value of its volume as a function of the desired concentration of substrate in the effluent. The design of the reactor can be made by using the mass balances for substrate and biomass. Under the assumptions explained above, the mass balances are particularly straightforward and are the same as for the aerobic packed bed reactor in Section 7.1.

Biomass:

$$\left(r_X + r_{\text{end}} \right) V = Q X_{\text{eff}} \quad \Rightarrow \quad (\mu - b) = \frac{QX_{\text{eff}}}{XV} = \frac{1}{\text{SRT}} = \frac{\delta}{\text{HRT}} \tag{7.3}$$

By solving for S, we obtain

$$S = \frac{\delta K_s + bK_s \text{HRT}}{\left(\mu_{\max} - b \right)\text{HRT} - \delta} = f(\text{HRT}) \tag{7.4}$$

which is essentially the same as the equation for the effluent substrate concentration in the activated sludge process.

The substrate balance is

$$Q(S_0 - S) = \frac{r_X V}{Y_{X/S}} = \frac{\mu X V}{Y_{X/S}} \quad \Rightarrow \quad \frac{XV}{Q} = \frac{\left(S_0 - S \right)Y_{X/S}\text{HRT}}{\delta + b\text{HRT}} = f(\text{HRT}) \tag{7.5}$$

Therefore, it is evident that the choice of the HRT determines both the substrate concentration in the effluent and the total mass of biomass in

TABLE 7.2   Values of the Surface Area and of the Biomass
Concentration for Packing Materials for RBRs

Disc Material	Surface Area (m²/m³)	Biomass Concentration (kg/m² Support)
HDPE	74	0.06–0.08
Polystyrene	144	0.011

the reactor per unit of the influent flow rate. It is an obvious consequence
of Equation 7.5 that the choice of the HRT also determines the value of the
biomass concentration in the reactor, that is,

$$X = \frac{(S_0 - S)Y_{X/S}}{\delta + b\text{HRT}} \quad (7.6)$$

RBRs are characterised by the specific area, that is, by the surface area of
the packing material per unit volume of the reactor, that is,

$$a_{surf}\left(\frac{m^2}{m^3}\right) = \frac{S_{pack}}{V} \quad (7.25)$$

where S_{pack} is the total surface area of the packing material in the submerged
volume of the RBR. The choice of the HRT determines the submerged
volume of the reactor

$$\frac{V}{Q} = \text{HRT} \quad (4.11)$$

Once the submerged volume is calculated, the biomass concentration
per unit surface area of the disc $X_{pack,surf}$ (kg biomass/m² support) can be
calculated

$$XV = a_{surf}VX_{pack,surf} \quad (7.26)$$

which gives

$$X_{pack,surf}\left(\frac{\text{kg biomass}}{m^2 \text{ support}}\right) = \frac{(S_0 - S)Y_{X/S}}{(\delta + b\text{HRT})a_{surf}} \quad (7.27)$$

Table 7.2 reports literature values for the surface area and biomass concentration in RBRs.

7.2.1 Choice of the Design Parameters

We have seen that for a given type of packing material characterised by the surface area a_{surf}, designing an RBR means choosing a value of the HRT. Of course, since the HRT determines the extent of substrate degradation, the HRT should be high enough to achieve the desired removal of the influent COD. Similarly as what was discussed for the aerobic packed bed reactor, also for the RBR, the HRT affects the effluent concentration S only if there are losses of biomass with the effluent, otherwise the SRT will be infinite for any values of the HRT and the removal of the substrate will not be dependent on the HRT. Once the minimum HRT for substrate degradation has been calculated, it might be necessary to increase the HRT even further. Increasing the HRT decreases the biomass concentration on the surface of the packing materials, $X_{pack,surf}$, and high HRT might be needed in order to maintain the value of $X_{pack,surf}$ below the maximum acceptable value for the disc. If the required value of the HRT in order to keep the value of $X_{pack,surf}$ below the maximum acceptable is too high, and the corresponding size of the disc would be too large, the use of multiple RBRs in parallel or in series will be required.

Example 7.3

A wastewater with a flow rate of 1000 m^3/day and a substrate concentration of 1 $kgCOD/m^3$ is to be treated with an RBR having a surface area a_{surf} of 100 m^2/m^3 and a maximum acceptable biomass concentration on the support $X_{pack,surf}$ of 0.04 kg biomass/m^2 support. The discs are available at a standard size with diameter $D = 3.5$ m and length $L = 8$ m. The submerged fraction of the disc is 0.40. Assume that the parameter δ is equal to 0.05. The effluent COD has to be less than 1 mgCOD/L. Determine the required HRT, the volume and number of discs. Use the kinetic parameters of Example 4.1.

Solution

The first criterion in the choice of the HRT is that the effluent COD meets the required value of 1 mgCOD/L or less. The effluent substrate profile as a function of the HRT is given by Equation 7.4 and it is qualitatively similar to what was reported in Figure 7.2. Using Equation 7.4 we find that we need an HRT of 0.05 day or higher in order to have an effluent concentration of 1 mgCOD/L or lower.

The second criterion in the choice of the HRT is the requirement that the biomass concentration on the support $X_{pack,surf}$ is lower than the maximum value possible. Using Equation 7.27 we find that for HRT $= 0.05$ day, $X_{packsurf} = 0.05$ kg/m², which is above the maximum acceptable value. Using Equation 7.27 again we find that a minimum HRT of 0.12 day is required in order to have $X_{pack,surf}$ of 0.04 kg/m² or lower. Choosing HRT $= 0.20$ day, we have $X_{pack,surf} = 0.03$ kg/m² and a required submerged volume of 200 m³. With the available geometry of the discs and submerged fraction, using Equation 7.24 we find that each disc has a submerged volume of approximately 31 m². Therefore, at least seven discs are required for the treatment.

7.3 FURTHER EXAMPLES ON ATTACHED GROWTH PROCESSES

Example 7.4

The same wastewater (flow rate 1000 m³/day) is treated in a conventional activated sludge process and in a packed bed reactor, both processes being aerobic for carbon removal only. The activated sludge process operates with an SRT of 22 days and the effluent substrate concentration is 2 mgCOD/L. The packed bed reactor has a volume of 500 m³ and is packed with a support material having a bulk density of 800 kg/m³. Measurements of the biomass concentration on the support give a value of 0.02 kg biomass/kg support. In the effluent, the biomass concentration is 50 mg/L and the substrate concentration is 10 mgCOD/L. Does the available data indicate that in the packed bed reactor there are mass transfer limitations (e.g. for the substrate or for oxygen)?

Solution

We need to calculate the SRT of the packed bed reactor. In the reactor, we have a mass of support material equal to

$$500 \text{ m}^3 \times 800 \text{ kg/m}^3 = 400 \text{ tonnes}$$

The value of the biomass concentration on the support is $X_{pack} = 0.02$ kg biomass/kg support, therefore in the reactor there are $400,000 \times 0.02 = 8000$ kg biomass, that is, a biomass concentration, referred to the entire volume of the reactor,

$$X = \frac{8000 \text{ kg}}{500 \text{ m}^3} = 16 \text{ kg/m}^3$$

The SRT for the packed bed reactor is

$$SRT = \frac{VX}{QX_{\text{eff}}} = \frac{8000 \text{ kg}}{1000 \text{ m}^3/\text{day} \cdot 0.05 \text{ kg/m}^3} = 160 \text{ day}$$

The SRT in the packed bed reactor is very high and considerably higher than for the activated sludge process, and yet the effluent concentration from the packed bed reactor is higher than for the activated sludge process. This indicates that probably the performance of the packed bed reactor is limited by mass transfer limitations, for substrate or oxygen transfer.

Example 7.5

A rotating biological reactor treats a wastewater with a flow rate of 200 m³/day and an influent COD of 800 mgCOD/L. The effluent has a substrate concentration of 5 mgCOD/L and a biomass concentration of 25 mg/L. Calculate the oxygen consumption rate by the microorganisms.

Solution

This problem can be solved with a simple application of the COD balance. The influent COD is

$$200 \text{ m}^3/\text{day} \cdot 0.8 \text{ kg COD/m}^3 = 160 \text{ kgCOD/day}$$

The effluent COD is

$$200 \text{ m}^3/\text{day} \cdot (0.005 + 0.025 \cdot 1.42) \text{ kgCOD/m}^3 = 8.1 \text{ kgCOD/day}$$

Therefore, the oxygen consumption by the microorganisms is

$$Q_{\text{O2biomass}} = 160 - 8.1 = 151.9 \text{ kgO}_2/\text{day}.$$

Example 7.6

An anaerobic packed bed reactor treats 10000 m³/day of a wastewater with an influent COD of 20 kgCOD/m³. The effluent of the reactor has a total biomass concentration of 80 mg/L and a soluble COD concentration of 0.2 kg COD/m³. Assuming that the gas phase only contains methane and carbon dioxide, which is the daily methane production in Nm³/day?

Solution

The influent COD is 10,000 m³/day · 20 kgCOD/m³ = 200,000 kgCOD/day. The effluent COD is

$$10,000\,\text{m}^3/\text{day} \cdot (0.2 + 0.080 \cdot 1.42)\,\text{kgCOD/m}^3 = 3136\,\text{kgCOD/day}$$

Therefore, the COD of the produced methane is

$$200,000 - 3,136 = 196,864\,\text{kgCOD/day}$$

which corresponds to 49,216 kg methane/day, that is, to 3076 kmol methane/day. Using ideal gas law, this corresponds to

$$Q_{methane} = \frac{3,076 \cdot 0.0821 \cdot 273.15}{1} = 68,981\,\text{Nm}^3\,\text{methane/day}.$$

7.4 KEY POINTS

- Attached growth processes have the advantage over dispersed growth processes of a higher biomass concentration, and therefore they can obtain the same efficiency of treatment in a reduced volume. Therefore, attached growth processes are particularly suitable for wastewater with high organic load;

- In attached growth processes, due to the dense and thick flocs, mass transfer limitations for the substrate and oxygen may occur inside the flocs. Therefore, not all the biomass may be active and the substrate removal rate per unit of biomass may be lower than with dispersed growth processes. However, the design procedure described here does not consider mass transfer limitations and therefore refers to 'ideal' attached growth processes where the rate of substrate removal per unit of biomass is the same as for suspended growth processes. Another important assumption made in this book is that the liquid phase in attached growth processes is completely mixed;

- With the described assumptions, the design of attached growth processes can be done with a similar approach as for suspended growth processes. The design parameters are the HRT and the SRT, which are linked via the solid losses with the effluent. For aerobic processes, increasing the HRT, and therefore the SRT, gives a higher treatment efficiency and a lower biomass concentration on the support. However, increasing the HRT causes the requirement for a larger volume and a larger mass of packing material, with a consequent increase in costs. In the design of attached growth processes, it needs to be taken care that the biomass concentration on the support does not exceed the maximum possible value that the support can hold;

- For anaerobic attached growth processes, with the assumption that endogenous metabolism generates inert biomass, the choice of the HRT affects the performance of the process in the same way as for aerobic processes (i.e. longer HRT gives better substrate degradation and higher gas production); however, differently than for aerobic processes, the HRT does not affect significantly the biomass concentration on the support;

- Rotating biological reactors can be designed using a similar approach than packed bed processes. The main design parameter is the HRT, which, under the assumptions made here, is linked to the SRT. Therefore increasing the HRT, and therefore the size of the reactor, leads to a better treatment efficiency. However, the main limitation of RBRs is that they can only be produced in relatively small sizes due to mechanical and physical constraints; therefore, they are suitable only for wastewaters with a relatively low flow rate.

Appendix A

Measurement of the Mass Transfer Coefficient for Oxygen in Water

M ANY PROCEDURES HAVE BEEN reported in the literature to measure $k_L a$. One of the easiest procedures to implement in a laboratory is the one that measures the $k_L a$ for the oxygen–water system. This experiment is described here.

The procedure requires an agitated vessel, where air can be either provided by diffusers or just by contact with the atmosphere (mechanical aeration). The experimental procedure is very simple:

1. Water is added to the vessel to the desired level and a sensor for dissolved oxygen is inserted into the liquid.

2. Nitrogen is sparged to the vessel in order to bring down the concentration of dissolved oxygen in water. Nitrogen sparging is stopped when oxygen concentration decreases below approximately 2 mg/L.

3. Air (or pure oxygen) at known flow rate is sparged into the vessel. Oxygen concentration starts to rise and oxygen concentration is recorded as a function of time.

4. The experiment is terminated when oxygen concentration reaches a value which is close to the saturation value with the inlet gas.

The $k_L a$ for the system can be calculated by applying the mass balance for oxygen. Indeed:

$$\frac{dC_{O_2}}{dt} = k_L a \cdot \left(C_{O_2}^* - C_{O_2} \right)$$

Integrating this equation between time 0, where the oxygen concentration is $C_{O_2,in}$ and time 0, with generic oxygen concentration C_{O_2}, we obtain:

$$\ln\left(C_{O_2}^* - C_{O_2} \right) = \ln\left(C_{O_2}^* - C_{O_2,in} \right) - k_L a \cdot t$$

Therefore, from a plot of $\ln\left(C_{O_2}^* - C_{O_2} \right)$ versus t the value of $k_L a$ can be calculated as the slope of the line.

It is important to observe that calculation of $k_L a$ based on this procedure is based on several assumptions:

- Both the gas and the liquid phase are perfectly mixed.

- The dynamics of the oxygen probe are negligible compared to the dynamics of oxygen transfer.

- If air is used as gas, the amount of oxygen that transfers to the liquid phase is negligible compared to the amount of oxygen that is fed to the vessel. This ensures that oxygen partial pressure in the bubbles in the vessel is the same as the in the atmosphere, and so $C_{O_2}^*$ is known and constant during the experiment.

An example of $k_L a$ determination using this method is shown in Figure A.1.

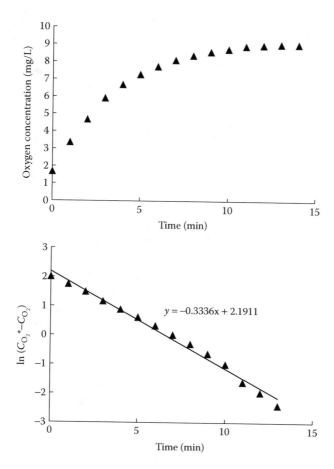

FIGURE A.1 Example of $k_L a$ determination. The value of $k_L a$ obtained is this example is 0.33 min^{-1}.

Appendix B

Solution of Differential Equations and Parameter Optimisation in Excel

MASS OR HEAT BALANCES carried out on batch systems usually give a set of differential equations, which need to be solved in order to obtain the time profiles of the desired variables. Similarly, differential equations may also be obtained by mass balances on systems where there is a concentration, or temperature, gradient in one or more directions, for example, plug-flow reactors and trickling filters. One of the reasons to solve differential equations is to find the optimum values of the parameters of a certain model, that is the values of the parameters that minimise the difference between the model simulations and the experimental data.

Many types of specialised software are available on the market to solve differential equations and to do parameter estimation. Here some simple procedures to do this using the widely available software Microsoft Excel are shown. All the calculations reported in this book have been carried out using Microsoft Excel.

B.1 SOLUTION OF DIFFERENTIAL EQUATIONS

Consider the generic differential equation

$$\frac{dx}{dt} = f(x) \tag{B.1}$$

with the initial condition $x(t=0) = x_0$.

Solving this differential equation means finding the function $x(t)$ that satisfies equation (B.1). A numerical solution of the equation can be found by noting that, if the time increment dt is very small, we have:

$$x\big|_{t+dt} = x\big|_t + \frac{dx}{dt}\bigg|_t dt = x\big|_t + f(x)\big|_t dt \qquad (B.2)$$

In this way, the values of the function $x(t)$ can be obtained as a series of data points starting from the initial condition $x(0) = x_0$, for example

$$x\big|_{dt} = x_0 + f(x)\big|_0 dt$$
$$x\big|_{2dt} = x_{dt} + f(x)\big|_{dt} dt \qquad (B.3)$$
$$\dots$$
$$x\big|_{n\cdot dt} = x_{(n-1)dt} + f(x)\big|_{(n-1)dt} dt$$

This numerical integration of the differential Equation B.1 gives correct results only if dt is sufficiently small. Equations B.3 can be easily coded into Excel.

Example B.1

Solve numerically the differential equation

$$\frac{dx}{dt} = k \cdot x \qquad (B.4)$$

where $k = 2$, over the time interval $t = \begin{bmatrix} 0 & 1 \end{bmatrix}$ with the initial condition $x(0) = 0.1$. The units of time k and x in this example are arbitrary.

Solution

In order to solve this equation numerically using Excel, we need to generate a column of time values using small increments dt, for example $dt = 0.01$. Then we need to create columns for x and for $(dx/dt) = k \cdot x$. The first value in the x column will be $x(0) = 0.1$ and the following values will be given by:

$$x(0.01) = x(0) + \frac{dx}{dt}\bigg|_0 \cdot 0.01$$

$$x(0.02) = x(0.01) + \frac{dx}{dt}\bigg|_{0.01} \cdot 0.01$$

| E3 | ▾ | ⋮ | ✕ | ✓ | *fx* | =E2+F2*(D3-D2) |

	A	B	C	D	E	F	G
1	Parameters			Time	x	dx/dt	
2	k	2		0	0.1	0.2	
3	x₀	0.1		0.01	0.102	0.204	
4				0.02	0.10404	0.20808	
5				0.03	0.106121	0.212242	
6				0.04	0.108243	0.216486	
7				0.05	0.110408	0.220816	
8				0.06	0.112616	0.225232	
9				0.07	0.114869	0.229737	
10				0.08	0.117166	0.234332	
11				0.09	0.119509	0.239019	
12				0.1	0.121899	0.243799	
13				0.11	0.124337	0.248675	
14				0.12	0.126824	0.253648	
15				0.13	0.129361	0.258721	
16				0.14	0.131948	0.263896	
17				0.15	0.134587	0.269174	
18				0.16	0.137279	0.274557	
19				0.17	0.140024	0.280048	
20				0.18	0.142825	0.285649	
21				0.19	0.145681	0.291362	
22				0.2	0.148595	0.297189	
23				0.21	0.151567	0.303133	
24				0.22	0.154598	0.309196	
25				0.23	0.15769	0.31538	
26				0.24	0.160844	0.321687	
27				0.25	0.164061	0.328121	

FIGURE B.1 Excel screenshot exemplifying the numerical solution of the differential equation in Example B.1.

and so on. This is shown in the screenshot above (Figure B.1).

The values of $x(t)$ obtained with this procedure are shown in Figure B.2.

The numerical solution obtained in this way can be compared to the analytical solution which, in this particular case, can be obtained very easily:

$$x = x_0 e^{(k \times t)}$$

The comparison of the analytical and numerical solution is shown in Figure B.3. It can be seen that the agreement between the numerical and the analytical solution is very good; this means that the numerical error is very small.

It is important to observe that the critical parameter that determines the accuracy of the numerical solution is the time interval

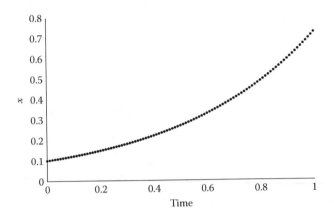

FIGURE B.2 Plot of $x(t)$ generated by numerical integration of the differential equation in Example B.1. Values of x obtained with $dt = 0.01$.

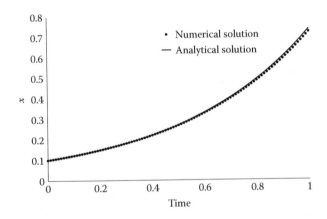

FIGURE B.3 Comparison of the numerical solution (obtained with $dt = 0.01$) with the analytical solution.

used for discretisation, dt, which needs to be as small as possible, compatibly with the computational power available. For example, let's see what happens if we choose a time interval $dt = 0.1$, that is 10 times larger than the dt value used so far. Figure B.4. compares the values of the function x with $dt = 0.1$, $dt = 0.01$ and the analytical solution.

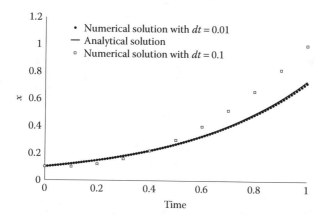

FIGURE B.4 Comparison of the numerical solution obtained with $dt = 0.1$ with the numerical solution obtained with $dt = 0.01$ and the analytical solution.

From Figure B.4, it is evident that the numerical solution with $dt = 0.1$ is largely incorrect (large numerical error) and this shows the need to choose a value of dt which is small enough to obtain the solution with the desired accuracy. The faster is the variable x changing with time, the smaller needs to be the value of dt to avoid numerical errors in the solution. This means that to obtain accurate solutions for high values of the derivative (dx/dt), which, in this example corresponds to high values of the parameter k, the value of dt needs to be smaller than for low values of the derivative.

B.2 PARAMETER OPTIMISATION

Parameter optimisation means finding the values of the model parameters that give the best possible fitting of the available experimental data. We assume that our mathematical model is given by a function $x(t)$. This function can be expressed explicitly, for example $x(t) = a \cdot t + b$, or it can be defined as the solution of a differential equation, for example the generic differential equation (B.1):

$$\frac{dx}{dt} = f(x) \qquad (B.1)$$

with the initial condition $x(t = 0) = x_0$.

We also assume that we have some experimental data which should correspond to the function $x(t)$. These experimental data are available at the times t_1, t_2, ..., t_n and we will call them $x_{exp}(t_1)$, $x_{exp}(t_2)$, ... $x_{exp}(t_n)$. Optimising the parameters of the function $x(t)$ means finding the values of the parameters which minimise of the objective function:

$$\sum_{i=1}^{n}\left(x_{exp}(t_i) - x(t_i)\right)^2$$

In practice this means finding the values of the parameters of the function $x(t)$ that make the time profile of $x(t)$ as close as possible to the experimental data. Parameter estimation can be done easily using Excel, as shown in the example below.

Example B.2

Consider the function of the example B.1:

$$\frac{dx}{dt} = k \cdot x \qquad (B.4)$$

where in this case k is unknown, over the time interval $t = \begin{bmatrix} 0 & 1 \end{bmatrix}$ with the initial condition $x(0) = 0.1$. The units of time k and x are arbitrary. We have a set the experimental data below which we want to use to find the optimum value of k.

Time	x_{exp}
0	0.1
0.1	0.11
0.2	0.16
0.3	0.19
0.4	0.2
0.5	0.24
0.6	0.37
0.7	0.4
0.8	0.49
0.9	0.57
1.0	0.7

Solution

The first step is solving the differential equation (B.1) for an arbitrary value of k. The differential equation can be solved numerically in Excel as described in Example B.1. This step generates a column of x values as a function of time. The values of x depend on the parameter k. A very important observation is that the dt used to calculate x in this step has nothing to do with the time intervals at which the experimental data are available (which is 0.1 in this example). The value of dt has to be as small as possible in order to minimise the numerical error due to discretisation, as shown in Section B.1.

The second step is to input in Excel the experimental data and to create an objective function:

$$\sum_{i=1}^{n}\left(x_{\exp}(t_i)-x(t_i)\right)^2=\left(0.1-x(t=0)\right)^2$$

$$+\left(0.11-x(t=0.1)\right)^2+\left(0.16-x(t=0.2)\right)^2+......$$

The final step is to use Solver to determine the value of the parameter k that gives the minimum value of the objective function. This is shown in Figure B.5.

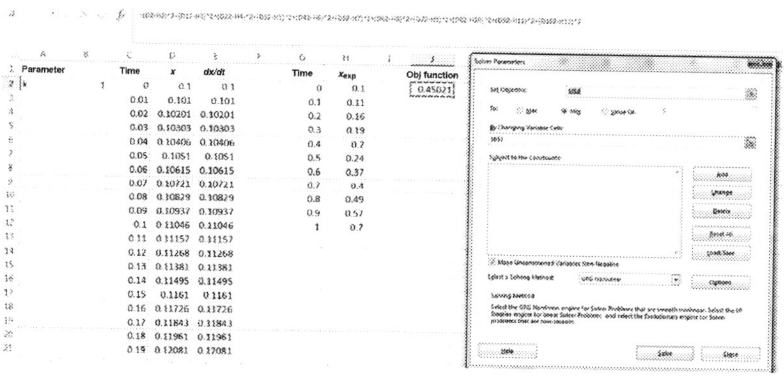

FIGURE B.5   Excel screenshot which shows the procedure to solve Example B.2.

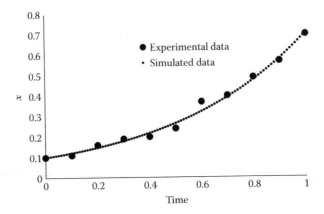

FIGURE B.6 Comparison of the experimental data with the best of the model, obtained with $k = 1.975$.

In this case, Solver gives a value of $k = 1.975$, which is the value of optimum value of k which minimises the objective function; that is it minimises the difference between the model and the experimental data. The comparison between the optimised model and the experimental data is shown in Figure B.6.

Appendix C

Solution of Systems of Equations in Excel

Consider a system of equations:

$$\begin{cases} f_1\left(x_1, x_2,..., x_n\right)=0 \\ f_2\left(x_1, x_2,..., x_n\right)=0 \\ ...... \\ f_n\left(x_1, x_2,..., x_n\right)=0 \end{cases}$$

(C.1)

This is a system of n equations in the n unknowns x_1, x_2, ..., x_n. Solving this system of equations means finding the values of $x_1, x_2,...x_n$ that satisfy all the equations (C.1). For simple systems, this problem can be solved by manual isolation of the unknowns and substitution. However, for more complicated system, the substitution method is impractical or unfeasible, but the system of equation can still be solved, in many cases, using Excel.

The procedure is quite simple. Firstly, initial guesses have to be given to the unknowns. Then the equations have to be written in the same form as for equations (C.1), that is, with the left hand side equal to 0. The left hand sides of the equations have be written into Excel, as a function of the given initial guesses for the unknowns. Excel will calculate a numerical value for each equation (or rather for the left hand side of each equation) which will be in general different than 0, because the initial guesses given are not the solution of the system of equations. We will call these calculated values 'residuals'. Then each residual is squared and all the squared residuals are added up together. At this point, Solver is used to find the minimum of the sum of the squared residuals, by manipulating the values of the unknowns. If Solver is successful and the sum of squared residuals is very close to 0, the values found for the unknown are the solutions of the system of equations.

Note that the procedure is quite sensitive (or very sensitive in some cases) to the given values of the initial guesses. Therefore, it is important to have an idea at least of the order of magnitude of the values of the unknowns. The procedure is shown in the example below.

Example C.1

Solve the following system of equations:

$$\begin{cases} x+y+z=2 \\ 3x+3y-z=6 \\ x-y+z=-1 \end{cases}$$

Solution

This is a system of three equations in the three unknowns x, y, z. This is a simple system that can be easily solved manually by substitution; however, we will use it as an example of the Excel method. First of all, initial guesses have to be given to x, y, z, we will give the value of 1 to all of them. Then we need to write the equations in Excel in the form:

$$\begin{cases} x+y+z-2 \\ 3x+3y-z-6 \\ x-y+z+1 \end{cases}$$

Excel will calculate a value for each equation (residual). We need to square each of the residuals and then add all the squared residuals together. Then use Solver to find the values of x, y, z than make the sum of the squared residuals minimum.

This procedure is shown in the screenshot below (Figure C.1).

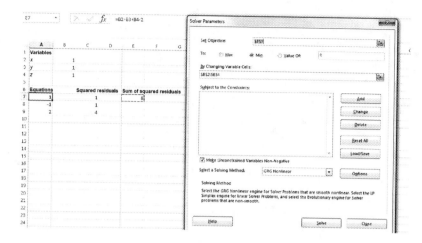

FIGURE C.1   Screenshot showing the procedure to solve Example C.1.

In this case, Solver gives the correct values of the unknowns:

$$\begin{cases} x = 0.5 \\ y = 1.5 \\ z = 0 \end{cases}$$

With these values, the sum of the squared residuals is of the order of 10^{-14}, that is a very low value which ensures we have found the solution of the system of equations.

Appendix D

Physical Properties Used in This Text

THE PHYSICAL PROPERTIES SHOWN in Table D.1 have been used in the calculation of the enthalpy balances. The reported values have been collected from various sources and are only meant to be suitable for example problems aimed at showing the application of the methods, but are not necessarily suitable for accurate calculations.

TABLE D.1   Physical Properties Used in the Enthalpy Calculations in This Book

Substance	Reference State	Standard Enthalpy of Formation (J/mol)	Heat of Dissolution (J/mol)
Glucose	Solid	$-1.28 \cdot 10^6$	$1.1 \cdot 10^4$
Oxygen	Gas	0	$-1.47 \cdot 10^4$
Ammonia	Liquid	$-8.1 \cdot 10^4$	
Biomass	Solid	$-7.7 \cdot 10^5$	
Carbon dioxide	Gas	$-3.9 \cdot 10^5$	
Water	Liquid	$-2.9 \cdot 10^5$	
Acetic acid	Liquid	$-4.8 \cdot 10^5$	$-1.5 \cdot 10^3$
Nitrate (NO_3^-)	Dissolved	$-2.7 \cdot 10^5$	
Sodium bicarbonate	Dissolved	$-6.9 \cdot 10^5$	

Appendix E

Problems and Questions Solutions

1.1 a. Propionic acid:

$$CH_3CH_2COOH + 3.5O_2 \rightarrow 3CO_2 + 3H_2O$$

$$\Rightarrow 1.51\,gCOD/g \text{ propionic acid}$$

b. Benzene:

$$C_6H_6 + 7.5O_2 \rightarrow 6CO_2 + 3H_2O \Rightarrow 3.08\,gCOD/g \text{ benzene}$$

c. Oleic acid:

$$C_{18}H_{34}O_2 + 25.5O_2 \rightarrow 18CO_2 + 17H_2O \Rightarrow 2.89\,gCOD/g \text{ oleic acid}$$

d. Xylose:

$$C_5H_{10}O_5 + 5O_2 \rightarrow 5CO_2 + 5H_2O \Rightarrow 1.067\,gCOD/g \text{ xylose}$$

e. Alanine:

$$C_3H_7O_2N + 3O_2 \rightarrow 3CO_2 + 2H_2O + NH_3 \Rightarrow 1.079\,gCOD/g \text{ alanine}$$

1.2 In this BOD test, 130 mg COD/L of substrates are removed and 50 mg/L of microorganisms are produced, and this corresponds to $50 \cdot 1.42 = 71$ mg COD/L of microorganisms. Therefore, the oxygen consumed is $130 - 71 = 59$ mg O_2/L and this is the BOD of the sample.

This example shows that, even for a wastewater where most of the COD is biodegradable, the BOD can be significantly lower than the COD, because of the production of microorganisms.

1.3 The conversion factor for the microorganisms into COD can be calculated immediately:

$$C_5H_7O_2N + 5O_2 \rightarrow 5CO_2 + 2H_2O + NH_3 \Rightarrow 1.42 \, gCOD/g \, biomass$$

Therefore, in the BOD the COD produced as microorganisms is $30 \cdot 1.42 = 42 \, mgCOD/L$

Using the COD balance:

$$(-\Delta S_{COD}) = \Delta X_{COD} + (-\Delta O_2) = 42 \, mgCOD/L + 60 \, mgCOD/L$$

$$= 102 \, mgCOD/L$$

This is the difference between the initial and final COD of the sample. The final COD is 10 mg COD/L, so the initial COD was 112 mg COD/L.

1.4 From the reduction reactions of dichromate and oxygen:

$$K_2Cr_2O_7 + 6e^- + 6H_3O^+ \rightarrow Cr_2O_3 + 2KOH + 8H_2O$$

$$O_2 + 4e^- + 4H_3O^+ \rightarrow 6H_2O$$

We get that 1 mol of $K_2Cr_2O_7$ consumed corresponds to 1 mol of Cr_2O_3 formed and to 1.5 mol of oxygen, that is to 1.5 mol of COD.

At the end of the experiment, the concentration of chromic anhydride is 100 mg/L. Since the total volume at the end of the experiment is 6 mL, the mass of chromic anhydride produced is 600 mg, that is 3.95 mmol. This corresponds to 5.92 mmol of oxygen, that is to 189 mg of COD. Since the volume of wastewater sample was 1 mL, the COD concentration in the sample is

$$\frac{189 \cdot 10^{-6} \, g}{10^{-3} \, L} = 189 \, mgCOD/L$$

1.5 During this test 190 mg COD/L of substrate are removed and 40 mg/L of microorganisms are produced, which correspond to

57 mg COD/L. Therefore, the removed COD is $190 - 57 = 133$ mg COD/L, which corresponds to concentration of removed nitrate of $133/2.86 = 46.5$ mg N-NO$_3$/L.

1.6 The wastewater with low COD loading is better suited for aerobic treatment, and the wastewater with high COD loading is better suited for anaerobic treatment.

With low COD loading, oxygen consumption and therefore aeration costs are likely to be low. Also, the amount of methane, and therefore of energy, that can be obtained with anaerobic treatment of wastewater with low COD loading is low.

However, if the wastewater has high COD loading oxygen consumption and aeration costs are likely to be high, while the amount of methane that can be obtained using anaerobic treatment is potentially high.

1.7 The HRT is the ratio between the reactor volume and the influent flow rate which in this case is 4000 m^3/10,000 m^3 = 0.4 day.

The SRT is the ratio between the mass of microorganisms in the reactor and their mass flow rate leaving the system. The mass of microorganisms in the reactor is 4000 m$^3 \cdot$ 1.5 kg/m^3 = 6000 kg. The mass flow rate of microorganisms leaving the reactor is 200 m^3/day $\cdot$ 3 kg/m^3 = 600 kg/day. The SRT is therefore 6000 kg/600 kg/day = 10 day.

The OLR is the ratio between the mass flow rate of COD entering the system and the reactor volume. The mass flow rate of COD entering the system is 10,000 m^3/day $\cdot$ 0.3 kg COD/m^3 = 3000 kg COD/day. The OLR is 3000 kg COD/day/4000 m^3 = 0.75 kg COD/m^3.day.

2.1 a. From the general stoichiometry in Table 2.3:

$$C_w H_x O_y N_z + \left(w + \frac{x}{4} - \frac{y}{2} - \frac{3}{4}z - 5\frac{Y_{X/S} MW_{substrate}}{MW_{biomass}} \right) O_2$$

$$+ \left(\frac{Y_{X/S} MW_{substrate}}{MW_{biomass}} - z \right) NH_3 \rightarrow \frac{Y_{X/S} MW_{substrate}}{MW_{biomass}} C_5 H_7 O_2 N$$

$$+ \left(w - 5\frac{Y_{X/S} MW_{substrate}}{MW_{biomass}} \right) CO_2 + \left(\frac{x}{2} - \frac{3}{2}z - 2\frac{Y_{X/S} MW_{substrate}}{MW_{biomass}} \right) H_2 O$$

Adapted for glutamic acid ($w = 5$, $x = 9$, $y = 4$, $z = 1$), which has a MW equal to 147 (biomass has a MW equal to 113), we obtain:

$$C_5H_9O_4N + (4.5 - 6.50Y_{X/S})O_2 + (1.3Y_{X/S} - 1)NH_3$$
$$\rightarrow 1.3Y_{X/S}C_5H_7O_2N + (5 - 6.50Y_{X/S})CO_2 + (3 - 2.60Y_{X/S})H_2O$$

From this formula, the maximum value of $Y_{X/S}$ is the one for which the stoichiometric coefficient of oxygen is 0, that is $Y_{X/S} = 0.69$ kg biomass/kg glutamic acid;

b. If $Y_{X/S} = 0.2$ kg biomass/kg glutamic acid, the stoichiometric coefficient for ammonia is −0.74. This means that ammonia is generated in the reaction and not consumed. This is because glutamic acid contains nitrogen and therefore ammonia addition is not required. The stoichiometric coefficient for carbon dioxide is 3.7 and for biomass is 0.26. Therefore, the rates of production of ammonia and carbon dioxide are:

$$r_{NH3} = 2\frac{kg}{m^3.day}\cdot\frac{1}{113\frac{kg}{kmol}}\cdot\frac{0.74}{0.26}\cdot 14\frac{kg}{kmol} = 0.71\frac{kgN-NH_3}{m^3 day}$$

$$r_{CO_2} = 2\frac{kg}{m^3.day}\cdot\frac{1}{113\frac{kg}{kmol}}\cdot 3.7\cdot 44\frac{kg}{kmol} = 2.88\frac{kgCO_2}{m^3 day}$$

2.2 a. In the hydrolysis of the protein, one molecule of water is added per molecule of alanine. The MW of alanine is 89 and of its monomer in the protein is 71. Therefore, the rate of alanine production is:

$$r_{alaine} = 100\cdot\frac{89}{7L}\frac{g}{m^3 day} = 125.3\frac{g}{m^3 day}$$

b. In each fat molecule, there are three molecules of stearic acid, each of them with the formula $C_{18}H_{35}O$ ($MW = 267$, each fatty acid loses one OH group when it combines to form a fat), and one molecule of glycerol, with the formula $C_3H_5O_3$ ($MW = 89$, glycerol loses three H atoms when it combines to form a fat). The molecular weight of the fat is therefore 890. When the fat hydrolyses, from one molecule of fat we generate three molecules of stearic acid $C_{18}H_{36}O_2$ ($MW = 284$) and one molecule of glycerol

$C_3H_8O_3$ ($MW = 92$). Per 100 g of fat molecule there are $100 \cdot 267 \cdot 3/890 = 90$ g of $C_{18}H_{35}O$ and $100 \cdot 89/890 = 10$ g of $C_3H_5O_3$. When the fat hydrolyses, 90 g of $C_{18}H_{35}O$ generate $90 \cdot 284/267 = 95.7$ g of stearic acid and 10 g of $C_3H_5O_3$ generate $10 \cdot 92/89 = 10.3$ g of glycerol. Therefore, if the rate of fat hydrolysis is 100 g/m³.day, the rate of stearic acid production will be 95.7 g/m³.day and the rate of glycerol production will be 10.3 g/m³.day.

2.3 It has to be

$$\frac{O_2}{K_{O_2}+O_2}=0.9 \Rightarrow O_2=\frac{0.9}{0.1}K_{O_2}=0.9 \text{ mg/L}$$

Therefore, a minimum concentration of at least 0.9 mg/L is required to have a growth rate which is at least 90% of the maximum growth rate in the absence of any oxygen limitation.

2.4 The charge balance for this problem is:

$$10^{-pH}+\left[Na^+\right]=\frac{\left[CH_3COOH_{tot}\right]}{1+\frac{10^{-pH}}{K_{CH_3COOH}}}$$

$$+\frac{\left[H_2CO_{3tot}\right]}{1+K_{CO_2}+\frac{K_{H_2CO_3}}{10^{-pH}}+\frac{K_{HCO_3}K_{H_2CO_3}}{10^{-2pH}}}\left(\frac{K_{H_2CO_3}}{10^{-pH}}+\frac{K_{HCO_3}K_{H_2CO_3}}{10^{-2pH}}\right)$$

$$+\frac{K_W}{10^{-pH}}$$

In this case [Na⁺] = [H_2CO_{3tot}] = [CH_3COOH_{tot}] = 0.1 M. Solving for pH we obtain pH = 5.56.

3.1 The amount of COD fed to the reactor per unit time is:

$$300\frac{mgCOD}{l}\cdot 0.1\frac{1}{hr}=30\frac{mgCOD}{hr}$$

The flow rate in the effluent will be the same as in the influent so the amount of COD going out of the reactor per unit time is:

$$\left(20\frac{mg\ COD}{l}+50\frac{mg}{l}\cdot 1.42\frac{mg\ COD}{mg}\right)\cdot 0.1\frac{1}{hr}=9.1\frac{mg\ COD}{hr}$$

Therefore, from the COD balance the oxygen consumption rate in the reactor is:

$$30 \text{ mg COD/hr} - 9.1 \text{ mg COD/hr} = 20.9 \text{ mg COD/hr}$$

$$= 0.502 \text{ g oxygen/day}$$

3.2 Since we assume no changes in the growth yield and kinetic parameters, the rate of oxygen consumption rate has to be equal to the rate of nitrate consumption expressed as COD. From the reduction reactions of nitrate and oxygen:

$$O_2 + 4e^- + 4H_3O^+ \rightarrow 6H_2O$$

$$HNO_3 + 10e^- + 10H_3O^+ \rightarrow N_2 + 16H_2O$$

We see that 1 g of N-NO$_3$ corresponds to 2.86 g of oxygen.

Therefore, we need to provide $0.502/2.86 = 0.175 \text{ g N} - \text{NO}_3/\text{day}$. With a feed flow rate of 100 mL/hr = 2.4 L/day, the nitrate concentration in the feed has to be equal to 72.75 mgN-NO$_3$/L = 442 mg NaNO$_3$/L.

4.1 a. The oxygen consumption rate by the biomass can be calculated by the COD balance in the reactor:

$$\text{Oxygen consumption} \left(\frac{\text{kg O}_2}{\text{day}} \right) = Q(S_0 - S) - Q_W X_R \cdot 1.42$$

where the conversion factor 1.42 is used to convert biomass into COD, according to the oxidation reaction:

$$C_5H_7O_2N + 5O_2 \rightarrow 5CO_2 + 2H_2O + NH_3$$

Q_W is not given so it needs to be calculated from the given data.

The mass balances for the biomass in the biological reactor and in the whole system are respectively:

$$(\mu - b)XV + Q_R X_R = (Q + Q_R)X$$

$$(\mu - b)XV = Q_W X_R$$

By combining them we obtain

$$Q_w X_R + Q_R X_R = (Q + Q_R) X$$

from which we calculate Q_w:

$$Q_w = \frac{(Q + Q_R) X - Q_R X_R}{X_R}$$

Using the numerical values given:

$$Q_w = \frac{1800 \text{ m}^3/\text{day} \cdot 0.9 \text{ kg biomass/m}^3 - 800 \text{ m}^3/\text{day} \cdot 1.8 \text{ kg biomass/m}^3}{1.8 \text{ kg biomass/m}^3}$$

$$= 100 \text{ m}^3/\text{day}$$

Therefore, the oxygen consumption rate by the biomass is given by:

$$\text{Oxygen consumption} \left(\frac{\text{kg O}_2}{\text{day}} \right) = 1000 \text{m}^3/\text{day} \cdot 0.27 \text{ kgCOD/m}^3$$

$$-10 \text{ m}^3/\text{day} \cdot 1.8 \text{ kg biomass/m}^3 \cdot 1.42 = 14.4 \text{ kgO2/day}$$

b. The oxygen concentration in the biological reactor can be calculated by the mass balance for oxygen in the liquid phase (neglecting the oxygen concentration in the feed and in the recycle stream):

$$k_L a (k \cdot p_{O_2} - C_{O_2}) V = Q_{O_2\text{biomass}} + (1 + R) Q C_{O_2}$$

Since pure oxygen is used and we assume atmospheric pressure: $P_{O_2} = 1$ atm

Therefore, we can calculate C_{O_2} by re-arranging the oxygen mass balance:

$$C_{O_2} = \frac{k_L a \cdot k \cdot p_{O_2} V - Q_{O_2\text{biomass}}}{Q + Q_R + k_L a V}$$

$$= \frac{10 \text{ days}^{-1} \cdot 0.043 \text{ kg/m}^3/\text{atm} \cdot 1 \text{ atm} \cdot 1000 \text{ m}^3 - 14.4 \text{ kgO}_2/\text{day}}{1800 \text{ m}^3/\text{day} + 10 \text{ days}^{-1} \cdot 1000 \text{ m}^3}$$

$$= 0.035 \text{ kgO}_2/\text{m}^3$$

4.2 The k_La can be calculated from the mass balance for oxygen in the aeration tank:

$$QC_{O_20} + Q_R C_{O_2} + k_La(C^*_{O_2} - C_{O_2})V = Q_{O_2biom} + (Q + Q_R)C_{O_2}$$

The oxygen consumption rate by the microorganisms, $Q_{O2biomass}$ can be calculated from a COD balance around the whole system:

$$Q_{O2biomass}\left(\frac{kg\ O_2}{day}\right) = Q(S_0 - S) - Q_W X_R \cdot 1.42$$

Everything in this equation is given except X_R, which can be calculated from a mass balance around the settling tank:

$$(Q + Q_R)X = (Q_W + Q_R)X_R \Rightarrow X_R = \frac{(Q + Q_R)X}{(Q_W + Q_R)} = 3.93\ g/L$$

So $Q_{O2biomass}$ is:

$$Q_{O_2biomass}\left(\frac{kg\ O_2}{day}\right) = Q(S_0 - S) - Q_W X_R \cdot 1.42 = 12,105\ kg\ oxygen/day$$

And so:

$$k_La = \frac{Q_{O_2biom} + Q(C_{O_2} - C_{O_20})}{(C^*_{O_2} - C_{O_2})V} = 29.93\ day^{-1} \cong 30\ day^{-1}$$

4.3 Nitrogen balance in the whole system:

$$QNH_{30} = QNH_3 + Q_W X_R \cdot 0.12$$

To find X_R:

$$(Q + Q_R)X = (Q_W + Q_R)X_R \Rightarrow X_R = \frac{(Q + Q_R)X_R}{(Q_W + Q_R)} = 5.2\ kg/m^3$$

And:

$$NH_3 = NH_{30} - \frac{Q_W X_R \cdot 0.12}{Q} = 18.75\frac{mgN - NH_3}{1}$$

4.4 In the original design, the biomass balance in the whole system can be written as usual:

$$(\mu - b) XV = Q_W X_R$$

In the new design (subscript 'new'), which assumes there are solid losses with the effluent, the biomass balance has to be written as:

$$(\mu - b) XV = Q_W X_{Rnew} + (Q - Q_W) X_{eff}$$

Note that in the new design Q_w and X will not change while X_R will change. The reason for X being the same is that in the new design both the SRT and the HRT will stay the same (the influent flow rate and the reactor volume don't change) and X depends only on these two parameters. Another way of seeing that X has to stay constant is writing the substrate balance in the aeration tank:

$$QS_0 + Q_R S = \frac{\mu XV}{Y} + (Q + Q_R)S \Rightarrow Q(S_0 - S) = \frac{\mu XV}{Y}$$

where it can be observed that all the values, including the term μ/Y, do not change with the new design, and so X cannot change too.

$\mu - b$ will also stay constant, since the substrate concentration in the effluent (i.e. in the reactor) will have to stay the same. Therefore, we can write:

$$(\mu - b) = \frac{1}{SRT} = \left(\frac{Q_W X_{Rnew} + (Q - Q_W) X_{eff}}{XV} \right)_{new\,design}$$

$$= \left(\frac{Q_W X_R}{XV} \right)_{original\,design}$$

Note that this means that the solids residence time has to be the same under the new design as in the original design, as expected since S does not change. In the original design X is not given, but it can be calculated immediately:

$$X = \frac{(Q_W + Q_R) X}{(Q + Q_R)} = 2.71\,g/L$$

The SRT, to be held constant, is equal to 30 days.

The condition of constant SRT can be used to calculate the new X_R:

$$X_{Rnew} = \frac{XV}{Q_W SRT} - \frac{(Q - Q_W) X_{eff}}{Q_W} = 3.5 \, kg/m^3$$

And the new required Q_R is, from the biomass balance on the settling tank under the new design conditions:

$$(Q + Q_{Rnew}) X = (Q_W + Q_{Rnew}) X_{Rnew} + (Q - Q_W) X_{eff}$$

From we obtain:

$$Q_{Rnew} = 68,000 \, m^3/day$$

4.5 The nitrate concentration in the effluent, NO_{32}, can be calculated from a mass balance for nitrogen on the aerobic reaction tank:

$$(Q + Q_R + Q_1) NH_{31} + (Q + Q_R + Q_1) X_{TOT1} \cdot 0.12 + (Q + Q_R + Q_1) NO_{31}$$

$$= (Q + Q_R + Q_1) NH_{32} + (Q + Q_R + Q_1) X_{TOT2} \cdot 0.12 + (Q + Q_R + Q_1) NO_{32}$$

In this equation, we don't know X_{TOT1} and NH_{31}, the biomass and ammonia concentration in the anoxic reactor. We need to write the nitrogen balance in the anoxic tank:

$$QNH_{30} + Q_R (NH_{32} + NO_{32}) + Q_1 (NH_{32} + NO_{32}) + Q_R X_{TOTR} \cdot 0.12$$

$$+ Q_1 X_{TOT2} \cdot 0.12 = (Q + Q_R + Q_1) NH_{31} + (Q + Q_R + Q_1) X_{TOT1} \cdot 0.12$$

$$+ (Q + Q_R + Q_1) NO_{31} + \dot{N}_{2generated}$$

where $\dot{N}_{2generated}$ is the rate of nitrogen gas generation (in kgN/day) in the anoxic reactor. $\dot{N}_{2generated}$ can be calculated from the mass balance for nitrate in the anoxic tank:

$$(Q_R + Q_1) NO_{32} = (Q + Q_R + Q_1) NO_{31} + \dot{N}_{2generated}$$

Eliminating $\dot{N}_{2generated}$ from the two last equations we obtain:

$$QNH_{30} + Q_R NH_{32} + Q_I NH_{32} + Q_R X_{TOTR} \cdot 0.12 + Q_{IXTOT2} \cdot 0.12$$
$$= (Q + Q_R + Q_I)NH_{32} + (Q + Q_R + Q_I)X_{TOT1} \cdot 0.12$$

This equation can be combined with the nitrogen balance on the aerobic tank:

$$QNH_{30} + Q_R X_{TOTR} \cdot 0.12 + (Q + Q_R + Q_I)NO_{31}$$
$$= QNH_{32} + (Q + Q_R)X_{TOT2} \cdot 0.12 + (Q + Q_R + Q_I)NO_{32}$$

which can be rearranged to give NO_{32}:

$$NO_{32} = NO_{31} + \frac{Q}{Q + Q_R + Q_I}(NH_{30} - NH_{32}) + \frac{Q_R}{Q + Q_R + Q_I}X_{TOTR} \cdot 0.12$$
$$- \frac{(Q + Q_R)}{Q + Q_R + Q_I}X_{TOT2} \cdot 0.12$$

The microorganisms' concentration in the recycle stream is not known but can be easily calculated with a biomass balance on the settling tank (assuming perfect settling):

$$(Q + Q_R)X_{TOT2} = (Q_R + Q_W)X_{TOTR} \implies X_{TOTR} = \frac{(Q + Q_R)}{Q_R + Q_W}X_{TOT2}$$

Using this equation, the nitrate concentration in the effluent NO_{32} can be calculated as:

$$NO_{32} = NO_{31} + \frac{Q}{Q + Q_R + Q_I}(NH_{30} - NH_{32}) - \frac{X_{TOT2}Q_W(Q + Q_R) \cdot 0.12}{(Q_R + Q_W)(Q + Q_R + Q_I)}$$
$$= 15.0 \frac{kgN}{m^3}$$

Another method to solve this problem is to use the overall nitrogen balance in the whole system (two reactors and settling tank):

$$QNH_{30} = Q_W X_{TOTR} \cdot 0.12 + Q(NH_{32} + NO_{32}) + \dot{N}_{2generated}$$

$\dot{N}_{2\text{generated}}$ can be calculated as shown previously and we obtain:

$$Q\text{NH}_{30} = Q_\text{W} X_{\text{TOTR}} \cdot 0.12 + Q(\text{NH}_{32} + \text{NO}_{32})$$

$$+ (Q_1 + Q_\text{R})\text{NO}_{32} - (Q + Q_\text{R} + Q_1)\text{NO}_{31}$$

which, after rearrangements and combination with the biomass balance in the settling tank gives the same equation for NO_{32} found with the previous method.

4.6 It can be solved with a nitrogen balance in the anoxic reactor:

$$Q\text{NH}_{30} + Q_\text{R}(\text{NH}_{32} + \text{NO}_{32}) + Q_1(\text{NH}_{32} + \text{NO}_{32}) + Q_\text{R} X_{\text{TOTR}} \cdot 0.12$$

$$+ Q_1 X_{\text{TOT2}} \cdot 0.12 = (Q + Q_\text{R} + Q_1)\text{NH}_{31} + (Q + Q_\text{R} + Q_1) X_{\text{TOT1}} \cdot 0.12$$

$$+ (Q + Q_\text{R} + Q_1)\text{NO}_{31} + \dot{N}_{2\text{generated}}$$

where $\dot{N}_{2\text{generated}}$ is the rate of nitrogen gas generation (in kgN/day) in the anoxic reactor. $\dot{N}_{2\text{generated}}$ can be calculated from the mass balance for nitrate in the anoxic tank:

$$(Q_\text{R} + Q_1)\text{NO}_{32} = (Q + Q_\text{R} + Q_1)\text{NO}_{31} + \dot{N}_{2\text{generated}}$$

Eliminating $\dot{N}_{2\text{generated}}$ from the two last equations we obtain:

$$Q\text{NH}_{30} + Q_\text{R}\text{NH}_{32} + Q_1\text{NH}_{32} + Q_\text{R} X_{\text{TOTR}} \cdot 0.12 + Q_1 X_{\text{TOT2}} \cdot 0.12$$

$$= (Q + Q_\text{R} + Q_1)\text{NH}_{31} + (Q + Q_\text{R} + Q_1) X_{\text{TOT1}} \cdot 0.12$$

So:

$$\text{NH}_{31} = \frac{Q\text{NH}_{30} + Q_\text{R}\text{NH}_{32} + Q_1\text{NH}_{32} + Q_\text{R} X_{\text{TOTR}} \cdot 0.12 + Q_1 X_{\text{TOT2}} \cdot 0.12}{(Q + Q_\text{R} + Q_1)}$$

$$- X_{\text{TOT1}} \cdot 0.12$$

To find X_{TOT2}, we use the biomass balance around the settling tank:

$$\left(Q_R + Q\right)X_{TOT2} = \left(Q_R + Q_W\right)X_{TOTR} + \left(Q - Q_W\right)X_{EFF} \Rightarrow X_{TOT2}$$

$$= \frac{\left(Q_R + Q_W\right)X_{TOTR} + \left(Q - Q_W\right)X_{EFF}}{\left(Q_R + Q\right)}$$

$$X_{TOT2} \cong 0.9\,g/L$$

So we obtain:

$$NH_{31} = 7\,mgN/L$$

4.7 The nitrogen balance in the whole system is:

$$QNH_{30} = Q_W X_{TOTR} \cdot 0.12 + Q(NH_{32} + NO_{32}) + \dot{N}_{2generated}$$

Eliminating $\dot{N}_{2generated}$ as done in previous problems and re-arranging we obtain an expression for NO_{32}:

$$NO_{32} = NO_{31} + \frac{Q\left(NH_{30} - NH_{32}\right) - Q_W X_R \cdot 0.12}{\left(Q + Q_R + Q_I\right)}$$

Without changing any of the flow rates, improving nitrogen removal, that is decreasing NO_{32}, can be obtained by decreasing NO_{31}, and this can be obtained by adding external COD, in this case methanol, to the process.

The concentration of NO_{31} in the current process, without methanol addition, can be calculated by rearranging the previous equation:

$$NO_{31} = NO_{32} - \frac{Q\left(NH_{30} - NH_{32}\right) - Q_W X_R \cdot 0.12}{\left(Q + Q_R + Q_I\right)} = 10.8\,mgN/L$$

Ideally, in a process for nitrate removal with pre-denitrification, it is expected that the nitrate concentration in the effluent from the anoxic reactor is close to 0, because all the nitrate should be consumed by denitrification. In this case, this does not happen, due to the limited COD available with the feed. This is the reason for adding methanol, which should provide additional COD and reduce NO_{31}. The amount of nitrate required to reduce NO_{31} to 0, and therefore to reduce NO_{32} in the effluent is:

$$\text{Methanol required} = \frac{(Q+Q_R+Q_I)NO_{31}}{0.2\dfrac{\text{kgN}-\text{NO}_3}{\text{kg methanol}}} = 24,300 \text{ kg methanol/day}$$

This amount of methanol will reduce NO_{31} to 0 (or close to it), and so will reduce NO_{32}. A possible disadvantage of the methanol addition, apart from the cost, is that it will increase biomass production and biomass concentration in the system, with possible overloading problems in the settling tank and reduction in the $k_L a$ for the aerobic reactor.

5.1 The mass of COD fed to the reactor per day is:

$$10\frac{\text{kg COD}}{\text{m}^3} \cdot 100\frac{\text{m}^3}{\text{day}} = 1000\frac{\text{kg COD}}{\text{day}}$$

The mass of COD which leaves the reactor with the liquid phase per day is:

$$\left(\begin{array}{l} 1\dfrac{\text{kg acetic acid}}{\text{m}^3} \cdot 1.067\dfrac{\text{kgCOD}}{\text{kg acetic acid}} \\[4mm] +0.3\dfrac{\text{kg propionic acid}}{\text{m}^3} \cdot 1.51\dfrac{\text{kgCOD}}{\text{kg propionic acid}} \\[4mm] +0.5\dfrac{\text{kg microorganisms}}{\text{m}^3} \cdot 1.42\dfrac{\text{kgCOD}}{\text{kg microorganisms}} \end{array}\right) \cdot 100\frac{\text{m}^3}{\text{day}}$$

$$= 223\frac{\text{kgCOD}}{\text{day}}$$

Therefore, assuming there are no other products in the effluent, the methane production rate per day corresponds to $1000 - 223 = 777$ kg COD/day. This corresponds to:

$$\frac{777\dfrac{\text{kg COD}}{\text{day}}}{4\dfrac{\text{kg COD}}{\text{kg methane}}} = 194.25\frac{\text{kg methane}}{\text{day}} = 12.1\frac{\text{kmol methane}}{\text{day}}$$

5.2 This problem is easily solved with a COD balance:

$$COD_{in} = 20\frac{\text{kg COD}}{\text{m}^3} - 1000\frac{\text{m}^3}{\text{day}} = 20{,}000\frac{\text{kg COD}}{\text{day}}$$

The COD that leaves the system with the liquid–solid phase:

$$COD_{out} = 1.5\frac{\text{kg COD}}{\text{m}^3} \cdot 1000\frac{\text{m}^3}{\text{day}} = 1500\frac{\text{kg COD}}{\text{day}}$$

Therefore, from the COD balance:

$$COD_{methane} = COD_{in} - COD_{out}$$

$$= 18{,}500\frac{\text{kg COD(methane)}}{\text{day}} = \frac{18{,}500\dfrac{\text{kg COD(methane)}}{\text{day}}}{4\dfrac{\text{kg COD}}{\text{kg methane}}}$$

$$= 4625\frac{\text{kg methane}}{\text{day}} = 289\frac{\text{kmol methane}}{\text{day}}$$

$$= 289\frac{\text{kmol methane}}{\text{day}} \times 22.4\frac{\text{Nm}^3}{\text{kg methane}} = 6475\frac{\text{Nm}^3}{\text{day}}$$

5.3 The COD of the solids is 1.5 g COD/g solid and the waste contains 20% solids, so the COD per unit mass of waste is:

$$1.5\frac{\text{g COD}}{\text{g solid}} \times 0.2\frac{\text{g solid}}{\text{g waste}} = 0.3\text{ g COD/g waste}$$

About 90% of the COD of the waste is converted into methane, so we generate:

$$0.3 \text{ kg COD/kg waste} \cdot 0.9 \frac{\text{kg methane(COD)}}{\text{kg waste(COD)}}$$

$$= 0.27 \frac{\text{kg methane(COD)}}{\text{kg waste}} = \frac{0.27 \dfrac{\text{kg methane(COD)}}{\text{kg waste}}}{4 \dfrac{\text{kg COD}}{\text{kg methane}}}$$

$$= 0.0675 \frac{\text{kg methane}}{\text{kg waste}} = 0.0042 \frac{\text{kmol methane}}{\text{kg waste}}$$

This is the amount of methane generated per unit mass of waste. The calorific heat of methane is 890 kJ/mol = 890 MJ/kmol and 50% of the energy generated is converted into electricity, therefore the electrical energy generated is:

$$0.0042 \frac{\text{kmol methane}}{\text{kg waste}} \cdot 890 \frac{\text{MJ}}{\text{kmol}} \cdot 0.5 = 1.875 \frac{\text{MJ}}{\text{kg waste}}$$

Since:

$$1 \text{kWh} = 1000 \frac{\text{J}}{\text{s}} \cdot 3600 \text{s} = 3.6 \text{ MJ}$$

The anaerobic digester generates

$$\frac{1.875 \dfrac{\text{MJ}}{\text{kg waste}}}{3.6 \dfrac{\text{MJ}}{\text{kWh}}} = 0.52 \frac{\text{kWh}}{\text{kg waste}}$$

We need to generate 200,000 kWh/year, so the amount of waste to be fed to the digester is:

$$\frac{200,000 \dfrac{\text{kW/hr}}{\text{year}}}{0.52 \dfrac{\text{kWh}}{\text{kg waste}}} = 384.6 \frac{\text{ton waste}}{\text{year}}$$

5.4 a. The feed flow rate is 5 ton/hr and the organic matter flow rate is:

$$5\frac{\text{ton feed}}{\text{hr}}\cdot0.05\frac{\text{ton organic matter}}{\text{ton feed}}=0.25\frac{\text{ton organic matter}}{\text{hr}}$$

The flow rate of carbon with the feed is:

$$0.25\frac{\text{ton organic matter}}{\text{hr}}\cdot0.4\frac{\text{ton C}}{\text{ton organic matter}}=0.1\frac{\text{ton C}}{\text{hr}}$$

About 45% of the carbon in the feed is converted to methane and so the methane production rate is:

$$0.1\frac{\text{ton C feed}}{\text{hr}}\cdot0.45\frac{\text{ton C methane}}{\text{ton C feed}}=0.045\frac{\text{ton C methane}}{\text{hr}}$$

$$=0.045\frac{\text{ton C methane}}{\text{hr}}\cdot16\frac{\text{ton methane}}{\text{ton C methane}}=0.06\frac{\text{ton methane}}{\text{hr}}$$

Therefore, the volume of methane produced is:

$$\frac{0.06\dfrac{\text{ton methane}}{\text{hr}}\cdot10^3\dfrac{\text{kg}}{\text{ton}}}{16\dfrac{\text{kg methane}}{\text{kmol}}}=3.75\frac{\text{kmol methane}}{\text{hr}}$$

Using ideal gas law at 1 atm and 0°C we obtain the volume of methane produced per hr:

$$V=\frac{nRT}{P}=\frac{3.75\cdot10^3\dfrac{\text{mol}}{\text{hr}}\cdot0.0821\dfrac{\text{l}\cdot\text{atm}}{\text{mol}\cdot\text{K}}\cdot273.15\text{K}}{1\,\text{atm}}=84.1\frac{\text{Nm}^3}{\text{hr}}$$

b. The calorific value of methane is 890 kJ/mol, so the total energy that can be generated by the combustion of methane is:

$$890\frac{\text{kJ}}{\text{mol}}\cdot3750\frac{\text{mol}}{\text{hr}}=927\,\text{kW}$$

Only 60% of this is converted to electricity so the electrical power generated is:

$$927\,kW \cdot 0.6 = 556\,kW$$

The digester operates for 8000 hrs per year so the electric energy generated in one year is:

$$556\,kW \cdot 8000\,\frac{hr}{year} = 4.44 \cdot 10^6\,\frac{kWh}{year}$$

c. The digester generates a liquid–solid slurry which is used as fertiliser. The solids in the fertiliser are composed of the microorganisms generated. We know that 10% of the carbon in the feed is converted to microorganisms, so the microorganisms generated are:

$$0.1\frac{ton\ C\ feed}{hr} \cdot 0.1\frac{ton\ C\ microorganisms}{ton\ C\ feed}$$

$$= 0.01\frac{ton\ C\ microorganisms}{hr}$$

Assuming microorganisms are $C_5H_7O_2N$ (1.88 g microorganisms/g C):

$$0.01\frac{ton\ C\ microorganisms}{hr} \cdot 1.88\frac{ton\ microorganisms}{ton\ C\ microorganisms}$$

$$= 18.8\frac{kg\ microorganisms}{hr}$$

This is the amount of fertiliser (as dry weight) produced per unit of time in the digester.

5.5 a. The total flow rate to the digester is 40 m³/day. The residence time in the digester is 30 days so the digester volume is:

$$V = 40\,\frac{m^3}{day}\,30\,day = 1200\,m^3$$

b. The total volume of gas produced is 1000 Nm³/day, and the fraction of methane is 60% in volume so the methane produced is 600 Nm³/day. In terms of mol, this corresponds to (using ideal gas law) 26,770 mol/day, that is to 428 kg/day.

The total mass of organic matter in the feed is:

$$\left(0.05 \cdot 10{,}000 + 0.1 \cdot 10{,}000 + 0.03 \cdot 20{,}000\right)\frac{kg}{day} = 2170\,\frac{kg}{day}$$

Therefore, the fraction of organic matter in the feed that is converted to methane is:

$$\frac{428\,\dfrac{kg}{day}}{2170\,\dfrac{kg}{day}} = 20\%$$

Bibliography

J.S. Chang, K.S. Lee, P.J. Lin. Biohydrogen production with fixed-bed bioreactors. *International Journal of Hydrogen Energy* 27, 1167–1174, 2002.

D. Dionisi. Potential and limits of biodegradation processes for the removal of organic xenobiotics from tidewaters. *ChemBioEng Reviews* 1(2), 67–82, 2014.

D. Dionisi, A.A. Rasheed, A. Majumder. A new method to calculate the periodic steady state of sequencing batch reactors for biological wastewater treatment: model development and applications. *Journal of Environmental Chemical Engineering* 4(3), 3665–3680, 2016.

C.P.L. Grady Jr., G.T. Daigger, N.G. Love, C.D.M. Filipe. *Biological Wastewater Treatment*, 3rd edn. IWA Publishing, CRC Press, Boca Raton, FL, 2011.

M. Henze, M. van Loosdrecht, G. Ekama, D. Brdjanovic (Eds.). *Biological Wastewater Treatment: Principles, Modelling and Design*. IWA Publishing, London, UK, 2008.

IWA task group on mathematical modelling for design and operation of biological wastewater treatment. *Activated sludge models ASM1, ASM2, ASM2D and ASM3*. IWA Publishing, London, UK, 2007.

IWA task group on mathematical modelling of anaerobic wastewater treatment. *Anaerobic Digestion Model No.1 (ADM1)*. IWA Publishing, London, UK, 2008.

S.V. Kalyuzhnyi. Batch anaerobic digestion of glucose and its mathematical modeling. II. Description, verification and application of model. *Bioresource Technology* 59, 249–258, 1997.

V. Kubsad, S.K. Gupta, S. Chaudari. Treatment of petrochemical wastewater by rotating biological contactor. *Environmental Technology* 26, 1317–1326, 2005.

T. Ognean. Aspects concerning scale-up criteria for surface aerators. *Water Research* 27, 477–484, 1993.

N. Qureshi, B.A. Annous, T.C. Ezeji, P. Karcher, I.S. Maddox. Biofilm reactors for industrial bioconversion processes: Employing potential of enhanced reaction rates. *Microbial Cell Factories* 4, 24, 2005.

J.F. Richardson, J.H. Harker, J.R. Backhurst, J.M. Coulson. *Coulson and Richardson's Chemical Engineering. Volume 2, Particle Technology and Separation Processes*, 5th edn. Butterworth-Heinemann, London, UK, 2002.

P. Saravanan, K. Pakshirajan, P. Saha. Growth kinetics of an indigenous mixed microbial consortium during phenol degradation in a batch reactor. *Bioresource Technology* 99, 205–209, 2008.

T. Sekizawa, K. Fujie, H. Kubota, T. Kasakura, A. Mizuno. Air diffuser performance in activated sludge aeration tanks. *Journal* (*Water Pollution Control Federation*) 57(1), 53–59, 1985.

A. Tawfik, H. Temmink, G. Zeeman, B. Klapwijk. Sewage treatment in a rotating biological contactor (RBC) system. *Water, Air, and Soil Pollution* 175, 275–289, 2006.

E.I.P. Volcke, M.C.M. van Loosdrecht, P.A. Vanrolleghem. Controlling the nitrite: Ammonium ratio in a SHARON reactor in view of its coupling with an Anammox process. *Water Science and Technology* 53(4–5), 45–54, 2006.

P.A. Wilderer, R.L. Irvine. *Sequencing Batch Reactor Technology*. IWA Publishing, London, UK, 2000.

Index

Note: Page numbers followed by f and t refer to figures and tables, respectively.

Printed in the United States
by Baker & Taylor Publisher Services